T0329525

ESSENTIAL
COMPUTATIONAL
FLUID DYNAMICS

ESSENTIAL COMPUTATIONAL FLUID DYNAMICS

Second Edition

Oleg Zikanov

University of Michigan–Dearborn
MI, USA

Registered Office
John Wiley & Sons, Inc., 111 River Street, Hoboken, NJ 07030, USA

Editorial Office
111 River Street, Hoboken, NJ 07030, USA

For details of our global editorial offices, customer services, and more information about Wiley products visit us at www.wiley.com.

Wiley also publishes its books in a variety of electronic formats and by print-on-demand. Some content that appears in standard print versions of this book may not be available in other formats.

Library of Congress Cataloging-in-Publication Data

Names: Zikanov, Oleg, author.
Title: Essential computational fluid dynamics / Oleg Zikanov, University of
    Michigan, Dearborn.
Description: Second edition. | Hoboken, NJ : John Wiley & Sons, Inc., [2019]
    | Includes bibliographical references and index. |
Identifiers: LCCN 2019011596 (print) | LCCN 2019016821 (ebook) | ISBN
    9781119474784 (Adobe PDF) | ISBN 9781119474814 (ePub) | ISBN 9781119474623
    (hardcover)
Subjects: LCSH: Fluid dynamics – Mathematics.
Classification: LCC QA911 (ebook) | LCC QA911 .Z55 2019 (print) | DDC
    532/.0501515 – dc23
LC record available at https://lccn.loc.gov/2019011596

Cover design: Wiley
Cover image: Courtesy of ANSYS

To Elena

CONTENTS

3 Partial Different Equations   37

PREFACE

This second edition is a result of major extension and restructuring of the book. The new material concerns such topics as the techniques of interpolation, finite volume discretization on unstructured grids, projection methods for steady-state problems, RANS modeling, and so on. A large number of new end-of-chapter problems are added. The text has been thoroughly edited, both to improve clarity and to reflect the recent changes in the practice of the computational fluid dynamics and heat transfer (commonly abbreviated as CFD).

The modifications, however, do not change the basic nature of the book. It remains a complete and self-contained introduction into the CFD. The subject is addressed on the very basic level suitable for a first course taught to beginning graduate or senior undergraduate students. No prior knowledge is assumed on the part of the reader.

To appreciate the purpose and flavor of the book, we have to consider the major shift that has recently occurred in the scope and character of CFD applications. From being primarily a research discipline, CFD has transformed into a tool of everyday engineering practice. It would be safe to say that, worldwide, tens of thousands of engineers are directly employed to run CFD computations at companies or consulting firms. Many others encounter CFD at some stages of their work.

Unlike solution of research problems, CFD analysis in an industrial environment does not, typically, involve development of new algorithms. Instead, one of the general-purpose codes is used. Such codes, nowadays, tend to provide a fusion of all the necessary tools: equation solver, mesh generator, turbulence and multiphysics models, and modules for post-processing and parallel computations. Two key factors contribute to the success in applying such codes: (i) understanding of physical and engineering aspects of the analyzed process and (ii) ability to conduct the CFD analysis properly, in a way that guarantees an accurate and efficient solution.

I recognized the need for a new textbook while teaching the graduate and senior undergraduate courses in CFD at the Department of Mechanical Engineering of the University of Michigan–Dearborn. Many of our graduate students are either working engineers or plan to enter the engineering field with a masters degree. Many undergraduate students pursue employment with industry after graduation. Potential future exposure of our students to CFD is often limited to the use of general-purpose codes. To respond to their needs, the instruction is focused on two areas: the fundamentals of the method (what we call the *essential CFD*) and the correct way of conducting the analysis using readily available software. A survey of the existing texts on CFD, although showing many excellent research-oriented texts, does not reveal a book that fully corresponds to this concept.

A comment is in order regarding the bias of the text. All CFD texts are, to some degree, biased in correspondence to the chosen audience and research interests of the authors. More weight is given to some of the methods (finite difference, finite element, spectral, etc.) and to some of the fields of application (heat transfer, incompressible fluid dynamics, or gas dynamics). The choices made in this book reflect the assumption of mechanical, chemical, and civil engineering students as the target audience rather than aerospace engineering students and the intended use of the text for applied CFD instruction. The focus is on the finite difference and finite volume methods. The finite element and spectral techniques are introduced only briefly. Also, somewhat more attention is given to numerical methods for incompressible fluid dynamics and heat transfer than for compressible flows.

The text can be used in combination with exercises in practical CFD analysis. As an example, our course at the University of Michigan–Dearborn is divided into two parts. The first part (about 60 percent of the total course time) is reserved for classroom instruction of the basic methods of CFD. It covers Part I, "Fundamentals," and Part II, "Methods," which includes a simple programming project (solving a one-dimensional nonlinear partial differential equation). The remainder of the course includes exercises with a CFD software and parallel discussion of the topics of Part III, "Art of CFD," dealing with turbulence modeling, computational grids, and rules of good CFD practice. This part is conducted in a computer laboratory and includes a project in which students perform a full-scale CFD analysis.

Acknowledgments: It is a pleasure to record my gratitude to many people who made writing the first and second editions of this book possible. This includes generations of students at the University of Michigan–Dearborn, who suffered through the first iterations of the text and provided priceless feedback. I wish to thank friends and colleagues who read the original manuscript and gave their insightful and constructive suggestions: Thomas Boeck, Dmitry Krasnov, Svetlana Poroseva, Tariq Shamim, Olga Shishkina, Sergey Smolentsev, Axelle Viré, and Anatoly Vorobev. The first serious attempt to write the book was undertaken during a sabbatical stay at the Ilmenau University of Technology. I appreciate the hospitality of André Thess and support by the German Science Foundation (DFG) that made this possible. Finally, and above all, I would like to thank my wife, Elena, and my, now fully grown, children Kirill and Sophia, for their understanding and support during the many hours it took to complete this book.

Oleg Zikanov
Dearborn, January 2019

ABOUT THE COMPANION WEBSITE

This book is accompanied by a companion website:

www.wiley.com/go/zikanov/essential

The website includes:

- Solution Manuals
- Power point slides

1

WHAT IS CFD?

1.1 INTRODUCTION

We start with a definition

CFD (Computational Fluid Dynamics) is a set of numerical methods applied to obtain approximate solutions of problems of fluid dynamics and heat transfer.

According to this definition, CFD is not a science on its own but a way to apply the methods of one discipline (numerical analysis) to another (heat and mass transfer). We will deal with details later. Right now, a brief discussion is in order of why exactly we need CFD.

A distinctive feature of the science of fluid flow and heat and mass transfer is the approach it takes toward description of physical processes. Instead of bulk properties, such as momentum or angular momentum of a body in mechanics or total energy or entropy of a system in thermodynamics, the analysis focuses on *distributed properties*. We try to determine the entire

Essential Computational Fluid Dynamics, Second Edition. Oleg Zikanov.
© 2019 John Wiley & Sons, Inc. Published 2019 by John Wiley & Sons, Inc.
Companion Website: www.wiley.com/go/zikanov/essential

fields such as temperature $T(x, t)$, velocity $v(x, t)$, density $\rho(x, t)$, etc.[1] Even when an integral characteristic, such as the friction coefficient or the net rate of heat transfer, is the ultimate goal of the analysis, it is derived from distributed fields.

The approach is very attractive by virtue of the level of details it provides. Evolution of the entire temperature distribution within a body can be determined. The effect of internal processes of a fluid flow such as motion, rotation, and deformation of minuscule fluid particles can be taken into account. Of course, the opportunities come at a price, most notably in the form of dramatically increased complexity of the governing equations. Except for a few strongly simplified models, the equations for distributed properties are *partial differential equations*, often nonlinear.

As an example of complexity, let us consider a seemingly simple task of mixing and dissolving sugar in a cup of hot coffee. An innocent question of how long would it take to completely dissolve the sugar leads to a very complex physical problem that includes a possibly turbulent two-phase (coffee and sugar particles) flow with variable physical properties and a chemical reaction (dissolving). Heat transfer (within the cup and between the cup and surroundings) is also of importance because temperature strongly affects the rate of the reaction. No simple solution of the problem exists. Of course, we can rely on the experience acquired after repeating the process daily (perhaps more than once) for many years. We can also add a couple of extra, possibly unnecessary, stirs. If, however, the task in question is more serious – for example, optimizing an oil refinery or designing a new aircraft – relying on everyday experience or excessive effort is not an option. We must find a way to *understand* and *predict* the process.

Generally, we can distinguish between three approaches to solving fluid flow and heat transfer problems:

1. *Theoretical approach*: Finding analytical solutions of governing equations or arriving to conclusions on the basis of some theoretical considerations.
2. *Experimental approach*: Staging an experiment using a model of the real object.
3. *Numerical approach*: Using computational procedures to find a solution of the governing equations.

---

[1]Throughout the book, we will use $x = (x, y, z)$ or $x = (x_1, x_2, x_3)$ for the vector of space coordinates and t for time.

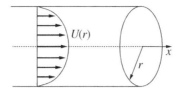

Figure 1.1   Laminar flow in an infinite pipe.

Let us look at these approaches in more detail.

Theoretical Approach. The approach has a crucial advantage of providing exact solutions. Among the disadvantages, the most important is that analytical solutions are only possible for a very limited class of problems, typically formulated in an artificial, idealized way. One example is the Hagen–Poiseuille solution for a flow in an infinitely long pipe (see Figure 1.1). The steady-state laminar velocity profile is

$$U(r) = \frac{r^2 - R^2}{4\mu}\frac{dp}{dx},$$

where U is the velocity, R is the pipe radius, dp/dx is the constant pressure gradient that drives the flow, and μ is the dynamic viscosity of the fluid. The solution is, indeed, simple and gives insight into the nature of flows in pipes and ducts, so its inclusion into all textbooks of fluid dynamics is not surprising. At the same time, the solution is correct only if the pipe is infinitely long,[2] temperature is constant, and the fluid is perfectly incompressible. Furthermore, even if we were able to build such a pipe and find some use for it, the solution would be correct only at Reynolds numbers $Re = UR\rho/\mu$ (ρ is the density of the fluid) below, approximately, 1200. Above this limit, the flow would take fully three-dimensional and time-dependent transitional or turbulent form, for which no analytical solution is possible.

It can also be noted that derivation of analytical solutions often requires substantial mathematical skills, which are not among the strongest traits of many modern engineers and scientists, especially if compared to the situation of 30 or 40 years ago. Several reasons can be named for the deterioration of such skills, one, no doubts, being development of computers and numerical methods, including the CFD.

---

[2]In practice, the solution is considered to be a good approximation for laminar flows in pipes at sufficiently large distance (dependent on the Reynolds number but at least few tens of diameters) from the entrance.

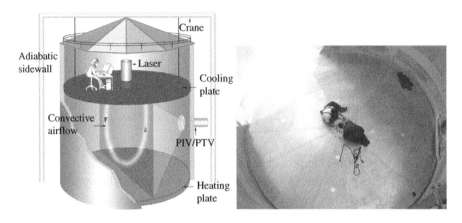

Figure 1.2   The experiment for studying thermal convection at the Ilmenau University of Technology, Germany. Turbulent convection similar to the convection observed in the atmosphere of Earth or Sun is simulated by air motion within a large barrel with thermally insulated walls and uniformly heated bottom. Source: Courtesy of André Thess.

Experimental approach: Well-known examples are the wind tunnel experiments, which help to design and optimize the external shapes of airplanes (also of ships, cars, buildings, and other objects). Another example is illustrated in Figure 1.2. The main disadvantages of the experimental approach are the technical difficulty (sometimes it takes several years before an experiment is set up and all technical problems are resolved) and high cost.

Numerical (computational) approach: Here, again, we employ our ability to describe almost any fluid flow or heat transfer process as a solution of a set of partial differential equations. An approximation to this solution is found by a computer executing an algorithm. This approach is not problem-free either. We will discuss the problems throughout the book. The computational approach, however, beats the analytical and experimental methods in some very important aspects: universality, flexibility, accuracy, and cost.

1.2   BRIEF HISTORY OF CFD

The history of CFD is a fascinating subject, which, unfortunately, we can only touch in passing. The idea to calculate approximate solutions of differential equations describing fluid flows and heat transfer is relatively old, definitely older than computers themselves. Development of numerical methods for solving ordinary and partial differential equations started in the first half of the twentieth century. The computations at that time required the use of calculation tables and dull mechanical work of tens, if not hundreds, of people. No wonder that only the apparently most important (often military-related) problems were addressed and only simple one-dimensional equations were solved.

Invention and subsequent fast development of computers opened a wonderful possibility of performing millions and, then, millions of millions of arithmetic operations in a matter of seconds. Together with the rapid development of the algorithms of numerical mathematics, this has led to impressive growth of speed and abilities of CFD analysis. First simulations of realistic two-dimensional flows were performed in the late 1960s, while three-dimensional flows could be seriously approached since the 1980s. Again, military tasks, such as modeling shock waves from an explosion or a flow past a hypersonic jet aircraft, were addressed first. In fact, development of faster and bigger computers until 1980s was largely motivated by the demands of the military-related CFD.

Over the last several decades, the field of CFD has changed profoundly. From a scientific discipline, in which researchers worked on unique projects using specially developed codes, it has transformed into *an everyday tool of engineering design and scientific research*. In engineering, the simulations are routinely used as "virtual experiments" replacing or complementing prototyping and other design techniques. The problem-specific codes are still developed for scientific purposes, but the practice has almost entirely switched to the use of commercial or open-source CFD codes. The market is largely divided between few major brands, such as ANSYS, STAR-CCM, OpenFOAM, or COMSOL. Standard codes are widely used in other areas of active CFD applications, such as meteorology, oceanography, or astrophysics. The codes differ in appearance and capabilities but are all essentially numerical solvers of partial differential equations with attached physical and turbulence models and algorithms for grid generation and post-processing of results.

1.3 OUTLINE OF THE BOOK

This book is intended as a brief but complete introduction into CFD. The focus is not on development of algorithms but on the fundamental principles, formulation of CFD problems, the most basic and common computational techniques, and essentials of a good CFD analysis. The book's main task is to prepare the reader to make educated choices while using one of the available CFD codes. A reader seeking deeper and more detailed understanding of specific computational methods is encouraged to use more advanced and more specialized texts, references to some of which are presented at the end of each chapter.

A comment is in order regarding the bias of the text. All CFD texts are, to some degree, biased in correspondence with the chosen audience and personal research interests of the authors. More weight is given to some of the methods (finite difference, finite element, spectral, etc.) and some of the fields of application (heat transfer, incompressible fluid dynamics, or gas dynamics). The preferences made in this book reflect the choice of mechanical, chemical, biomedical, and civil engineers as the target audience and the intended use for applied CFD instruction. The focus is on the finite difference and finite volume methods. The finite element and spectral techniques are introduced, but only briefly. Also, more attention is given to numerical methods for incompressible fluid dynamics and heat transfer than for compressible sub- and supersonic flows.

The book contains 13 chapters. We are already at the end of Chapter 1. The remaining chapters are separated into three parts: "Fundamentals," "Methods," and "Art of CFD." Part I deals with the basic concepts of numerical solution of partial differential equations. It starts with Chapter 2 introducing the equations we are most likely to solve: the governing equations of fluid flows and heat transfer. We consider various forms of the equations used in CFD and review common boundary conditions. Necessary mathematical background and the concept of numerical approximation are presented in Chapter 3. Chapter 4 discusses the basics of the finite difference method. We also introduce the key concepts associated with all CFD methods, such as the truncation error and consistency of numerical approximation. The principles and main tools of the finite volume method are presented in Chapter 5. Chapter 6 is devoted to the concept of stability of numerical time integration. Some popular and important (both historically and didactically) schemes for one-dimensional model equations are presented in Chapter 7. This material summarizes the discussion of the fundamental concepts and can be used for a midterm programming project.

Part II, which includes Chapters 8–10, contains a compact description of some of the most important and commonly used CFD techniques. Methods of solution of systems of algebraic equations appearing in the result of the CFD approximation are discussed in Chapter 8. Chapter 9 presents some schemes used for unsteady heat conduction and compressible flows. The discussion is deliberately brief for such voluminous subjects. It is expected that a reader with particular interest in any of them will refer to other, more specialized, texts. Significantly more attention is given to the methods developed for computation of flows of incompressible fluids. Chapter 10 provides a relatively broad explanation of the issues, presents the projection method, and introduces some popular algorithms.

Part III consists of Chapters 11–13 and deals with subjects that are not directly related to the numerical solution of partial differential equations, but nevertheless are irreplaceable in practical CFD analysis. They all belong to a somewhat imprecise science in the sense that the approach is often decided on the basis of knowledge and experience rather than exact knowledge alone. The subjects in question are the turbulence modeling (Chapter 11), design and quality of computational grids (Chapter 12), and the questions arising in the course of CFD analysis, such as uncertainty and validation of results (Chapter 13). The discussion is, by necessity, brief. A reader willing to acquire truly adequate understanding of these difficult but fascinating topics should consult the books listed at the end of each chapter.

BIBLIOGRAPHY

A rich source of information on CFD, online since 1994: books, links, discussion forums, jobs, etc. http://www.cfd-online.com/.

Official web site of the TOP500 project providing reliable and detailed information on the world most powerful supercomputers. http://www.top500.org/.

Part I

FUNDAMENTALS

GOVERNING EQUATIONS OF FLUID DYNAMICS AND HEAT TRANSFER

The methods of CFD can, at least in theory, be applied to any set of partial differential equations. The main area of application, however, has always been the solution of the equations describing processes of fluid flow and heat transfer. This chapter provides a brief description of the equations and can be skipped by a reader familiar with the matter. The material is included for the sake of completeness and is not intended as a replacement of the detailed and thorough account found in comprehensive texts on fluid dynamics and heat and mass transfer. Several such texts are listed at the end of the chapter.

2.1 PRELIMINARY CONCEPTS

From the physical viewpoint, the equations describing fluid flows and heat and mass transfer are simply versions of the conservation laws of classical physics, namely:

- Conservation of chemical species (law of conservation of mass).
- Conservation of momentum (Newton's second law of motion).
- Conservation of energy (first law of thermodynamics).

Essential Computational Fluid Dynamics, Second Edition. Oleg Zikanov.
© 2019 John Wiley & Sons, Inc. Published 2019 by John Wiley & Sons, Inc.
Companion Website: www.wiley.com/go/zikanov/essential

In some cases, additional equations are needed to account for other phenomena, such as entropy transport (expressing the second law of thermodynamics), chemical reactions, phase change, or effect of electromagnetic fields.

Our starting point is the concept of the *continuous* media (solid or liquid) consisting of *elementary volumes* that are infinitesimal from the macroscopic viewpoint but sufficiently large in comparison with the typical distance between molecules so they can themselves be considered as continua. In the case of a fluid flow, the elementary volumes, also called *fluid elements*, are defined as consisting of the same molecules at all times. They move, rotate, and deform under the action of the forces acting in the flow.

The conservation laws must be satisfied by any such fluid element. This can be mathematically expressed in two different ways. We can follow the so-called Lagrangian approach, where the equations are formulated directly in terms of properties of a given fluid element moving in space. This approach is rarely used in CFD. Much more common is the Eulerian approach, in which the conservation principles applied to a fluid element are reformulated in terms of distributed properties, such as density $\rho(x, t)$, temperature $T(x, t)$, or velocity $V(x, t)$, considered as scalar or vector functions of space x and time t.

Our next step is to introduce the *material derivative*. Let us consider a fluid with a variable scalar property, for example, density $\rho(x, t)$. There is a flow that moves the fluid element, so its position vector in the Cartesian coordinates varies with time as $R(t) = (x(t), y(t), z(t))$ (see Figure 2.1).

Differentiation of ρ with respect to time gives the rate of change of density *within the element*:

$$\frac{\partial \rho}{\partial t} + \frac{\partial \rho}{\partial x}\frac{dx(t)}{dt} + \frac{\partial \rho}{\partial y}\frac{dy(t)}{dt} + \frac{\partial \rho}{\partial z}\frac{dz(t)}{dt} = \frac{\partial \rho}{\partial t} + u\frac{\partial \rho}{\partial x} + v\frac{\partial \rho}{\partial y} + w\frac{\partial \rho}{\partial z}, \quad (2.1)$$

where we have used the obvious fact that the time derivatives of the components of the position vector are the respective components of the local velocity $V = u(x, t)i + v(x, t)j + w(x, t)k$. The right-hand side of the equation bears the name of the *material derivative* and has special notation

$$\frac{D\rho}{Dt} \equiv \frac{\partial \rho}{\partial t} + u\frac{\partial \rho}{\partial x} + v\frac{\partial \rho}{\partial y} + w\frac{\partial \rho}{\partial z} = \frac{\partial \rho}{\partial t} + V \cdot \nabla \rho. \quad (2.2)$$

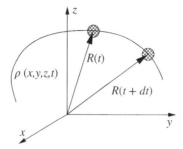

Figure 2.1 Elementary volume of fluid (fluid element) moving in a variable density environment.

Similarly, the rate of change of temperature within a moving fluid element is given by

$$\frac{DT}{Dt} \equiv \frac{\partial T}{\partial t} + \boldsymbol{V} \cdot \nabla T,$$

while for the velocity component u we have

$$\frac{Du}{Dt} \equiv \frac{\partial u}{\partial t} + \boldsymbol{V} \cdot \nabla u.$$

In vector form, the combination of the material derivatives of all three velocity components is denoted as

$$\frac{D\boldsymbol{V}}{Dt} \equiv \frac{\partial \boldsymbol{V}}{\partial t} + (\boldsymbol{V} \cdot \nabla)\boldsymbol{V}. \tag{2.3}$$

The formulas clearly show that the rate of change of any distributed property consists of two parts, one due to the time variation of the property at a given location and another due to the motion of the element in a spatially variable field of this property.

Another important concept is associated with the fact that, while the mass of an element is conserved, its volume continuously changes as it moves and transforms in the flow. It can be viewed as the change of volume that occurs because the velocity field is space dependent, and so the velocity values at opposite sides of the element are different. Let us consider the two-dimensional situation illustrated in Figure 2.2. The element has the sizes dx and L and volume $\delta V = L \, dx$. The velocity field

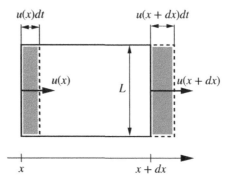

Figure 2.2 Change of the volume of fluid element because of spatial variability of velocity.

is purely one-dimensional $V = u\boldsymbol{i}$ but x-dependent with $u = u(x)$. During the time interval dt, the right-hand side boundary moves together with fluid molecules by the distance $u(x + dx)dt$. The corresponding increase of volume is $L\,dtu(x + dx)$. At the same time, the volume decreases by $L\,dtu(x)$ due to the motion of the left-hand side boundary. The rate of change of volume taken per unit volume is

$$\frac{1}{\delta\mathcal{V}}\frac{d(\delta\mathcal{V})}{dt} = \frac{1}{Ldx}\lim_{dt\to 0}\frac{Lu(x+dx)dt - Lu(x)dt}{dt}$$
$$= \frac{(Lu(x+dx) - Lu(x))}{Ldx} = \frac{u(x+dx) - u(x)}{dx}.$$

Taking the limit of an infinitely small element $dx \to 0$, we find

$$\frac{1}{\delta\mathcal{V}}\frac{d(\delta\mathcal{V})}{dt} = \frac{du}{dx}.$$

In the general case of a three-dimensional velocity field $V = u\boldsymbol{i} + v\boldsymbol{j} + w\boldsymbol{k}$, this formula generalizes to

$$\frac{1}{\delta\mathcal{V}}\frac{d(\delta\mathcal{V})}{dt} = \frac{\partial u}{\partial x} + \frac{\partial v}{\partial y} + \frac{\partial w}{\partial z} \equiv \nabla \cdot V. \qquad (2.4)$$

2.2 CONSERVATION LAWS

We are now prepared to write down the physical conservation laws in the way they are used in CFD. For each law, we will start with its application to a fluid element and then rewrite it in terms of distributed properties using the material derivative and other mathematical devices.

2.2.1 Conservation of Mass

Let us consider a fluid element of volume δV moving in a flow with density $\rho(x, t)$ and velocity $V(x, t)$. Since, according to the definition, the element consists of the same molecules at all times, its mass $\delta m = \rho \delta V$ must remain constant:

$$\frac{d(\rho \delta V)}{dt} = \delta V \frac{D\rho}{Dt} + \rho \frac{d(\delta V)}{dt} = 0.$$

Note that the material derivative is used to represent the rate of change of density within a moving fluid element. Dividing by δV and applying (2.4), we obtain the *continuity equation*

$$\frac{D\rho}{Dt} + \rho \nabla \cdot V = 0, \tag{2.5}$$

which can be rewritten using (2.2) as

$$\frac{\partial \rho}{\partial t} + V \cdot \nabla \rho + \rho \nabla \cdot V = \frac{\partial \rho}{\partial t} + \nabla \cdot (\rho V) = 0. \tag{2.6}$$

In many situations, the compressibility of the fluid can be ignored (fluid dynamics books provide the exact criteria). If this is the case, we can assume that ρ =const. and reduce (2.5) or (2.6) to the *incompressibility equation*

$$\nabla \cdot V = 0. \tag{2.7}$$

2.2.2 Conservation of Chemical Species

Let us now assume that the fluid is a composition of several chemical species, which can transform into each other by chemical reactions. A good example is the flow in a combustion chamber, where a mixture of a hydrocarbon fuel and air is burned to produce exhaust products and energy. The law of conservation of mass still holds, of course, but Eqs. (2.5) and (2.6) have to be modified to account for chemical reactions and interspecies diffusion.

The diffusion is, to put it simply, a process of spontaneous transport of chemical species from the location where their relative concentration is high to the location where the concentration is lower (see the books listed at the end of the chapter for an appropriately detailed and rigorous description). The transport is quantified by the vector field $J_i(x, t)$ of the *flux of a species i*, which denotes the direction and the rate of the mass flux of the species through a unit area at the point x. In the same manner as in the derivation of (2.4), we can find that the rate of change by diffusion of the mass content of species i in a fluid element of unit volume is $\nabla \cdot J_i$.

The concentration of species can be expressed in terms of the *mass fraction* $m_i(x, t)$, which is the ratio of the mass of species i to the total mass of the mixture in the same small volume. Another possibility is to use the mass concentration of species $C_i = m_i \rho$ defined as the mass of species i per unit volume. The conservation law is

$$\frac{\partial}{\partial t}(\rho m_i) + \nabla \cdot (\rho m_i V + J_i) = R_i, \qquad (2.8)$$

where we introduced the source term R_i that accounts for the production/consumption of the species by chemical reactions.

Our next step is to apply the Fick's law of diffusion. This is the simplest model of diffusion processes sufficiently accurate in many practical situations when variations of concentrations are not very strong:

$$J_i = -\Gamma_i \nabla \rho m_i. \qquad (2.9)$$

Here, Γ_i is the diffusion coefficient.

The conservation equation becomes

$$\frac{\partial}{\partial t}(\rho m_i) + \nabla \cdot (\rho m_i V) = R_i + \nabla \cdot (\Gamma_i \nabla \rho m_i). \qquad (2.10)$$

If the Fick diffusion coefficients Γ_i are approximated as constants, the equation simplifies to

$$\frac{\partial}{\partial t}(\rho m_i) + \nabla \cdot (\rho m_i V) = R_i + \Gamma_i \nabla^2 \rho m_i. \qquad (2.11)$$

2.2.3   Conservation of Momentum

The underlying physical principle is the Newton's second law, which states that the rate of change of momentum of a body is equal to the net force acting on it:

$$m \frac{d}{dt}(V) = F. \qquad (2.12)$$

For a fluid element of volume δV moving in a flow, we express its mass as $m = \rho \delta V$ and use the material derivative (2.3) for the rate of change of velocity. Dividing the equation by δV, we obtain

$$\rho \left[\frac{\partial V}{\partial t} + (V \cdot \nabla)V \right] = \rho f, \tag{2.13}$$

where $f = F/\rho \delta V$ is the net force acting on the fluid element taken per unit mass of the fluid.

In the Cartesian coordinates, (2.13) is

$$\rho \frac{Du}{Dt} = \rho \left(\frac{\partial u}{\partial t} + u \frac{\partial u}{\partial x} + v \frac{\partial u}{\partial y} + w \frac{\partial u}{\partial z} \right) = \rho f_x$$

$$\rho \frac{Dv}{Dt} = \rho \left(\frac{\partial v}{\partial t} + u \frac{\partial v}{\partial x} + v \frac{\partial v}{\partial y} + w \frac{\partial v}{\partial z} \right) = \rho f_y$$

$$\rho \frac{Dw}{Dt} = \rho \left(\frac{\partial w}{\partial t} + u \frac{\partial w}{\partial x} + v \frac{\partial w}{\partial y} + w \frac{\partial w}{\partial z} \right) = \rho f_z. \tag{2.14}$$

We can distinguish between two kinds of forces acting on a fluid element:

Body forces: They act directly on the mass of the fluid and originate from a remote source. Their cumulative strength is proportional to the fluid's mass. The examples are the gravity and electric (Coulomb), magnetic, and Lorentz forces. The fictitious centrifugal and Coriolis forces, which appear when the flow is described in a rotating reference frame, also belong to this list. For the purposes of this book, it is sufficient to ignore the physical nature of the force and simply consider the net body force per unit mass f as we have already done in (2.13) and (2.14).

Surface forces: They are the pressure and friction forces acting between neighboring fluid elements and between a fluid element and an adjacent wall. It is shown in the fluid dynamics books that the vector field of surface forces can be represented as divergence of a symmetric 3×3 tensor called the *stress tensor* τ. Its component τ_{ij} can be seen as the i-component of the surface force acting on a unit area surface, which is normal to the j-axis of the Cartesian coordinate system.[1] The diagonal

---

[1]Here and in the rest of the book, we assume that the values 1, 2, and 3 of indices i and j correspond to the Cartesian coordinates x, y, and z. Similarly, the notation u_i with $i = 1, 2, 3$ is used for the velocity components u, v, w.

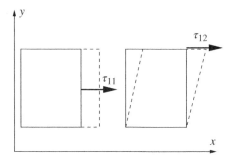

Figure 2.3   Illustration of normal (left) and shear (right) stresses acting on a fluid element.

elements τ_{ii} cause extension/contraction of the fluid element, while the off-diagonal elements are responsible for its deformation by the shear (see Figure 2.3).

The law of conservation of momentum can now be written as

$$
\begin{aligned}
\rho \frac{Du}{Dt} &= \rho f_x + \frac{\partial \tau_{xx}}{\partial x} + \frac{\partial \tau_{yx}}{\partial y} + \frac{\partial \tau_{zx}}{\partial z} \\
\rho \frac{Dv}{Dt} &= \rho f_y + \frac{\partial \tau_{xy}}{\partial x} + \frac{\partial \tau_{yy}}{\partial y} + \frac{\partial \tau_{zy}}{\partial z} \\
\rho \frac{Dw}{Dt} &= \rho f_z + \frac{\partial \tau_{xz}}{\partial x} + \frac{\partial \tau_{yz}}{\partial y} + \frac{\partial \tau_{zz}}{\partial z}.
\end{aligned}
\tag{2.15}
$$

The stress tensor can be separated into the isotropic pressure part, which is always present, and the viscous (friction) part, which exists only in flowing fluid and must be zero if the fluid is at rest:

$$
\tau_{ij} = -p\delta_{ij} + \sigma_{ij},
\tag{2.16}
$$

where

$$
\delta_{ij} = \begin{cases} 1 & \text{if } i = j \\ 0 & \text{if } i \neq j \end{cases}
\tag{2.17}
$$

is the Kronecker delta tensor.

For the equations to fully describe the flow, a model for the viscous stresses σ_{ij} has to be introduced. The most widely used model is

$$
\sigma_{ij} = \lambda \delta_{ij} (\nabla \cdot V) + \mu \left(\frac{\partial u_i}{\partial x_j} + \frac{\partial u_j}{\partial x_i} \right),
\tag{2.18}
$$

where μ and λ are the first and second viscosity coefficients. Note that in an incompressible fluid with $\nabla \cdot V = 0$, the term with the second viscosity disappears. For compressible fluids, it is generally believed that $\lambda = -\frac{2}{3}\mu$ is an accurate approximation except for special situations such as shock waves in hypersonic flows or attenuation of acoustic waves.

The model (2.18) does not have a fully satisfactory theoretical justification. It has, however, being validated in experiments and simply in everyday practice of applying the resulting equations. The fluids whose behavior satisfies the model are called Newtonian. There are non-Newtonian fluids that behave quite differently (e.g. polymer melts and solutions, human blood at high shear stress, etc.).

After substituting (2.18) into (2.14) and using the second viscosity assumption, we obtain the final form of the momentum conservation equations, the Navier–Stokes equations

$$\rho\frac{Du}{Dt} = \rho f_x - \frac{\partial p}{\partial x} + \frac{\partial}{\partial x}\left[\mu\left(-\frac{2}{3}\nabla \cdot V + 2\frac{\partial u}{\partial x}\right)\right]$$
$$+ \frac{\partial}{\partial y}\left[\mu\left(\frac{\partial v}{\partial x} + \frac{\partial u}{\partial y}\right)\right] + \frac{\partial}{\partial z}\left[\mu\left(\frac{\partial w}{\partial x} + \frac{\partial u}{\partial z}\right)\right]$$

$$\rho\frac{Dv}{Dt} = \rho f_y - \frac{\partial p}{\partial y} + \frac{\partial}{\partial y}\left[\mu\left(-\frac{2}{3}\nabla \cdot V + 2\frac{\partial v}{\partial y}\right)\right] \qquad (2.19)$$
$$+ \frac{\partial}{\partial x}\left[\mu\left(\frac{\partial v}{\partial x} + \frac{\partial u}{\partial y}\right)\right] + \frac{\partial}{\partial z}\left[\mu\left(\frac{\partial w}{\partial y} + \frac{\partial v}{\partial z}\right)\right]$$

$$\rho\frac{Dw}{Dt} = \rho f_z - \frac{\partial p}{\partial z} + \frac{\partial}{\partial z}\left[\mu\left(-\frac{2}{3}\nabla \cdot V + 2\frac{\partial w}{\partial z}\right)\right]$$
$$+ \frac{\partial}{\partial x}\left[\mu\left(\frac{\partial w}{\partial x} + \frac{\partial u}{\partial z}\right)\right] + \frac{\partial}{\partial y}\left[\mu\left(\frac{\partial w}{\partial y} + \frac{\partial v}{\partial z}\right)\right].$$

The equations can be written in a shorter form if we introduce the rate of strain tensor with components

$$S_{ij} \equiv \frac{1}{2}\left(\frac{\partial u_i}{\partial x_j} + \frac{\partial u_j}{\partial x_i}\right) \qquad (2.20)$$

and use the Einstein summation convention, according to which repeated indices in a term imply summation over all their possible values (1, 2, 3 in our case):

$$\rho\frac{Du_i}{Dt} = \rho f_i - \frac{\partial p}{\partial x_i} + \frac{\partial}{\partial x_j}\left[2\mu S_{ij} - \frac{2}{3}\mu(\nabla \cdot V)\delta_{ij}\right]. \qquad (2.21)$$

For the special case of an incompressible fluid with constant viscosity coefficient μ, the Navier–Stokes equations become

$$\rho\frac{DV}{Dt} = -\nabla p + \mu\nabla^2 V + \rho f. \tag{2.22}$$

Another special case is the asymptotic limit of an inviscid fluid with $\mu = \lambda = 0$, for which the so-called Euler equations are valid:

$$\rho\frac{DV}{Dt} = -\nabla p + \rho f. \tag{2.23}$$

Both (2.22) and (2.23) must be understood as idealizations, strictly speaking, achievable only as asymptotic limits of flows with very low compressibility and very weak viscosity effect, respectively. This does not prevent them from being widely used as approximations of real flow behavior.

2.2.4   Conservation of Energy

The energy conservation principle can be formulated for a fluid element in the manner similar to the mass and momentum conservation (see the books listed at the end of the chapter for a derivation) as

$$\rho\frac{De}{Dt} = -\nabla \cdot q - p(\nabla \cdot V) + \dot{Q}, \tag{2.24}$$

where $e(x, t)$ is the internal energy per unit mass, $q(x, t)$ is the vector field of the heat flux by thermal conduction, and $\dot{Q}$ is the rate of internal heat generation by the effects such as viscous friction, radiation, chemical reactions, or Joule dissipation. The conduction heat flux can be described by the Fourier conduction law

$$q = -\kappa\nabla T, \tag{2.25}$$

where $T(x, t)$ is the temperature field and κ is the coefficient of thermal conductivity of the fluid.

The energy conservation equation can also be written in the enthalpy form

$$\rho\frac{Dh}{Dt} = \frac{Dp}{Dt} + \dot{Q} - \nabla \cdot q, \tag{2.26}$$

where $h = e + p/\rho$ is the specific enthalpy. Yet another possibility is the equation for the total (internal plus mechanical) energy E

$$\rho\frac{DE}{Dt} = -\nabla \cdot q - \nabla \cdot (pV) + \dot{Q} + \rho f \cdot V. \tag{2.27}$$

In some cases, the equation can be brought into a simpler form. This is, in particular, true when the fluid can be approximately considered as incompressible (the Boussinesq approximation).[2] For such fluids the specific internal energy is approximated as $e = CT$, where $C = C_p = C_v$ is the specific heat. The energy equation becomes

$$\rho C \frac{DT}{Dt} = -\nabla \cdot \boldsymbol{q} + \dot{Q}. \tag{2.28}$$

Substituting (2.25), we obtain the equation of convection heat transfer

$$\rho C \frac{DT}{Dt} = \rho C \left(\frac{\partial T}{\partial t} + \boldsymbol{V} \cdot \nabla T \right) = \nabla \cdot (\kappa \nabla T) + \dot{Q}, \tag{2.29}$$

which in the case of a quiescent fluid and constant conduction coefficient κ reduces to the classical heat conduction equation

$$\rho C \frac{\partial T}{\partial t} = \kappa \nabla^2 T + \dot{Q} \tag{2.30}$$

or, when the internal heat generation is ignored, to

$$\rho C \frac{\partial T}{\partial t} = \kappa \nabla^2 T. \tag{2.31}$$

Evidently, Eqs. (2.30) and (2.31) or their versions with variable thermal conductivity κ also describe the conduction heat transfer in solids.

2.3 EQUATION OF STATE

To close the system of governing equations, we have to add an equation of state, which connects the thermodynamic variables p, ρ, and T. We also need an expression for the internal energy in terms of the thermodynamic variables. The simplest and most widely used are the ideal gas model

$$p/\rho = RT, \quad e = e(T), \tag{2.32}$$

and the already used model of an incompressible fluid

$$\rho = \text{const.}, \quad e = CT, \tag{2.33}$$

although many other models are possible and often necessary.

[2]To be precise, in the Boussinesq approximation density is assumed constant in all terms of the governing equations except the gravity force term in the momentum equation, where it is modeled as a linear function of temperature.

If the physical coefficients, such as viscosity μ or conductivity κ, are not assumed constant, we have to include formulas giving values of these coefficients as functions of temperature and other variables.

2.4 EQUATIONS OF INTEGRAL FORM

A different approach to the derivation of governing equations can be taken, in which the conservation principles are applied not to a fluid element moving with the flow but to a control volume fixed in space. Instead of analyzing the effect of moving boundaries as, for example, in Figure 2.2, we have, in this case, to take into account the fluid flow through the boundaries of the volume and the associated transport of conserved quantities.

Let us start with the conservation of mass. The total mass of fluid in a control volume Ω (see Figure 2.4) is $M = \int_{\Omega} \rho \, d\Omega$. By virtue of conservation of mass, M can only change because of the transport of mass into or out of the control volume by the flow. The correct term is the *flux* of mass. We are only interested in the component of the flux normal to the boundary S of Ω, since the tangential component does not change the mass inside Ω. The magnitude of this component per unit time and unit surface area is $V \cdot n\rho$, where n is the normal vector of unit length shown in Figure 2.4. It is conventional to use an outward-facing normal, so a positive flux means mass flow out of the control volume. Integrating over the boundary, we obtain the equation for the net mass balance

$$\frac{d}{dt}\int_{\Omega} \rho \, d\Omega + \int_{S} \rho V \cdot n \, dS = 0. \tag{2.34}$$

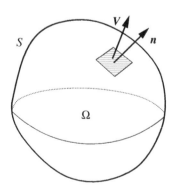

Figure 2.4 Control volume for derivation of equations in integral form.

This is the mass conservation equation in the integral form. It is important to realize that the control volume Ω can be of arbitrary size and shape. For example, it can be the entire flow domain or a small elementary cell of a finite volume grid (see Chapter 5).

The integral equation of conservation of momentum is derived similarly to (2.34). The fluxes are defined for the components of the momentum as $\rho u_i V$. Formulating the balance, we have to include the body forces within the control volume Ω and the surface forces at the surface S. The equation for the ith component is

$$\frac{d}{dt} \int_\Omega \rho u_i \, d\Omega + \int_S \rho u_i V \cdot n \, dS = \int_S t_i \cdot n \, dS + \int_\Omega \rho f_i \, d\Omega. \qquad (2.35)$$

Here t_i is the vector corresponding to the ith row of the stress tensor, such as the first row in the Cartesian coordinates $t_x = \tau_{xx} i + \tau_{xy} j + \tau_{xz} k$. The stresses should be expressed through the velocity components. For the Navier–Stokes model (2.16), (2.18), we obtain, for the i-component of momentum,

$$\frac{d}{dt} \int_\Omega \rho u_i \, d\Omega + \int_S \rho u_i V \cdot n \, dS =$$

$$\int_S \left[(-p + \lambda \nabla \cdot V) n_i + \sum_j \mu \left(\frac{\partial u_i}{\partial x_j} + \frac{\partial u_j}{\partial x_i} \right) n_j \right] dS + \int_\Omega \rho f_i \, d\Omega. \, (2.36)$$

For the energy balance, we consider that energy movement through the boundary S of Ω can be accomplished in two ways: by heat conduction and by motion of matter through the boundary. Accordingly, we consider the normal conduction and convection fluxes $q \cdot n$ and $\rho E V \cdot n$ at the boundary. Taking into account the work by body and surface forces, the integral analog of (2.27) is formulated as

$$\frac{d}{dt} \int_\Omega \rho E \, d\Omega + \int_S q \cdot n \, dS + \int_S \rho E V \cdot n \, dS = \int_S -pn \cdot V \, dS + \int_\Omega \dot{Q} \, d\Omega$$

$$+ \int_\Omega \rho f \cdot V \, d\Omega. \qquad (2.37)$$

When the fluid is incompressible and the energy generation by internal sources can be ignored, we obtain the integral analog of the equation of convection heat transfer (2.29)

$$\frac{d}{dt} \int_\Omega \rho C T \, d\Omega + \int_S \rho C T V \cdot n \, dS = \int_S \kappa \nabla T \cdot n \, dS. \qquad (2.38)$$

The integral equations (2.34)–(2.38) all have similar mathematical structure, which reflects conceptual similarity of the physical processes they describe. Every equation has a term with time derivative of a volume integral, which gives the rate of change of the amount of the conserved quantity within the control volume. Each of the other terms represented by a volume or surface integral corresponds to a certain factor responsible for the change. The integrals are of the following types:

Convective flux integrals: The surface integrals that do not contain derivatives of the conserved field but contain the velocity V of the flow. They represent the velocity transport of the conserved quantity through the boundary of the control volume. The examples are $\int_S \rho V \cdot n \, dS$ in (2.34), $\int_S \rho u_i V \cdot n \, dS$ in (2.36), $\int_S \rho C T V \cdot n \, dS$ in (2.38), etc.

Diffusive flux integrals: The surface integrals that contain first derivatives of the conserved field. These integrals represent the transport through the boundary by diffusion, heat conduction, or viscosity. The derivatives are typically multiplied by the corresponding transport coefficients. The examples are $\int_S \mu (\partial u_i/\partial x_j + \partial u_j/\partial x_i) n_j \, dS$ in (2.36) and $\int_S \kappa \nabla T \cdot n \, dS$ in (2.38).

Volumetric source integrals: The volume integrals corresponding to distributed sources or sinks of the conserved quantity within the control volume, such as $\int_\Omega \rho f_i \, d\Omega$ in (2.36) or $\int_\Omega \dot{Q} \, d\Omega$ and $\int_\Omega \rho f \cdot V \, d\Omega$ in (2.37).

Surface force integrals:The surface integrals representing the work by normal surface forces on the boundary of the control volume. Only one example is present in our equations, the pressure term $\int_S (-p) n \, dS$ in (2.36).

It is convenient for future use, especially for development of finite volume schemes (see Chapter 5), to leave the pressure term for separate consideration and combine the other three types in a formal integral conservation equation for an arbitrary scalar field Φ:

$$\underbrace{\frac{d}{dt} \int_\Omega \Phi \, d\Omega}_{\text{Rate of change}} + \underbrace{\int_S \Phi V \cdot n \, dS}_{\text{Convective flux}} = \underbrace{\int_S \chi \nabla \Phi \cdot n \, dS}_{\text{Diffusive flux}} + \underbrace{\int_\Omega Q \, d\Omega}_{\text{Volumetric source}}.$$

$$(2.39)$$

2.5 EQUATIONS IN CONSERVATION FORM

Since the integral equations describe the same physical processes as the differential governing equations such as (2.5), (2.14), and (2.27), they have to be equivalent mathematically. Let us try to derive the differential equations from the integral ones. We will do it for the formal conservation law (2.39). The procedure is very simple and consists of two steps. First, we convert the surface integrals into volume integrals by using the divergence theorem

$$\int_S \Phi V \cdot n \, dS = \int_\Omega \nabla \cdot (\Phi V) \, d\Omega, \quad \int_S \chi \nabla \Phi \cdot n \, dS = \int_\Omega \nabla \cdot (\chi \nabla \Phi) \, d\Omega,$$

$$(2.40)$$

which is true for any vector field (ΦV or $\chi \nabla \Phi$ in our case) with continuous first derivatives and any volume Ω with a piecewise smooth boundary S. Equation (2.39) can be rewritten as

$$\int_\Omega \left(\frac{\partial \Phi}{\partial t} + \nabla \cdot (\Phi V) - \nabla \cdot (\chi \nabla \Phi) - Q \right) d\Omega = 0.$$

We now remember that the equation is valid for an arbitrary volume Ω, which is only possible if the integrand itself is zero:

$$\frac{\partial \Phi}{\partial t} + \nabla \cdot (\Phi V) = Q + \nabla \cdot (\chi \nabla \Phi). \qquad (2.41)$$

Similar procedures applied to the mass and momentum conservation equations (2.34) and (2.35) result in

$$\frac{\partial \rho}{\partial t} + \nabla \cdot (\rho V) = 0, \qquad (2.42)$$

$$\frac{\partial (\rho u_i)}{\partial t} + \nabla \cdot (\rho u_i V) = \rho f_i + \nabla \cdot t_i. \qquad (2.43)$$

The integral equation of energy conservation (2.37) transforms into

$$\frac{\partial (\rho E)}{\partial t} + \nabla \cdot (\rho E V) = -\nabla \cdot q - \nabla \cdot (p V) + \dot{Q} + \rho f \cdot V. \qquad (2.44)$$

Equations (2.42)–(2.44) are in the *conservation form* also called *conservation law form*, *conservative form*, or *divergence form*. Their defining property is that each term corresponds directly to a term of the integral equation: the time derivative to the rate of change of the conserved quantity in a fixed control volume, nonderivative terms to the volume integrals of sources, and divergence terms to the surface integrals of fluxes. The derivative terms have coefficients, which are either constant or, if variable, not appearing elsewhere in the equations under a derivative sign.

It can be easily shown that the equations in the conservation form (2.42)–(2.44) are mathematically equivalent to the original equations (2.5), (2.14), and (2.27). When, however, the equations are solved numerically on a computational grid, the approximations derived from the two sets of the equations are not necessarily equivalent, and the results of calculations can be different.

An important feature of the numerical schemes based on the approximation of the equations in their conservation form is that such schemes, if properly arranged, conserve the quantities (mass, momentum, energy, etc.) exactly and in the global sense, that is for the entire computational domain.

2.6 EQUATIONS IN VECTOR FORM

It is sometimes convenient for development and analysis of computational algorithms to present the governing equations in a compact vector form. We can easily do this for the equations in conservation form (2.42)–(2.44). Let us introduce the vector fields

$$
U = \begin{bmatrix} \rho \\ \rho u \\ \rho v \\ \rho w \\ \rho E \end{bmatrix}, \quad
Q = \begin{bmatrix} 0 \\ \rho f_x \\ \rho f_y \\ \rho f_z \\ \dot{Q} + \rho (f \cdot V) \end{bmatrix}, \quad
A = \begin{bmatrix} \rho u \\ \rho u^2 + p - \sigma_{xx} \\ \rho u v - \sigma_{xy} \\ \rho u w - \sigma_{xz} \\ (\rho E + p)u + q_x \end{bmatrix}, \quad (2.45)
$$

$$
B = \begin{bmatrix} \rho v \\ \rho u v - \sigma_{xy} \\ \rho v^2 + p - \sigma_{yy} \\ \rho v w - \sigma_{yz} \\ (\rho E + p)v + q_y \end{bmatrix}, \quad
C = \begin{bmatrix} \rho w \\ \rho u w - \sigma_{xz} \\ \rho v w - \sigma_{yz} \\ \rho w^2 + p - \sigma_{zz} \\ (\rho E + p)w + q_z \end{bmatrix}. \quad (2.46)
$$

The system (2.42)–(2.44) abbreviates to

$$
\frac{\partial U}{\partial t} + \frac{\partial A}{\partial x} + \frac{\partial B}{\partial y} + \frac{\partial C}{\partial z} = Q \quad (2.47)
$$

or, when the body forces and internal heat sources are negligible, to

$$\frac{\partial U}{\partial t} + \frac{\partial A}{\partial x} + \frac{\partial B}{\partial y} + \frac{\partial C}{\partial z} = 0. \tag{2.48}$$

2.7 BOUNDARY CONDITIONS

In principle, one can say that all parts of the universe are connected to each other by fluxes of heat and mass and, thus, must be included into a good CFD solution. Since such an enterprise is hardly feasible, we have to compromise and formulate CFD problems for *finite* domains limited by *boundaries*. Such boundaries often appear naturally. For example, they can follow rigid walls. Sometimes, however, the choice is, by necessity, artificial. Several examples of such artificial boundaries are considered in Sections 2.7.2 and 2.7.3. Whatever the case, *a correctly formulated CFD problem should include a set of proper boundary conditions for velocity, temperature, and other variables.*

The importance of appropriate boundary conditions should not be underestimated. No correct CFD solution can be obtained without them. In practical CFD, when one of the general-purpose codes is used, setting the boundary conditions is one of the key "creative" acts performed by the user.

The exact meaning and detailed discussion of physical boundary conditions for specific problems can be found in the books on fluid dynamics and heat transfer. Here, we give only a brief review of the most common types using simple examples. We also touch the boundary conditions that do not follow physics but have to be introduced in a computational solution.

In the first example shown in Figure 2.5, a car is moving within a tunnel. The task is to calculate the airflow and temperature distribution around the car. The flow in the entire tunnel is impossible to simulate (the reasons of which will be discussed throughout the book), so we use the computational

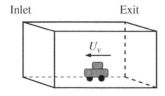

Figure 2.5 Example of setting solid wall, inlet, and exit boundary conditions.

domain in the form of a tunnel segment extending few meters ahead and behind the vehicle.[3]

2.7.1   Rigid Wall Boundary Conditions

At rigid walls, the velocity boundary conditions are different for viscous ($\mu \neq 0$) and inviscid ($\mu = 0$) flows. For viscous flows, the no-slip conditions are applied:

$$V = U_{wall} \text{ at the wall.} \tag{2.49}$$

In our example, if we use a reference frame moving with the car, the conditions are

$V = 0$ at the surface of the vehicle

$V = -U_V$ at the walls of the tunnel.

For inviscid flows, the impermeable wall conditions are applied, according to which only the velocity component normal to the wall is required to match the corresponding component of the wall velocity. The tangential component can slip:

$$V \cdot n = U_{wall} \cdot n \text{ at the wall.} \tag{2.50}$$

We will always assume that the normal n faces outward with respect to the solution domain. For example, in analysis of a fluid flow, n is directed into the wall.

For temperature, two asymptotic limits can be used. One is the condition of known wall temperature T_{wall} (imagine a wall in the form of a large copper slab kept at this temperature):

$$T = T_{wall} \text{ at the wall.} \tag{2.51}$$

Another is the condition of known normal heat flux into the wall q_{wall}:

$$\frac{\partial T}{\partial n} = \nabla T \cdot n = -\frac{1}{\kappa} q_{wall} \text{ at the wall.} \tag{2.52}$$

---

[3]The nature of this particular flow requires a moderate (few meters) extension before the vehicle, while the extension behind it has to be larger if we want to analyze the effect of the wake behind the car.

The special case of the latter is a perfectly insulating wall:

$$\frac{\partial T}{\partial n} = \nabla T \cdot \boldsymbol{n} = 0 \text{ at the wall.} \tag{2.53}$$

The Newton's cooling law can be used as a more realistic boundary condition when neither of the two asymptotic limits is acceptable. The heat flux is taken to be proportional to the difference between the temperatures on two sides of the boundary:

$$q_{wall} = h(T - T_{wall}), \tag{2.54}$$

where h is an empirical cooling constant.[4] A combination of (2.52) and (2.54) results in the boundary condition

$$\kappa \nabla T \cdot \boldsymbol{n} + h(T - T_{wall}) = 0. \tag{2.55}$$

For our example of a car in a tunnel, the tunnel walls can be assumed perfectly insulating (2.53) while (2.52) or (2.55) can be used for the vehicle surface.

2.7.2 Inlet and Exit Boundary Conditions

If the computational domain has open boundaries, such as the inlet and exit in our example, special boundary conditions must be set at them.

The common choice for the inlet is to prescribe velocity and temperature:

$$V = U_{inlet}, \quad T = T_{inlet}, \quad \text{at the inlet.} \tag{2.56}$$

Parameters of turbulent fluctuations should also be prescribed if the flow is turbulent and a turbulence model is used (see Chapter 11).

At the exit, special exit or outflow conditions must be imposed. Typically, the values of the solution variables at this part of the boundary are not a priori known. For example, in the car–tunnel problem in Figure 2.5, we artificially cut off a part of the wake generated behind the car, and there is no possibility to determine how this part affects the flow within the solution domain. Fortunately, if the flow at every point of the exit boundary is directed outward, the fluid elements located at the boundary are swept away from the solution

[4]A common problem with this approximation is that the coefficient h, which is determined by the properties of an often turbulent thermal boundary layer, cannot be given a reliable and accurate quantitative estimate.

domain at the next time moment. This means that any error in the values of velocity and temperature of these elements is unlikely to have a significant effect on the solution within the domain.

These considerations are valid under two conditions. One of them is that the flow is actually directed outward at the entire exit boundary. Typically, this means that the boundary is drawn sufficiently far behind the area of interest and does not cut through any recirculating vortices. Another condition is that the magnitude of the outward velocity is sufficiently high, so the resulting convective fluxes of momentum and energy are stronger than the diffusive fluxes due to viscosity and heat conduction. This condition can be formalized as the demand that the flow's Reynolds and Peclet numbers are much larger than 1.

When the conditions are satisfied, the exit boundary conditions, which are still required mathematically, can be taken in one of the commonly used forms, such as that of zero streamwise gradient (gradient in the direction of the flow x):

$$\frac{\partial V}{\partial x} = 0, \quad \frac{\partial T}{\partial x} = 0, \quad \text{at the exit.} \tag{2.57}$$

2.7.3   Other Boundary Conditions

In many situations, we can make plausible assumptions about the nature of the solution even before it is actually computed. This can be utilized to simplify the task at hand. An example illustrating this approach is given in Figure 2.6. A flow in a circular pipe with a series of equidistant ringlike obstructions is calculated. Two assumptions can be made, especially if our interest is in the mean (average) state of a turbulent flow: that the flow is axially symmetric and that its structure is periodic, repeating itself in every groove between the obstructions.

Relying on the first assumption allows us to consider a two-dimensional solution with all variables depending on the axial z-coordinate and radial r-coordinate of the cylindrical coordinate system instead of the general three-dimensional solution. The computational domain lies in the r–z plane and is limited by the solid walls and the symmetry axis. Special boundary conditions that guarantee regularity of solution have to be imposed at $r = 0$. The engineering CFD codes usually provide such conditions as an option.

The assumption of periodicity allows us to reduce the computational domain in the axial direction. Since the flows in the grooves are identical, only one of them needs to be computed. We can introduce periodic (cyclic)

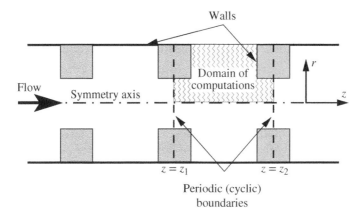

Figure 2.6 Flow in a circular pipe with periodic obstructions as an example of setting symmetry and cyclic boundary conditions.

boundaries as shown in Figure 2.6 and require that the solution variables are reproduced on such boundaries periodically:

$$V(r, z_1) = V(r, z_2), \quad T(r, z_1) = T(r, z_2). \tag{2.58}$$

The example illustrates the fact that the geometry-based simplifying assumptions can be very useful. They allow us to reduce the size of the computational domain and to consider two-dimensional flows instead of three-dimensional. As a result, solutions can be obtained more accurately, on a finer computational grid, and at lower computational cost. The assumptions should, however, be used with caution. The actual flow structure does not always follow the symmetries suggested by the geometry. For example, hydrodynamic instabilities and other effects would, in many cases, make the flow in Figure 2.6 three-dimensional and nonperiodic.

2.8 DIMENSIONALITY AND TIME DEPENDENCE

The vector and scalar fields that describe fluid flows and heat transfer are, in general, functions of three spatial variables and time. For example, one can use the Cartesian coordinates $x = (x, y, z)$ and write the temperature and velocity as

$$T = T(x, t) = T(x, y, z, t), \quad u = u(x, t) = u(x, y, z, t). \tag{2.59}$$

As we show in the following chapters, the computational cost of a CFD solution falls dramatically when the number of independent variables is reduced. The situations when this is justified are discussed in this section.

2.8.1 Two- and One-Dimensional Problems

One can save computational resources or solve an otherwise unattainable problem by assuming that the solution fields depend, for example, on x and y, but not z (reducing the problem from three- to two-dimensional):

$$T = T(x, y, t), \quad \boldsymbol{u} = \boldsymbol{u}(x, y, t), \tag{2.60}$$

or reducing further to a one-dimensional problem:

$$T = T(x, t), \quad \boldsymbol{u} = \boldsymbol{u}(x, t). \tag{2.61}$$

The reduction of dimensionality may come with a serious or even unacceptable penalty. If the solution variables, for example, T or $\boldsymbol{u}$, vary significantly along the z-coordinate, the approximations (2.60) and (2.61) are inaccurate. A CFD solution based on them would most likely result in incorrect predictions. It is, therefore, one of the key responsibilities of the person doing the analysis to decide whether or not the dimensionality of the problem can be reduced without excessive loss of accuracy. Sometimes, the decision can be made on the basis of a priori understanding of the process in question. In other cases, validation of the CFD model via comparison with an experiment is necessary (see Chapter 13).

Often the reduction of dimensionality can be achieved by changing the coordinate system. For example, when we are looking for an axisymmetric solution, it is natural to use polar cylindrical coordinates (r, θ, z) and solve a two-dimensional problem for

$$T = T(r, z, t), \quad \boldsymbol{u} = \boldsymbol{u}(r, z, t). \tag{2.62}$$

Such a simplification was already used in Section 2.7.3.

The reduction of dimensionality is also possible when we decide to solve not for the real three-dimensional flow variables, but for the variables averaged in a certain direction or directions. This is routinely done in the analysis of turbulent flows, where we solve equations for the mean flow, letting the averaging take care of the three-dimensional fluctuations (see Chapter 11).

2.8.2   Equilibrium and Marching Problems

Great savings of computational efforts are also obtained if one assumes that the problem is steady state, i.e. the variables do not depend on time:

$$T = T(x), \quad u = u(x). \tag{2.63}$$

Only one distribution of the variables in the spatial solution domain has to be found instead of many such distributions presenting the evolution of the system.

As we will see in the following chapters, there are differences in the numerical methods applied to the steady-state and transient equations. This reflects the specialty of time as a one-way coordinate. As most of us believe, the events of the present are affected by the events of the past but not by those of the future. This asymmetry certainly applies to the known solutions for fluid flows and heat transfer. The situation is more complex for spatial coordinates, which can be two-way (the value of a variable at a given point if affected by the values on both sides of this point) or one-way. We will clarify this statement in Chapter 3 when we discuss the classification of partial differential equations.

In order to account for the special role of time, the CFD problems are classified into two groups: *equilibrium*, in which the solution is assumed time-independent, and *marching*, in which no such assumption is made.

The choice is determined by the nature of the process and by the purpose of the analysis. For example, equilibrium problems arise when we want to know the air resistance coefficient of an airplane cruising with constant speed and latitude or the temperature distribution within a bioreactor operating in a steady-state mode. As an approximation, we assume that the solution variables are functions of space but not time and replace the time derivatives in the governing equations by zeros.

In other cases, the evolution toward the equilibrium state is of interest, or the equilibrium state does not exist even as an approximation. Returning to our examples, this would correspond to an airplane taking off or landing or to a change in the regime of operation of a bioreactor. Full transient equations should be solved in such cases.

As explained in Chapter 11, replacing a marching problem by an equilibrium one is also possible in the analysis of turbulent flows. This is achieved by averaging and justified in the situations when the variation of flow variables with time is limited to turbulent fluctuations, while the mean flow is steady state.

CFD is applied to both kinds of processes, but differently. For equilibrium problems, the equations are solved numerically only once to determine an approximation of the solution variables, such as T and u in (2.63) in the spatial solution domain. For marching problems, we start with initial conditions, which give the state of the system at some initial time moment $t = t_0$ and "march" the solution forward in time to determine the approximations of T and $u(x, t)$ at $t > t_0$.

BIBLIOGRAPHY

Kundu, P.K., Cohen, I.M., and Dowling, D.R. (2015). *Fluid Mechanics*. London: Elsevier Academic Press.

Landau, L.D. and Lifshitz, E.M. (1987). *Fluid Mechanics, Course of Theoretical Physics*, vol. **6**. Woburn, MA: Butterworth-Heinemann.

White, F.M. (2005). *Viscous Fluid Flow*. New York: McGraw-Hill.

PROBLEMS

1. Write the formula for the material derivative of the specific internal energy e of a flowing fluid. What does it represent?

2. What is Newtonian fluid?

3. What are the models of incompressible fluid and ideal gas?

4. Verify that the Navier–Stokes equations (2.19) reduce to (2.22) in the case of a flow with constant density and viscosity.

5. Write the integral equation for a fixed control volume Ω (see Section 2.4) that expresses the principle of conservation of chemical species described in Section 2.2.2.

6. Following the procedure described in Section 2.5 derive the continuity equation (2.6) from the integral mass conservation equation (2.34).

7. Verify that the Hagen–Poiseuille solution for a laminar flow in an infinitely long circular pipe discussed in Chapter 1 satisfies the mass and momentum conservation equations for a steady flow of an incompressible Newtonian fluid. Use fluid dynamics books for the equations expressed in cylindrical coordinates.

8. Consider the following problems of conduction heat transfer within a solid body. For each problem, write the entire system of governing equations and boundary conditions. Assume constant physical properties in (a) and (b), but not in (c). Do not make any assumptions about the nature of the solution, i.e. write it for the general three-dimensional time-dependent case. In each problem, write the solution twice: in a coordinate-free form and using an appropriate coordinate system.

   a) A cylindrical metal rod, the ends of which are maintained at constant temperatures T_1 and T_2 and the sidewall is perfectly thermally insulated.

   b) A metal cuboid with volumetric Joule heating $\dot{Q}(x, t)$ generated by electric currents. The walls are exposed to air of temperature T_{air}. The heat transfer coefficient h is a known constant.

   c) A plastic spherical shell with time-variable spatially uniform heat flux per unit area $\dot{q}(t)$ applied at the inner surface. The outer surface is cooled by water of constant temperature T_w. The heat transfer coefficient h at the outer surface is a known constant. In this problem, assume that the density and specific heat of the plastic are constant but the thermal conductivity varies significantly as a function of temperature $\kappa = \kappa(T)$.

9. Define the computational domain, and write the full system of governing equations and boundary conditions for the following situations. In each of them, consider a long straight duct with smooth walls and uniformly distributed circular pipes crossing the duct in the direction perpendicular to the duct axis and parallel to one set of walls (see Figure 2.7).

   a) There is a flow of air along the duct. Air can be assumed incompressible and having constant temperature equal to the temperature of the duct walls and pipes.

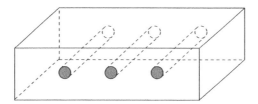

Figure 2.7 Schematic for Problem 2.8.9.

b) The same as in (a), but now temperature varies. The cylinders are maintained at constant temperature T_c, which is significantly higher than the air temperature T_i at the duct inlet. Air is still assumed incompressible.

c) The duct is now filled with a solid material of density ρ, specific heat C, and conductivity κ. Temperature of the cylinders is T_c, and the temperature of the walls is T_w.

3

PARTIAL DIFFERENT EQUATIONS

From the mathematical viewpoint, the equations of fluid flows and heat transfer are partial differential equations (PDEs). Before we actually start solving them and face the difficulties, we need to consider their mathematical properties. The reasons for that will be illustrated in this and the following chapters. Certain properties of the equations have profound effect on the behavior of their solutions and, significantly for us, on the choice of numerical method.

The full governing equations, such as (2.42)–(2.44), are complex. Exact analytical solutions can only be found in a few configurations typically describing strongly idealized physical systems. It should not, therefore, be surprising that the analysis of the mathematical issues and the initial development of the numerical methods are usually carried out for the simple model equations, for which analytical solutions are available and which possess principal mathematical properties of the original equations.

We will follow this approach and start by presenting model PDEs for a scalar field $u(x, t)$. These equations will be used throughout the Part I of the book to illustrate the basic principles of numerical solution. In this chapter, the model equations will help us to present the elements of a well-posed PDE problem and to discuss the mathematical classification of PDEs, its

Essential Computational Fluid Dynamics, Second Edition. Oleg Zikanov.
© 2019 John Wiley & Sons, Inc. Published 2019 by John Wiley & Sons, Inc.
Companion Website: www.wiley.com/go/zikanov/essential

consequences for the solution properties, and its relevance to the physics of fluid flows and heat transfer. At the end of the chapter, we will introduce the concept of numerical discretization of a PDE problem and review the main discretization techniques.

3.1 MODEL EQUATIONS: FORMULATION OF A PDE PROBLEM

3.1.1 Model Equations

The model equations used in our discussion are introduced in this section.

Heat Equation: The heat equation was derived in Section 2.2.4. It expresses the energy conservation principle in the case of conduction heat transfer with constant physical properties and absent sources of internal heat generation:

$$\frac{\partial u}{\partial t} = a^2 \nabla^2 u, \tag{3.1}$$

where $u(x, t)$ is the temperature field and $a^2 = \kappa/\rho C$ is the temperature diffusivity coefficient. In fact, the same equation can be used to describe many other processes, such as diffusion of an admixture in a quiescent fluid or evolution of an initially sharp velocity gradient in a viscous flow. In the one-dimensional case, the equation reduces to

$$\frac{\partial u}{\partial t} = a^2 \frac{\partial^2 u}{\partial x^2}. \tag{3.2}$$

Wave Equation: The wave equation

$$\frac{\partial^2 u}{\partial t^2} = a^2 \nabla^2 u \tag{3.3}$$

describes wavelike phenomena such as sound propagation or oscillations of a string or membrane. In the one-dimensional case, the equation is

$$\frac{\partial^2 u}{\partial t^2} = a^2 \frac{\partial^2 u}{\partial x^2}. \tag{3.4}$$

Linear Convection Equation: Another, even simpler, equation can be used as a representative of the equations with wavelike solutions. This is the so-called linear convection equation

$$\frac{\partial u}{\partial t} + c\frac{\partial u}{\partial x} = 0, \tag{3.5}$$

where c is a positive constant.

Laplace and Poisson Equations: The Laplace equation

$$\nabla^2 u = 0 \tag{3.6}$$

can be considered as a steady-state version of the heat equation (3.1) obtained by setting $\partial u/\partial t = 0$. An important generalization is the Poisson equation

$$\nabla^2 u = f(\boldsymbol{x}), \tag{3.7}$$

where f is a known function of spatial coordinates. The simplest PDE forms of (3.6) and (3.7) are for the two-dimensional case $u = u(x, y)$, for example,

$$\frac{\partial^2 u}{\partial x^2} + \frac{\partial^2 u}{\partial y^2} = f(x, y). \tag{3.8}$$

Burgers and Generic Transport Equations: The Burgers equation is

$$\frac{\partial u}{\partial t} + u\frac{\partial u}{\partial x} = \mu\frac{\partial^2 u}{\partial x^2}, \tag{3.9}$$

where $u = u(x, t)$ and $\mu \geq 0$ is a constant coefficient. The equation was suggested by J.M. Burgers in the 1940s as a one-dimensional model of the Navier–Stokes dynamics and, presumably, of turbulence in fluid flows. The terms of the equation can be considered as counterparts of unsteady, convective, and viscous terms of the momentum conservation equation of the Navier–Stokes system.

A modification of (3.9) often considered in CFD literature is the one-dimensional generic transport equation

$$\frac{\partial \phi}{\partial t} + u\frac{\partial \phi}{\partial x} = \mu\frac{\partial^2 \phi}{\partial x^2}, \tag{3.10}$$

where ϕ is a transported and diffused scalar field (e.g. temperature) and $u(x, t)$ is a known function acting as a velocity-like transporting agent.

It has become clear with time that turbulence is an essentially three-dimensional phenomenon and cannot be modeled by (3.9). Similarly, (3.10) is not a good model for the majority of heat and mass transfer processes, which are either two- or three-dimensional. It has also become clear that Eqs. (3.9) and (3.10) serve as excellent benchmarks for development and testing of the CFD methods.

3.1.2 Domain, Boundary and Initial Conditions, and Well-Posed PDE Problem

The subject of the PDE analysis is not an equation itself but a complete PDE problem consisting of the equation, domain of solution, boundary conditions, and initial conditions. The problem has to be solved in a *spatial domain* Ω and, in the case of time dependency, in a *time interval* between t_0 and t_{end} (see Figure 3.1). t_{end} can be a finite number or infinity. Similarly, the domain Ω may have finite size or extend to infinity in one or several directions. In numerical simulations, the infinite limits of the spatial or time domain are replaced by sufficiently large finite numbers.

Boundary conditions have to be imposed at the entire boundary $\partial\Omega$ of the spatial domain. This is necessary not only to account for the effect of real physical boundaries but also from a purely mathematical viewpoint.

According to our discussion in Chapter 2, the physical boundary conditions are usually expressed in terms of the boundary values of the unknown

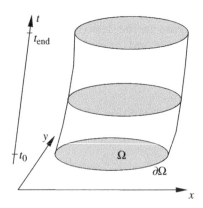

Figure 3.1 Solution domain of a PDE problem.

field u or its normal derivative. Mathematically, the possibilities are the Dirichlet boundary condition

$$u(\boldsymbol{x}, t)|_{\partial\Omega} = g \quad \text{at } t_0 < t < t_{end}, \tag{3.11}$$

the Neumann boundary condition on the normal derivative

$$\left.\frac{\partial u(\boldsymbol{x}, t)}{\partial n}\right|_{\partial\Omega} = g \quad \text{at } t_0 < t < t_{end}, \tag{3.12}$$

the Robin (mixed) boundary condition

$$\left(a_1 \frac{\partial u(\boldsymbol{x}, t)}{\partial n} + a_2 u(\boldsymbol{x}, t)\right)\Bigg|_{\partial\Omega} = g \quad \text{at } t_0 < t < t_{end}, \tag{3.13}$$

and the periodicity condition (here in the x-direction)

$$u(\boldsymbol{x}, t)|_{x_0} = u(\boldsymbol{x}, t)|_{x_0 + L} \quad \text{at } t_0 < t < t_{end}. \tag{3.14}$$

In (3.11)–(3.14), g is a known function of space and time defined at the boundary, and L is the length of periodicity. It is allowed mathematically and sometimes required by the physics that the boundary conditions of different types are applied at different parts of the boundary.

If the domain Ω is infinite, special boundary conditions have to be applied at infinity. For example, if the domain extends to ∞ in the x-direction, the conditions may be

$$u \to 0 \text{ or } u \to A = \text{const.} \quad \text{or} \quad u \text{ is bounded} \quad \text{at } x \to \infty. \tag{3.15}$$

In the marching problems, *initial conditions* have to be imposed at $t = t_0$. Depending on the type of the equation, one or two conditions are necessary. The most common situations are when the solution itself is known

$$u(\boldsymbol{x}, t_0) = h(\boldsymbol{x}) \quad \text{in } \Omega \tag{3.16}$$

and when its first time derivative is known

$$\frac{\partial u}{\partial t}(\boldsymbol{x}, t_0) = f(\boldsymbol{x}) \quad \text{in } \Omega. \tag{3.17}$$

Among our model equations, the heat equation (3.2), linear convection equation (3.5), Burgers equation (3.9), and generic transport equation (3.10) require the initial condition (3.16), while the wave equation (3.4) needs both (3.16) and (3.17).

Only a problem with properly set boundary and initial conditions is *well posed*, i.e. (i) has a solution, which is (ii) unique and (iii) changes continuously with changing parameters, such as the boundary and initial conditions and the inhomogeneous terms. We note that well-posedness is not just for mathematicians. While some ill-posed problems can be approached computationally, a CFD solution in the classical sense of this term used in this book is only possible when the problem is well posed. The importance of existence and uniqueness of the solution approximated in the CFD analysis is evident. The third criterion of the well-posedness is essential to avoid the situations when a small error in the approximation of the initial and boundary conditions or the inhomogeneous terms, which inevitably appears in the course of CFD analysis (see Chapters 4 and 5), causes unacceptably large errors in the solution.

3.1.3   Examples

Let us illustrate the setting of PDE problems using simple examples.

One-Dimensional Heat Equation: We invoke the classical example and consider (3.2) as an equation describing the temperature distribution in a thin long rod with thermally insulated sidewalls (see Figure 3.2). In this

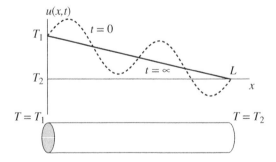

Figure 3.2   One-dimensional temperature field in a long rod as an example of the physical field described by the heat equation (3.2). Initial and asymptotic final distributions of temperature are shown for a problem with Dirichlet boundary conditions at both ends.

case, we disregard temperature variations across the rod and assume that the temperature is a function of the coordinate x and time t.

The solution domain consists of the space interval $[0, x]$ and the time interval, which can be finite $[t_0, t_{end}]$ or extend to infinity $[t_0, \infty)$. Different kinds of boundary conditions are possible. The situation when the ends of the rod are kept at constant temperature corresponds to the Dirichlet boundary conditions

$$u(0, t) = T_1, \quad u(L, t) = T_2 \text{ at } t > t_0. \tag{3.18}$$

The situation of known heat flux at the ends is described by the Neumann boundary conditions

$$\frac{\partial u}{\partial x}(0, t) = a_1, \quad \frac{\partial u}{\partial x}(L, t) = a_2 \text{ at } t > t_0. \tag{3.19}$$

The Newton's cooling law (2.54) can also be applied. This results in the Robin boundary conditions (here at the end $x = 0$)

$$\frac{\partial u}{\partial x}(0, t) + b_1 u(0, t) = b_2. \tag{3.20}$$

The values of the parameters, such as $T_1, T_2, a_1, a_2, b_1,$ or b_2 in (3.18)–(3.20) can be constants or functions of time.

Periodic boundary conditions are also possible:

$$u(0, t) = u(L, t) \quad \text{at } t > t_0. \tag{3.21}$$

Initial temperature distribution $u_0(x)$ is used for the initial conditions

$$u(x, t_0) = u_0(x) \quad \text{at } 0 < x < L. \tag{3.22}$$

The complete PDE problem is of marching type and includes the PDE (3.2); the computational domain; one boundary condition, such as (3.18), (3.19), (3.20), or (3.21), on each end; and the initial condition (3.22).

Laplace Equation: The Laplace equation (3.6) can be obtained as an equation that describes a steady-state temperature distribution in a domain Ω. For example, let us assume that we consider a heat transfer problem in a body with known constant boundary temperature or known constant boundary heat flux and are not interested in transient behavior. We know that there must be the final equilibrium distribution of temperature and want to find it. Setting the time derivative of temperature to zero transforms (3.1) into (3.6). The boundary conditions can be $u|_{\partial\Omega} = g$, $\partial u/\partial n|_{\partial\Omega} = g$, or $u|_{\partial\Omega} + a\partial u/\partial n|_{\partial\Omega} = g$.

If internal heat sources of constant intensity $f(x)$ are present within Ω, the final steady-state temperature distribution is a solution of the Poisson equation (3.7).

Another situation described by the Laplace equation is the irrotational flow, in which velocity is a gradient of a scalar potential: $V = \nabla\phi(x, t)$. If the fluid is incompressible, the continuity equation becomes

$$\nabla \cdot V = \nabla^2 \phi = 0.$$

It should be stressed that, in this case, the Laplace equation does not imply the steady-state character of the process. An important example of a similar situation is the behavior of pressure in incompressible flows. As discussed in Chapter 10, the pressure field satisfies a Poisson equation with a time-dependent right-hand side.

Despite the fact that it can describe time-dependent behavior, the Laplace or Poisson PDE problem is formally of equilibrium type. It consists of Eq. (3.6), domain Ω, and one boundary condition (Dirichlet, Neumann, or Robin type) at each point of the boundary.

One-Dimensional Wave Equation: As derived in the textbooks of applied mathematics, the shape of a one-dimensional perfectly elastic string is a solution of the one-dimensional wave equation (3.4). The string motion is frictionless and limited to the x–y-plane. The displacement from the line $y = 0$ is defined as $y = u(x, t)$ (see Figure 3.3). The solution domain includes the space interval $[0, L]$ and the time interval $[t_0, t_{end}]$.

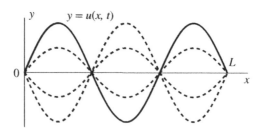

Figure 3.3   Oscillations of an one-dimensional elastic string described by the wave equation (3.4).

The boundary conditions at $x = 0$ and $x = L$ can be, for example, those of a fixed end

$$u(0, t) = u_0 \quad \text{at } t > t_0$$

or a freely moving end

$$\frac{\partial u}{\partial x}(L, t) = 0 \quad \text{at } t > t_0.$$

The initial conditions must include both the shape and velocity of the string at $t = t_0$:

$$u(x, 0) = f(x), \quad \frac{\partial u}{\partial t}(x, 0) = g(x) \text{ at } 0 < x < L. \tag{3.23}$$

This is a marching problem. A correct formulation includes Eq. (3.4), the solution domain, one boundary condition at each boundary, and the two initial conditions (3.23).

3.2 MATHEMATICAL CLASSIFICATION OF PDES OF SECOND ORDER

Partial differential equations of fluid dynamics and heat transfer belong to the class of *quasi-linear* PDEs, which means that they are linear in their highest-order derivatives, but perhaps not in other terms. Inspection of the governing equations in Chapter 2 and of the model equations in the previous section shows that this is, indeed, the case. The quasi-linear PDEs can be classified into three types according to the existence and form of their *characteristics*, the special lines in the solution domain. As usual, we leave the exact definition and detailed discussion to specialized texts, this time on the mathematical theory of PDEs. One aspect is, however, very important for us: the information in the solutions tends to propagate along the characteristics if they exist. This has deep implications not only for the mathematical properties of the solution but also for the choice of numerical methods. To put it simply, *different numerical methods must be used for equations of different types.*

3.2.1 Classification

The classification is applicable to a broad range of systems of quasi-linear PDE. For simplicity, we will limit the formal discussion to a single linear equation of second order for a function of two variables $\phi(x, y)$. The most general form of such an equation is

$$A\phi_{xx} + B\phi_{xy} + C\phi_{yy} + D\phi_x + E\phi_y + F\phi = G, \qquad (3.24)$$

where A, B, C, D, E, F, and G are known coefficients, which can be either constants or functions of x and y. Our choice of (3.24) is not as arbitrary as it may seem. Many equations of fluid dynamics and heat transfer are of the second order, for example, the Navier–Stokes equations (2.19) or the heat transfer equation (2.29). Their highest-order derivatives have the same general form as the highest-order derivatives of (3.24). The lower-order terms are quite different, but, as we will learn imminently, they do not directly affect the classification.

The characteristics of (3.24) can be defined as curves, on which the second derivatives ϕ_{xx}, ϕ_{xy}, and ϕ_{yy} are not uniquely determined by the other terms of the equation. It can be shown that, if a characteristic curve $y = y(x)$ exists, its slope is given by

$$h(x) = \frac{dy}{dx} = \frac{B \pm \sqrt{B^2 - 4AC}}{2A}. \qquad (3.25)$$

The classification is based on this relation, more specifically, on the combination $B^2 - 4AC$. Depending on its value at the point (x, y) there are three possibilities:

$B^2 - 4AC > 0$: There are two real (as in real versus imaginary numbers) characteristics intersecting at this point. The equation is called *hyperbolic*.

$B^2 - 4AC = 0$: There is one real characteristic. The equation is called *parabolic*.

$B^2 - 4AC < 0$: No real characteristics exist at this point. The equation is called *elliptic*.

If A, B, and C are constants, the classification holds in the entire domain Ω. If A, B, or C are functions of x and y, the classification must be done separately for each point (x, y). The equation may be of different types in different parts of Ω.

Examples: Let us determine the types of the model equations considered in the previous section. For the one-dimensional heat equation (3.2), we replace t by y and obtain $A = a^2$, $B = 0$, and $C = 0$, so $B^2 - 4AC = 0$ and the equation is parabolic. For the one-dimensional wave equation (3.4), the same substitution gives $A = a^2$, $B = 0$, and $C = -1$, which corresponds to $B^2 - 4AC = 4a^2 > 0$, so the equation is of hyperbolic type. The two-dimensional Laplace and Poisson equations (3.6) and (3.7) have $A = 1$, $B = 0$, $C = 1$, and $B^2 - 4AC = -4 < 0$. Both the equations are elliptic.

The situation is more complicated for the remaining model equations. The linear convection equation (3.5) is of the first order and does not belong to our classification. Nevertheless, as we will discuss in Section 3.2.2, it has essential features of a hyperbolic system. The classification of the Burgers equation (3.9) and the generic transport equation (3.10) depends on the value of the coefficient μ. If $\mu = 0$, they look as variations of the linear convection equation, in which the constant c is replaced by a variable, and can be considered hyperbolic. We will discuss this matter further when we consider the hyperbolic equations. At $\mu > 0$, the equations are formally parabolic, although the left-hand side implies elements of hyperbolic behavior.

Classification can also be applied to systems of linear or quasi-linear PDEs of the first order. The simplest example is the system

$$\frac{\partial \boldsymbol{q}}{\partial t} + \boldsymbol{R}\frac{\partial \boldsymbol{q}}{\partial x} = 0, \tag{3.26}$$

where $\boldsymbol{q}(x, t)$ is an unknown vector function and $\boldsymbol{R}(x, t)$ is a square matrix of known coefficients. The system is classified as hyperbolic if all the eigenvalues of $\boldsymbol{R}$ are real and distinct. The eigenvalues determine the slopes of the characteristic curves. Another identifiable case is when all the eigenvalues are complex. Such systems are called elliptic.

The classification of first-order systems plays an important role in gas dynamics, acoustics, and other fields, where the governing equations are expressed in the form of such systems. Further details can be found in the books listed at the end of this chapter. Here, we only mention that the two classifications produce identical results if they are applied to mathematically identical cases. For example, the one-dimensional wave equation (3.4) and the Laplace equation (3.6) can be rewritten as systems of two first-order equations (3.26). Analyzing the eigenvalues we find that the systems should be classified as hyperbolic and elliptic, respectively. This is in full agreement with the classification of the original PDEs of second order.

The formal classification can be extended to some PDEs or systems of PDEs with more than two independent variables. This requires more complex mathematics than we have allowed ourselves so far and does not always lead to a clear interpretation. For this reason, we skirt the subject and refer an interested reader to the books listed at the end of this chapter. As a more practical alternative, a discussion of the properties of the Navier–Stokes and heat equations is given in Section 3.2.5.

3.2.2   Hyperbolic Equations

To illustrate the typical properties of hyperbolic systems, we will analyze solutions of the one-dimensional wave equation (3.4). From (3.25), we find the slope of the characteristics as

$$h = \frac{dt}{dx} = \frac{0 \pm \sqrt{0 + 4a^2}}{2a^2} = \pm\frac{1}{a}.$$

There are two families of characteristics: *left-running* $x + at = $ const. and *right-running* $x - at = $ const..

It is shown by straightforward substitution that the general solution of (3.4) can be represented as

$$u(x,t) = F_1(x + at) + F_2(x - at), \tag{3.27}$$

where F_1 and F_2 are functions determined by initial and boundary conditions. If we ignore the effect of boundaries, the solution for the specified initial conditions $u(x,0) = f(x)$, $\partial_t u(x,0) = g(x)$ can be written in the d'Alembert form

$$u(x,t) = \frac{f(x + at) + f(x - at)}{2} + \frac{1}{2a} \int_{x-at}^{x+at} g(\tau)d\tau. \tag{3.28}$$

The first part of the solution can be interpreted using the illustration in Figure 3.4a. The initial perturbation of $u(x,0) = f(x)$ around the point x_0 (e.g. a localized deformation of a string) is split into halves, which propagate without changing their shape along the characteristics $x + at = x_0$ and $x - at = x_0$.

The second part of the solution (3.28) represents the response to the initial perturbation of $\partial_t u$ (e.g. initial velocity of the string). If, for example, it is given by the delta function $g(x) = \delta(x - x_0)$, the solution evolves as illustrated in Figure 3.4b. $u(x,t)$ is a constant equal to $1/2a$ within the cone between the left-running and right-running characteristics and zero outside this cone.

An important feature of the hyperbolic systems is illustrated by (3.28). The perturbations propagate in space with a finite speed. Let us consider the situation, when a source of perturbations suddenly appears at the time

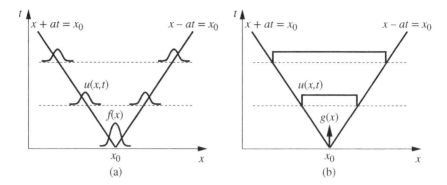

Figure 3.4 d'Alembert solution (3.28) of the hyperbolic equation (3.4). (a) Effect of the initial perturbation of u. (b) Effect of the initial perturbation of $\partial_t u$.

moment t_0 and space location x_0 (point P in Figure 3.5). An observer located at the distance L from the source will not notice the perturbations until the time $t_0 + L/a$. In general, the state of the solution at the point P only affects the solution within a cone between the left-running and right-running characteristics intersecting at P. The cone is called the *domain of influence*. Similarly, the solution at P itself is affected only by the solution within the *domain of dependence* (see Figure 3.5).

The behavior described by hyperbolic equations and, thus, determined by characteristics appears in many physical systems – for example, in

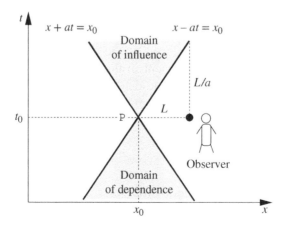

Figure 3.5 Characteristics and domains of influence and dependence of the hyperbolic equation (3.4).

supersonic flows, propagation of acoustic and electromagnetic waves, and flows in stratified systems such as waves on water surface of internal waves in the ocean. All these processes involve wavelike motions along the characteristics or discontinuities across them (e.g. shock waves in supersonic flows). The Navier–Stokes and energy conservation equations have some features of hyperbolic systems as we discuss in Sections 3.2.5 and 9.2.1.

We now return to the linear convection equation (3.5) and consider the line $x = ct + \text{const.}$. Along this line, $dx = c\, dt$, and the value of u remains constant because

$$du = \frac{\partial u}{\partial t}dt + \frac{\partial u}{\partial x}dx = \left(\frac{\partial u}{\partial t} + c\frac{\partial u}{\partial x}\right)dt = 0. \tag{3.29}$$

The line is, thus, a right-running characteristics of the PDE. Its existence is a good enough reason to classify the linear convection equation as hyperbolic. It is easy to check by direct substitution that the equation has the exact solution in the form of a wave:

$$u(x, t) = F(x - ct), \tag{3.30}$$

where F is a function determined by the initial and boundary conditions.

In the case of zero diffusivity coefficient μ, the Burgers and generic transport equations become

$$\frac{\partial u}{\partial t} + u\frac{\partial u}{\partial x} = 0 \tag{3.31}$$

and

$$\frac{\partial \phi}{\partial t} + u\frac{\partial \phi}{\partial x} = 0. \tag{3.32}$$

The only difference between them and the linear convection equation is that the constant coefficient c is replaced by the unknown solution in the case of (3.31) or by a variable coefficient in the case of (3.32). This does not change the hyperbolic, characteristic-determined character of the solution, except that now the characteristics are not straight lines but curves in the (x, t)-space, with the slope determined by the local value of $u(x, t)$.

3.2.3 Parabolic Equations

The parabolic equations of second order have only one family of real characteristics. For example, for the one-dimensional heat equation (3.1), the slope is

$$h = \frac{dt}{dx} = \frac{0 \pm \sqrt{0 + 0}}{2a^2} = 0.$$

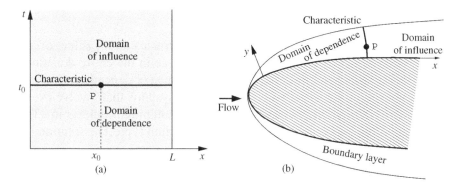

Figure 3.6   (a) Characteristics and domains of influence and dependence of the parabolic equation (3.2). (b) Parabolic behavior of viscous boundary layer.

The characteristics are lines t = const. (see Figure 3.6a). The perturbation that occurs at the space location x_0 and time moment t_0 (point P in Figure 3.6a) affects the solution in the entire space domain $0 < x < L$, although the effect becomes weaker with the distance to P. The domain of influence of the point P includes the domain $0 < x < L$ and time moments $t > t_0$. Accordingly, the domain of dependence of P includes all points $0 < x < L$ and all moments of time prior to the time of P.

In the solutions of parabolic equations, the interaction occurs at infinite speed but relaxes with distance. For example, in the problem of conduction heat transfer in a one-dimensional rod, increase of temperature at a single point is initially felt only slightly at other points (weaker signal for larger distances). With time, however, the effect gradually becomes stronger, and the nonuniformity of the temperature field is smoothed out.

The physical processes described by the parabolic equations can be characterized by the common term of *diffusion*. The term is used here with the general meaning of smoothing out perturbations of physical fields by pseudo-chaotic movement of molecules, turbulent eddies, of other pseudo-randomly moving objects, e.g. people in a dense crowd. This group of processes includes the actual diffusion of chemical species introduced in Section 2.2.2, heat conduction, and, as we discuss later in the book, the effect of viscous friction on flow velocity.

An important example of parabolic systems is the reduced Navier–Stokes equations describing the flow within a viscous boundary layer. The characteristics are perpendicular to the wall, and the flow evolves along the boundary layer similarly to the time evolution of solutions of other parabolic systems (see Figure 3.6b).

3.2.4   Elliptic Equations

The elliptic equations do not have real characteristics at all. Effect of any perturbation is felt immediately and to full degree in the entire domain of solution. There are no limited domains of influence or dependence.

As opposite to the hyperbolic and parabolic systems that involve evolution forward in time and have to be treated as marching problems in CFD, the elliptic PDE problems are always of equilibrium type. The solution has to be found at once in the entire domain Ω.

Physically, the elliptic systems describe equilibrium distributions of properties in spatial domains with boundary conditions. As examples, we name the steady-state heat transfer, electrostatics, and irrotational fluid flows. We will also learn in Chapter 10 that pressure field in incompressible flows is a solution of an elliptic Poisson equation.

3.2.5   Classification of Full Fluid Flow and Heat Transfer Equations

The three-dimensional governing equations of fluid flow and heat transfer introduced in Chapter 2 cannot be written in the simple form of (3.24) or (3.26) and classified by the simple methods outlined in Section 3.2.1. Furthermore, rigorous classification is, in many cases, impossible or misleading in the CFD context. The practical and useful approach fully sufficient for our purposes is to identify elements of hyperbolic, parabolic, or elliptic behavior based on the presence of certain terms in the equations. In addition to highlighting expected properties of the solution, such an analysis influences our choice of numerical methods.

The application of this approach to the Navier–Stokes equations is discussed in Chapters 9 and 10. Here, we will illustrate it on the example of the equation of convection heat transfer (see (2.29)) with constant conductivity coefficient and zero volumetric heat source:

$$\frac{\partial T}{\partial t} + V \cdot \nabla T = \frac{\kappa}{\rho C} \nabla^2 T. \tag{3.33}$$

We start with implications of the second-order derivatives in the right-hand side. They allow us to classify the equation as parabolic. In fact, keeping only these terms and the time derivative, we obtain the three-dimensional heat equation (3.1) with the diffusion coefficient $a^2 = \kappa/\rho C$. The reduced equation describes the process of diffusion of nonuniformities of temperature by heat conduction, which is still active in the processes described by the full equation.

The terms of the material derivative in the left-hand side, while of lower order and, therefore, not affecting the formal classification, are still important for understanding the behavior of the solution. Setting the right-hand side to zero and using the Cartesian coordinates, we find

$$\frac{\partial T}{\partial t} + u\frac{\partial T}{\partial x} + v\frac{\partial T}{\partial y} + w\frac{\partial T}{\partial z} = 0, \tag{3.34}$$

which is, evidently, a three-dimensional version of the zero diffusivity generic transport equation (3.32). The differential of temperature in the direction of the velocity vector $\boldsymbol{V}$ is

$$dT = \left(\frac{\partial T}{\partial t} + \frac{\partial T}{\partial x}\frac{dx}{dt} + \frac{\partial T}{\partial y}\frac{dy}{dt} + \frac{\partial T}{\partial z}\frac{dz}{dt} \right) dt$$

$$= \left(\frac{\partial T}{\partial t} + u\frac{\partial T}{\partial x} + v\frac{\partial T}{\partial y} + w\frac{\partial T}{\partial z} \right) dt = 0. \tag{3.35}$$

We see that a curve parallel at its every point to the local velocity $\boldsymbol{V}(\boldsymbol{x}, t)$ is a characteristic of (3.34). This implies that the solution of the entire equation (3.33) has an element of hyperbolic behavior in the sense that constant values of temperature are transported by velocity along the characteristic curves. The hyperbolicity is, of course, imperfect since the values are changed by the physical effects associated with the equation terms we have neglected in (3.34).

3.3 NUMERICAL DISCRETIZATION: DIFFERENT KINDS OF CFD

In the rest of the book, we shall assume that the analytical solution of a PDE is unavailable or cannot be used for some reason. The equation has to be solved numerically. This section introduces the key concept of the numerical solution – the concept of *discretization*. We also briefly review the main discretization techniques.

Discretization can be understood as replacement of an *exact* solution of a PDE or a system of PDEs in a *continuum* domain by an *approximate* numerical solution in a *discrete* domain. Instead of continuous distributions of solution variables, we find a finite set of numerical values that represents an approximation of the solution.

The discretization can be implemented in different ways. The majority of the methods used in the computational fluid dynamics and heat transfer follow one of the three general approaches: finite difference, finite element, or spectral.

3.3.1 Spectral Methods

Conceptually, the spectral methods are close to the technique of separation of variables used to solve PDEs analytically. The solution is represented by a series of linearly independent functions, which are, in many cases, mutually orthogonal with respect to the integral norm (cos, sin, Bessel functions, orthogonal polynomials, etc.). The coefficients of the series are assumed to be unknown functions of time. The series is infinite in the separation of variable method, where we try to find the coefficients such that the series converges to the exact solution of the PDE. In the spectral numerical method, the series is finite, and the coefficients are chosen so that they *minimize the error of approximation.*

Let us consider a simple example. The modified one-dimensional heat equation

$$\frac{\partial u}{\partial t} = a^2 \frac{\partial^2 u}{\partial x^2} + \sin 5x \qquad (3.36)$$

is solved at $0 < x < \pi$ and $0 < t < T$ with the boundary conditions $u(0, t) = u(\pi, t) = 0$ and initial conditions $u(x, 0) = x(\pi - x)$. We will use the Galerkin method, which is a version of the general spectral approach. First, we have to find a set of orthogonal functions that satisfies the boundary conditions (the *trial functions*). An obvious possibility is

$$\sin x, \sin 2x, \sin 3x, \dots .$$

The solution is sought in the form of a series of trial functions:

$$\tilde{u}(x, t) = \sum_{n=1}^{N} A_n(t) \sin nx, \qquad (3.37)$$

where we use the notation $\tilde{u}$ for the numerical approximation in order to distinguish it from the exact solution of the PDE u.

Substituting (3.37) into Eq. (3.36), we obtain the error of the approximation

$$\epsilon(x,t) = \frac{\partial \tilde{u}}{\partial t} - a^2\frac{\partial^2 \tilde{u}}{\partial x^2} - \sin 5x = -\sin 5x + \sum_{n=1}^{N}[A_n' + (na)^2 A_n]\sin nx.$$
(3.38)

In order to minimize ϵ, we require that its projection on the functional subspace spanned by N orthogonal functions (the *test functions*) is 0. In other words, we require that the inner product of the error with each test function is 0:

$$\langle \epsilon, \phi_m \rangle = \int_0^\pi \epsilon(x,t), \phi_m(x)dx = 0, \quad m = 1, \ldots, N.$$
(3.39)

In the Galerkin method, the same set of functions serves as a test and trial set (different approach may be taken by other spectral methods). Using the orthogonality

$$\int_0^\pi \sin nx \sin mx \, dx = \begin{cases} 0 & \text{if} \quad m \neq n, \\ \pi/2 & \text{if} \quad m = n, \end{cases}$$

we obtain, from (3.38) and (3.39), a system of N ordinary differential equations for $A_n(t)$:

$$A_n' = -(na)^2 A_n + \delta_{5n}, \quad n = 1, \ldots, N.$$
(3.40)

The initial conditions are found by multiplying each side of the expression $\tilde{u}(x,0) = x(\pi - x)$ by $\sin mx$ and integrating from 0 to π. The result is

$$A_m(0) = \frac{2}{\pi} \int_0^\pi x(\pi - x)\sin mx \, dx = -\frac{4(\cos(m\pi) - 1)}{\pi m^3}, \quad m = 1, \ldots, N.$$

The system can be solved numerically using any of the methods developed for ordinary differential equations – for example, the Runge–Kutta method.

The minimized approximation error ϵ decreases with growth of the number of trial and test functions N. This feature (accuracy improving with the total number of the degrees of freedom in the model) is a general feature of all correctly designed numerical discretization methods. The spectral methods, however, have a strong advantage here. Their rate of reduction is much faster than for the other methods. Typically, a moderate value of N is sufficient to achieve good accuracy.

Naturally, there are also bad news. One of them is that the series such as (3.37) can only be developed for a multidimensional PDE if the solution domain has simple geometry, such as a rectangle, sphere, or cylinder. This essentially precludes application of the method to practical engineering problems, which typically involve complex geometries. On the contrary, the fundamental research problems can often be simplified to model geometries. It is, therefore, not surprising that the spectral methods have found widespread use in the theoretical fluid mechanics, especially in investigations of turbulence. We will show an example in Section 11.2.

3.3.2   Finite Element Methods

The finite element methods are widely applied in many areas of engineering – for example, in structural analysis and conduction heat transfer. They are also used for simulating fluid flows, although not as widely as the finite difference and finite volume methods. The basic concept is similar to that of the spectral methods. The main difference is that the decomposition (3.37) is done not in the entire solution domain but within each of many small elements, into which the domain is subdivided. A small number of trial functions are used. The functions are chosen so that they are zero outside the element. Projecting the approximation errors in a manner similar to the Galerkin procedure and performing summation over all elements results in a system of ordinary differential equations not unlike (3.40).

3.3.3   Finite Difference and Finite Volume Methods

The focus of this book is on the finite difference approach (which also includes the finite volume methods in our classification). The discretization is done by approximating the continuous solution at discrete grid points and time layers of a *computational grid*.

The general strategy is summarized in Table 3.1 and illustrated in Figure 3.7. The spatial domain of solution Ω is covered by a grid of points with coordinates $(x, y, z)_i$, where the index i is used to number the points. If a marching problem is solved, the time range of solution $[0, t_{end}]$ should also be covered by discrete points usually called *time layers* t^n.

Our goal is to find the set of numbers u_i^n that approximates the exact solution $u(x, t)$ at points $(x, y, z)_i$ and layers t^n. This is achieved by approximating the original PDE using *finite differences* or *finite volumes* and solving the resulting system of algebraic equations.

Table 3.1   Relation between the elements of a mathematical PDE problem and its finite difference approximation.

PDE problem	Finite difference approximation
Partial differential equation	System of algebraic discretization equations
Exact analytical solution	Approximate finite difference solution
Domain Ω and time interval $[t_0, t_1]$	Computational grid – set of points $(x, y, z)_i$ and time layers t^n
Exact solution – function $u(x, y, z, t)$	Approximate solution – set of values u_i^n approximating u at $(x, y, z)_i$ and t^n

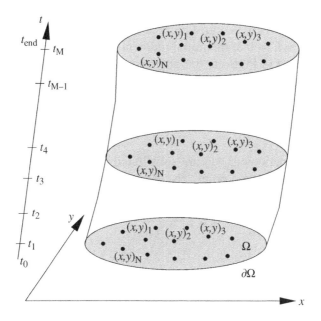

Figure 3.7   Finite difference discretization of the solution domain.

We will discuss in the following chapters how the grids are designed and how the equations are approximated by algebraic formulas. For now, as a preliminary example, we consider the same PDE problem as in the discussion of the spectral methods in Section 3.3.1:

$$\frac{\partial u}{\partial t} = a^2 \frac{\partial^2 u}{\partial x^2} + \sin 5x, \quad 0 \le x \le \pi, 0 < t < T,$$

$$u(0, t) = u(\pi, t) = 0, \quad u(x, 0) = x(\pi - x).$$

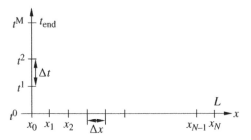

Figure 3.8 Computational grid used to solve the one-dimensional heat equation.

The computational grid is illustrated in Figure 3.8. It includes $N + 1$ equally spaced points $x_i = i\Delta x$ with $i = 0, 1, 2, \ldots, N$ and $\Delta x = L/N$ and $M + 1$ equally spaced time layers $t^n = n\Delta t$ with $n = 0, 1, 2, \ldots, M$ and $\Delta t = t_{end}/M$. Note that there is a grid point at each boundary $x_0 = 0$ and $x_N = \pi$ and a time layer $t^0 = 0$ at the initial moment of time.

The terms of the PDE can be approximated, for example, as (we will discuss these approximations in Chapter 4)

$$\left.\frac{\partial u}{\partial t}\right|_{x_i, t^n} \approx \frac{u_i^{n+1} - u_i^n}{\Delta t}, \quad \left.\frac{\partial^2 u}{\partial x^2}\right|_{x_i, t^n} \approx \frac{u_{i-1}^n - 2u_i^n + u_{i+1}^n}{(\Delta x)^2}. \tag{3.41}$$

The approximations (3.41) are substituted for the terms of the PDE at every inner discretization point. The result is

$$\frac{u_i^{n+1} - u_i^n}{\Delta t} = a^2 \frac{u_{i-1}^n - 2u_i^n + u_{i+1}^n}{(\Delta x)^2} + \sin 5x_i \tag{3.42}$$

for $i = 1, 2, \ldots, N - 1$ and $n = 0, 2, \ldots, M - 1$. The boundary and initial conditions are approximated as

$$u_0^n = 0, \quad u_N^n = 0 \text{ at } n = 1, 2, \ldots, M, \tag{3.43}$$

$$u_i^0 = x_i(\pi - x_i) \text{ at } i = 0, 1, \ldots, N. \tag{3.44}$$

Equations (3.42)–(3.44) represent a system of linear algebraic equations. It is important to realize that the equations are coupled with each other. For example, the finite difference equations (3.42) written for the four adjacent points (x_i, t^{n-1}), (x_i, t^n), (x_{i-1}, t^n), (x_{i+1}, t^n) have the common unknown u_i^n. The boundary and initial conditions are coupled with the finite difference equations at the internal points next to the boundary. The result of the coupling is that, in general, all the finite difference equations have to be solved simultaneously as an algebraic system. One of the tasks of CFD is to formulate such systems in the ways that allow efficient solution. This is easy in the case of (3.42)–(3.44). Regrouping the terms in (3.42) as

$$u_i^{n+1} = u_i^n + \frac{a^2 \Delta t}{(\Delta x)^2}(u_{i-1}^n - 2u_i^n + u_{i+1}^n)$$

and using the boundary conditions (3.43), we can calculate all the values u_i^{n+1} at the $(n + 1)$st layer provided the values u_i^n at the nth layer have already been found. We already have the values of u_i^0 at the zeroth time layer from (3.44). Starting with it, we can advance to the layer t^1, then t^2, and so on in a *marching procedure* and solve the problem in no time.

BIBLIOGRAPHY

Burgers, J.M. (1995). Hydrodynamics–Application of a model system to illustrate some points of the statistical theory of free turbulence. In *Selected Papers of JM Burgers* (pp. 390–400). Dordrecht: Springer.

Canuto, C., Hussaini, M.Y., Quarteroni, A., and Zang, T.A. (2006). *Spectral Methods: Fundamentals in Single Domains*. Berlin, Heidelberg: Springer-Verlag.

Fletcher, C.A.J. (1991). *Computational Techniques for Fluid Dynamics: Fundamental and General Techniques*, vol. **1**, 2e. Springer-Verlag: Berlin.

Greenberg, M. (1998). *Advanced Engineering Mathematics*, 2e. Englewood Cliffs, NJ: Prentice Hall.

Haberman, R. (2003). *Applied Partial Differential Equations*, 4e. Upper Saddle River, NJ: Prentice Hall.

Leveque, R.J. (2002). *Finite Volume Methods for Hyperbolic Problems*. Cambridge: Cambridge University Press.

Logan, D.L. (2006). *A First Course in the Finite Element Method*. Florence, KY: Cengage-Engineering.

Tannehill, J.C., Andersen, D.A., and Pletcher, R.H. (1997). *Computational Fluid Mechanics and Heat Transfer*, 2e. Philadelphia, PA: Taylor & Francis.

PROBLEMS

1. Formulate complete PDE problems (specify the equation, space domain, time interval, and boundary and initial conditions) for the following model situations.

a) Conduction heat transfer occurs in a thin rod of length L with insulated sidewalls (see Figure 3.2). Temperature is initially constant $T(0) = T_0$. We are asked to find the temperature distribution in the time period $0 < t < t_1$, during which the left end of the rod is kept at the temperature T_0 and the right end is subject to cooling with the constant heat transfer rate q.

b) Equilibrium temperature distribution is to be found in a rectangular metal plate $0 \leq x \leq L_x, 0 \leq y \leq L_y$. The boundaries $x = 0$ and $x = L_x$ are kept at constant temperatures T_1 and T_2. The boundaries $y = 0$ and $y = L_y$ are thermally insulated.

c) One-dimensional string stretched between the points $(x, y) = (0, 0)$ and $(x, y) = (L_x, 0)$ oscillates elastically during the time period $0 < t < t_1$ (see Figure 3.3). At the initial moment $t = 0$, the deviation of the string from the horizontal line and its velocity are given by the functions $f(x)$ and $g(x)$, respectively.

d) The situation is almost the same as in part (a) of this problem, but with two changes. The temperature of the left end of the rod is not constant but a function of time: $T_0 + \sin(\pi t)$. Also, there are internal heat sources of constant and uniform volumetric heat generation rate Q distributed along the rod.

e) The situation is almost the same as in part (b) of this problem, but the temperatures of the boundaries at $x = 0$ and $x = L$ are now functions of time: $T_1 = \sin(\pi t)$, $T_2 = \sin(2\pi t)$. The solution is to be found at $0 \leq t \leq t_1$. The temperature is equal to 0 in the entire plate at $t = 0$.

2. What are the characteristics of a quasi-linear equation of second order? How are they determined?

3. Classify (determine the type of) the following PDEs of second order:

$$\frac{\partial^2 u}{\partial x \partial y} = 0,$$

$$(x^2 - 1)\frac{\partial^2 u}{\partial x^2} + 2\frac{\partial^2 u}{\partial y^2} = 25(x^3 - 1)\frac{\partial u}{\partial x},$$

$$\frac{\partial^2 u}{\partial t^2} + \frac{\partial^2 u}{\partial x^2} + \frac{\partial u}{\partial x} = \cos(5t).$$

4. Determine the type (hyperbolic, parabolic, or elliptic) of each PDE in Problem 1.

5. Transform the one-dimensional wave equation (3.4) into a system of PDE of the first order, such as (3.26). Find the eigenvalues of matrix R, and determine the type of the system. Does it agree with the type of (3.4) as an equation of second order? *Hint*: Introduce new variables $v = \partial u/\partial t$ and $w = a\partial u/\partial x$.

6. Conduct the same analysis as in the previous problem for the Poisson equation (3.8).

7. Consider the equation

$$\frac{\partial u}{\partial t} + c\frac{\partial u}{\partial x} = \mu\frac{\partial^2 u}{\partial x^2}, \qquad (3.45)$$

where $u = u(x, t)$ and c and μ are positive constants. Would you expect hyperbolic (wavelike) or parabolic (diffusion-like) behavior of the solution?

8. Define the domain of dependence and domain of influence of a point P in a solution of a PDE? How are these domains determined for each type of the PDE of second order?

9. Verify coupling of the linear algebraic equations of the system (3.42)–(3.44) by writing the discretization formula (3.42) for the interior grid points (x_i, t^n), (x_{i+1}, t^n), (x_i, t^{n-1}) and the points (x_0, t^n) and (x_1, t^n) near the boundary.

Programming Exercise   Develop the algorithm for solution of the heat equation Problem (3.36) using the Galerkin spectral method. Use $a = 0.5$ and $T = 50$. Apply the Runge–Kutta scheme of the second order with the time step $\Delta t = 0.02$ to solve the ordinary differential equations for the coefficients (3.40). Compute the solution using 5, 10, 50, and 100 test and trial functions. Compare the results. Analyze how the amplitudes of the coefficients $A_n(t)$ change with n and t.

4

FINITE DIFFERENCE METHOD

The example at the end of the last chapter provides the first illustration of the finite difference method. There is much more to the method than a few simple formulas. We begin a thorough discussion in this chapter. The goal is to introduce the main concepts and answer the questions:

- How do we construct the finite difference schemes?
- How close is the numerical approximation to the exact solution?
- How can we reliably achieve desired accuracy of the approximation?

4.1 COMPUTATIONAL GRID

The first step toward a finite difference approximation is to cover the solution domain together with its boundaries by a computational grid.

4.1.1 Time Discretization

The grid covering the time interval $[t_0, t_{end}]$ of the solution consists of time layers $t^n = t_0 + n\Delta t$. They can be distributed *uniformly*, i.e. with constant

Essential Computational Fluid Dynamics, Second Edition. Oleg Zikanov.
© 2019 John Wiley & Sons, Inc. Published 2019 by John Wiley & Sons, Inc.
Companion Website: www.wiley.com/go/zikanov/essential

time step Δt, or *nonuniformly*, with varying Δt. The second approach is called the *variable time step* method and can be applied to accelerate solution of marching problems by increasing Δt, where possible. We stress the word *possible* since, as discussed in this and the following chapters, there are upper limits on Δt set by the requirements of accuracy and numerical stability.

4.1.2 Space Discretization

If the space domain Ω is one-dimensional (an interval), the computational grid is a one-dimensional set of points $x_i = x_0 + i\Delta x$, with Δx being either constant (*uniform* grids) or a function of x (*nonuniform*, also called *clustered* or *stretched* grids).

In the multidimensional case (Ω is a two-dimensional or three-dimensional domain), one can also choose between structured and unstructured grids. We review these options here but postpone the deep discussion until Chapter 12. In an *unstructured* grid, the nodes (grid points) are placed irregularly (as illustrated in Figure 3.7). Such grids are never used with the classical finite difference methods considered in this chapter. They, however, have become a powerful tool of modern practical CFD analysis in combination with the finite volume discretization, which we will discuss in Chapter 5 and later in Chapter 12. The main advantage of the unstructured grids is their geometric flexibility. Any geometric shape can be covered by the grid, with some nodes placed exactly at the boundaries.

The *structured* grids have traditionally been used with the finite difference discretization. They are simpler to work with than unstructured grids and, thus, represent a better tool for developing numerical schemes and for learning. In such grids, the nodes are placed at the intersections of the lines of a coordinate system. The nodes of a structured grid do not have to be numbered throughout by one index. Instead, two (in the two-dimensional case) or three (in the three-dimensional case) indices can be assigned to every node, so that each index represents the node's location along the corresponding coordinate. As an example, Figure 4.1 shows two-dimensional structured grids built along the lines of a Cartesian and a polar coordinate systems. The point A in Figure 4.1 numbered as (i, j) has Cartesian coordinates (x_i, y_j) and polar coordinates (θ_i, r_j). The point immediately to the right of A, which has coordinates $(x_i + dx, y_j)$ in Figure 4.1a, is identified by indices $(i + 1, j)$, etc.

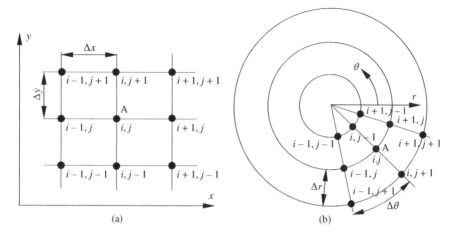

Figure 4.1 Structured discretization grids built along the lines of Cartesian (a) and polar (b) coordinate systems.

A multidimensional structured grid can be uniform with constant grid steps Δx, Δy, and Δz or nonuniform (clustered), in which the grid steps vary in space.

Throughout this chapter and in much of the following discussion, we predominantly use uniform structured grids. The two-dimensional rectangular Cartesian grid shown in Figure 4.1a will be our favorite. The approximation of the function u at the point (x_i, y_j) and time layer t^n will be written as

$$u_{i,j}^n \approx u(x_i, y_j, t^n) \tag{4.1}$$

or, if time is not involved, as

$$u_{i,j} \approx u(x_i, y_j). \tag{4.2}$$

4.2 FINITE DIFFERENCE APPROXIMATION

4.2.1 Approximation of $\partial u / \partial x$

The concept of a finite difference approximation of a partial derivative is, in fact, familiar to us from the elementary calculus. For example, the partial derivative of a function $u(x, y)$ with respect to x taken at the point (x_0, y_0) is defined as

$$\left. \frac{\partial u}{\partial x} \right|_{x_0, y_0} \equiv \lim_{\Delta x \to 0} \frac{u(x_0 + \Delta x, y_0) - u(x_0, y_0)}{\Delta x}. \tag{4.3}$$

In a numerical solution, we do not have access to the function $u(x, y)$ itself. Only the approximate values of u at the grid points (x_i, y_j) are available. Let $x_0 = x_i$, so $x_0 + \Delta x = x_{i+1}$ and $y_0 = y_j$. The closest available approximation of (4.3) is

$$\left. \frac{\partial u}{\partial x} \right|_{i,j} \approx \frac{u(x_{i+1}, y_j) - u(x_i, y_j)}{\Delta x} = \frac{u_{i+1,j} - u_{i,j}}{\Delta x}, \tag{4.4}$$

which, of course, makes sense only if Δx is "sufficiently small."

More rigorously, we have to prove that the expression in the right-hand side of (4.4) approximates the partial derivative and determine the error of the approximation. This can be easily done using the Taylor series expansion of $u(x, y)$. We shall assume that the function is sufficiently smooth, so the Taylor series converges, and use the one-dimensional expansion with respect to x at constant y:

$$u_{i+1,j} = u_{i,j} + \left. \frac{\partial u}{\partial x} \right|_{i,j} \Delta x + \left. \frac{\partial^2 u}{\partial x^2} \right|_{i,j} \frac{(\Delta x)^2}{2!} + \left. \frac{\partial^3 u}{\partial x^3} \right|_{i,j} \frac{(\Delta x)^3}{3!} + \cdots$$
$$+ \left. \frac{\partial^{n-1} u}{\partial x^{n-1}} \right|_{i,j} \frac{(\Delta x)^{n-1}}{(n-1)!} + \left. \frac{\partial^n u}{\partial x^n} \right|_{\zeta,j} \frac{(\Delta x)^n}{n!}, \tag{4.5}$$

where ζ is some point between x_i and x_{i+1}. We now rearrange the formula moving the term with the partial derivative $\partial u / \partial x$ into the left-hand side and everything else into the right-hand side and dividing by Δx:

$$\left. \frac{\partial u}{\partial x} \right|_{i,j} = \frac{u_{i+1,j} - u_{i,j}}{\Delta x} - \left. \frac{\partial^2 u}{\partial x^2} \right|_{i,j} \frac{(\Delta x)}{2!} - \cdots - \left. \frac{\partial^n u}{\partial x^n} \right|_{\zeta,j} \frac{(\Delta x)^{n-1}}{n!}. \tag{4.6}$$

The formula provides all the necessary information. We see that the finite difference approximation (4.4) includes only the first term of the expansion (4.6). All the other terms in the right-hand side are dropped. The error of the approximation is the sum of these dropped terms. We also see that the error decreases with decreasing Δx and vanishes in the limit $\Delta x \to 0$.

4.2.2 Truncation Error, Consistency, and Order of Approximation

The formula (4.6) can be rewritten as

$$\left. \frac{\partial u}{\partial x} \right|_{i,j} = \frac{u_{i+1,j} - u_{i,j}}{\Delta x} + \text{TE}, \tag{4.7}$$

where we used the notation TE for the *truncation error* of discretization of a partial derivative. It contains the rest of the series dropped in the finite difference approximation and, thus, shows the difference between the exact expression (4.6) and the finite difference approximation (4.4). The truncation error is the key characteristic of the accuracy of a finite difference scheme. Although it cannot be evaluated directly (since the function u is usually an unknown solution of a PDE problem), it provides vital information about *how fast the error of approximation decreases with decreasing grid step.*

At this point, we need to recall some facts of basic calculus. The notation $f = O(\alpha)$ (f is of the order of α) means that f remains smaller than $K\alpha$ with some constant K as $\alpha \to 0$. In other words, f decreases with α not slower than α itself.

In our case, the small parameter is Δx. Each term of the truncation error is of the order of a certain power of Δx:

$$\left.\frac{\partial^2 u}{\partial x^2}\right|_{i,j} \frac{(\Delta x)}{2!} = O(\Delta x), \quad \left.\frac{\partial^3 u}{\partial x^3}\right|_{i,j} \frac{(\Delta x)^2}{3!} = O((\Delta x)^2),$$

$$\dots, \quad \left.\frac{\partial^n u}{\partial x^n}\right|_{\zeta} \frac{(\Delta x)^{n-1}}{n!} = O\left((\Delta x)^{n-1}\right).$$

As Δx decreases, the higher-order terms become negligible in comparison with the first term, and the entire truncation error is evaluated as

$$\text{TE} = O(\Delta x).$$

We can rewrite (4.6) again, this time as

$$\left.\frac{\partial u}{\partial x}\right|_{i,j} = \frac{u_{i+1,j} - u_{i,j}}{\Delta x} + O(\Delta x). \tag{4.8}$$

This version illustrates the fact that only the lowest-order term of the truncation error is important when the grid step Δx is small. The order of this term is called *the order of truncation error or the order of approximation of the finite difference scheme*. It defines how fast the error of the approximation decreases with the grid step.

Another important concept is that of *consistency* of a finite difference approximation. It is obvious that the approximation makes sense only if the truncation error vanishes at $\Delta x \to 0$, that is, if the scheme has, at least, the first order of approximation. A scheme satisfying this requirement is called consistent.

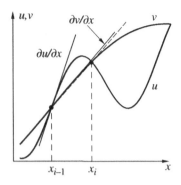

Figure 4.2 Effect of strong gradients of function on truncation error. The finite difference approximation (4.4) (the slope of the dashed line) accurately reproduces $\partial v/\partial x$ but not $\partial u/\partial x$.

One consideration should be taken into account in practical computations, where the limit $\Delta x \to 0$ is never attained. In addition to the order of approximation and the size of grid step, the magnitude of the truncation error is determined by the amplitudes of the derivatives of the approximated function: $\partial u/\partial x$, $\partial^2 u/\partial x^2$, etc. If the function has strong gradients, the amplitudes of derivatives are large, and the truncation error can be significant even when the consistency is satisfied and Δx is not particularly large.

A simple view on this phenomenon is illustrated in Figure 4.2. On the same computational grid, the derivative of function u characterized by strong variations over small distances is approximated with lower accuracy than the derivative of v. The figure also shows that the rather poorly defined term "small distances" should be understood as "small in comparison with the grid step." It is clear that accurate approximation of derivatives requires a grid with steps smaller – desirably several times smaller – than the smallest distance on which the function experiences significant variation.

We will give one more illustration of the effect by approximating the first derivative of $u(x) = e^{\lambda x}$, $\lambda = $const. (the y-dependency of u is omitted for brevity). The largest term of the truncation error of the finite difference approximation (4.4) is (see (4.6))

$$-\left.\frac{\partial^2 u}{\partial x^2}\right|_i \frac{\Delta x}{2} = -\frac{\lambda^2 e^{\lambda x_i}}{2}\Delta x,$$

so that the magnitude of the truncation error relative to the value of the function is $\sim \lambda^2 \Delta x$. We see that keeping the same magnitude of error for a faster changing function with larger λ requires that the grid step is reduced as $\Delta x \sim 1/\lambda^2$.

One may think that the discussion above is impractical since the solution is not known when we design a grid. In truth, the areas where the solution has sharp gradients, for example, the boundary layers in a fluid flow, can often be predicted in advance (see Chapter 12). In the situations when such areas appear where they are not expected, and there is a suspicion of unacceptably large truncation error, additional computations with reduced grid steps can be warranted. This matter is further discussed in Chapter 13.

4.2.3 Other Formulas for $\partial u/\partial x$: Evaluation of the Order of Approximation

The *forward difference* formula (4.4) is not the only possible approximation of $\partial u/\partial x$. In fact, infinitely many approximations can be developed. For example, we can use the *backward difference*

$$\left.\frac{\partial u}{\partial x}\right|_{i,j} \approx \frac{u_{i,j} - u_{i-1,j}}{\Delta x} \tag{4.9}$$

or the *central difference*

$$\left.\frac{\partial u}{\partial x}\right|_{i,j} \approx \frac{u_{i+1,j} - u_{i-1,j}}{2\Delta x}. \tag{4.10}$$

The formulas can be given a simple graphic interpretation as approximations of the slope of the curve $u(x, y = \text{const.})$ (see Figure 4.3). Such an

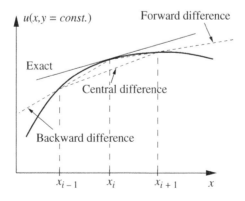

Figure 4.3 Approximation of the partial derivative $\partial u(x, y)/\partial x$ by finite difference formulas (4.4), (4.9), and (4.10).

interpretation, however, is impossible for many other schemes and does not provide information on the error of the approximation. A formal procedure of quantitative assessment of finite difference formulas is needed. The most commonly applied and simplest way is to apply the Taylor series expansions around the point (x_i, y_j) at which the derivative is approximated. For example, we can substitute

$$u_{i-1,j} = u_{i,j} - \frac{\partial u}{\partial x}\bigg|_{i,j} \Delta x + \frac{1}{2!}\frac{\partial^2 u}{\partial x^2}\bigg|_{i,j} (\Delta x)^2 - \frac{1}{3!}\frac{\partial^3 u}{\partial x^3}\bigg|_{i,j} (\Delta x)^3 + O\left((\Delta x)^4\right)$$

into (4.9) to obtain

$$\frac{\partial u}{\partial x}\bigg|_{i,j} = \frac{u_{i,j} - u_{i-1,j}}{\Delta x} + O(\Delta x). \tag{4.11}$$

Substitution of the same expression for $u_{i-1,j}$ and

$$u_{i+1,j} = u_{i,j} + \frac{\partial u}{\partial x}\bigg|_{i,j} \Delta x + \frac{1}{2!}\frac{\partial^2 u}{\partial x^2}\bigg|_{i,j} (\Delta x)^2 + \frac{1}{3!}\frac{\partial^3 u}{\partial x^3}\bigg|_{i,j} (\Delta x)^3 + O\left((\Delta x)^4\right)$$

into (4.10) shows that for the central difference there is incidental cancelation of the first-order terms (proportional to $(\partial^2 u/\partial x^2)|_{i,j}$), so

$$\frac{\partial u}{\partial x}\bigg|_{i,j} = \frac{u_{i+1,j} - u_{i-1,j}}{2\Delta x} + \frac{1}{3!}\frac{\partial^3 u}{\partial x^3}\bigg|_{i,j} (\Delta x)^2 + O\left((\Delta x)^3\right)$$

$$= \frac{u_{i+1,j} - u_{i-1,j}}{2\Delta x} + O\left((\Delta x)^2\right). \tag{4.12}$$

Let us see what we have. First, all three schemes – forward, backward, and central differences – are consistent since the truncation errors vanish as $\Delta x \rightarrow 0$. Second, the schemes have different orders of approximation: first order for the backward and forward differences (4.4) and (4.9) and second order for the central difference (4.10). This is of fundamental importance, since it shows that the TE of the central difference scheme decreases faster with Δx than the TE of the backward and forward schemes. For example, if Δx is reduced from 10^{-2} to 10^{-3}, the error decreases by a factor of 10 for a first-order approximation but by a factor of 100 for a second-order approximation. We can say that the central difference scheme (4.10) inevitably becomes more accurate than the forward and backward schemes (4.4) and (4.9) at sufficiently small grid steps Δx.

Two other finite difference formulas for the first derivative $\partial u / \partial x$ are

$$\frac{\partial u}{\partial x}\bigg|_{i,j} = \frac{-3u_{i,j} + 4u_{i+1,j} - u_{i+2,j}}{2\Delta x} + O\left((\Delta x)^2\right) \qquad (4.13)$$

and

$$\frac{\partial u}{\partial x}\bigg|_{i,j} = \frac{3u_{i,j} - 4u_{i-1,j} + u_{i-2,j}}{2\Delta x} + O\left((\Delta x)^2\right). \qquad (4.14)$$

They are often used at the boundaries, where the values of u on only one side of the (i,j) point are available, but the second order of accuracy is desired.

4.2.4 Schemes of Higher Order for First Derivative

Schemes of the order higher than second are possible and, sometimes, desirable. Such schemes are, evidently, more accurate than the lower-order schemes. At the same time, the solution algorithms based on higher-order approximations are more difficult to program and may require larger amount of computations. They also tend to have more stringent numerical stability requirements on the time step (the numerical stability is discussed in Chapter 6).

As an example, the central fourth-order scheme is

$$\frac{\partial u}{\partial x}\bigg|_{i,j} = \frac{-u_{i+2,j} + 8u_{i+1,j} - 8u_{i-1,j} + u_{i-2,j}}{12\Delta x} + O\left((\Delta x)^4\right). \qquad (4.15)$$

A separate class of finite difference approximations is the so-called Padé or compact schemes. In these schemes, the approximation of the derivative at the (i,j)-point is not explicitly expressed in terms of the values of u at neighboring nodes, but is found as a part of the solution of a system of coupled linear algebraic equations. As an example, we give the fourth-order scheme, where the values of $\partial u / \partial x$ at different grid points are found as a solution of

$$\frac{1}{3}\frac{\partial u}{\partial x}\bigg|_{i+1,j} + \frac{4}{3}\frac{\partial u}{\partial x}\bigg|_{i,j} + \frac{1}{3}\frac{\partial u}{\partial x}\bigg|_{i-1,j} = \frac{u_{i+1,j} - u_{i-1,j}}{\Delta x}, \quad i = 1, \dots, N. \quad (4.16)$$

4.2.5 Higher-Order Derivatives

Numerous finite difference approximations exist for the higher-order derivatives. The examples below are given for the x-derivatives. The same formulas can be applied for the derivatives with respect to y, z, and t after trivial substitution of, say, y and j for x and i.

Particularly important are the three-point forward, backward, and central schemes for the second derivative:

$$\left.\frac{\partial^2 u}{\partial x^2}\right|_{i,j} = \frac{u_{i,j} - 2u_{i+1,j} + u_{i+2,j}}{(\Delta x)^2} + O(\Delta x) \tag{4.17}$$

$$\left.\frac{\partial^2 u}{\partial x^2}\right|_{i,j} = \frac{u_{i,j} - 2u_{i-1,j} + u_{i-2,j}}{(\Delta x)^2} + O(\Delta x) \tag{4.18}$$

$$\left.\frac{\partial^2 u}{\partial x^2}\right|_{i,j} = \frac{u_{i-1,j} - 2u_{i,j} + u_{i+1,j}}{(\Delta x)^2} + O\left((\Delta x)^2\right). \tag{4.19}$$

Some other schemes are

$$\left.\frac{\partial^2 u}{\partial x^2}\right|_{i,j} = \frac{-u_{i+3,j} + 4u_{i+2,j} - 5u_{i+1,j} + 2u_{i,j}}{(\Delta x)^2} + O\left((\Delta x)^2\right) \tag{4.20}$$

$$\left.\frac{\partial^3 u}{\partial x^3}\right|_{i,j} = \frac{u_{i+2,j} - 2u_{i+1,j} + 2u_{i-1,j} - u_{i-2,j}}{2(\Delta x)^3} + O\left((\Delta x)^2\right) \tag{4.21}$$

$$\left.\frac{\partial^4 u}{\partial x^4}\right|_{i,j} = \frac{u_{i+2,j} - 4u_{i+1,j} + 6u_{i,j} - 4u_{i-1,j} + u_{i-2,j}}{(\Delta x)^4} + O\left((\Delta x)^2\right) \tag{4.22}$$

$$\left.\frac{\partial^2 u}{\partial x^2}\right|_{i,j} = \frac{-u_{i+2,j} + 16u_{i+1,j} - 30u_{i,j} + 16u_{i-1,j} - u_{i-2,j}}{12(\Delta x)^2} + O\left((\Delta x)^4\right). \tag{4.23}$$

The consistency and the order of approximation of the schemes can be verified in the same way as in the previous sections: by developing the Taylor series expansions around (x_i, y_j) and substituting them into the right-hand sides of (4.17)–(4.23).

The finite difference formulas for higher-order derivatives can be considered as results of repeated application of simpler schemes, such as the elementary forward, backward, and central formulas (4.4), (4.9), and (4.10) applied to the first derivatives. As an example, let us derive the central difference approximation of the second derivative (4.19) from the central scheme (4.10). Using the shorthand notation $f = \partial u/\partial x$, we write the second derivative as $\partial^2 u/\partial x^2 = \partial/\partial x(\partial u/\partial x) = \partial f/\partial x$ and apply (4.10) to approximate $\partial f/\partial x$ at (x_i, y_j). The formula is modified a little. We use the half step $\Delta x/2$ instead of the full step Δx and the values of f at the half-integer points

$x_{i-1/2} = x_i - \Delta x/2$ and $x_{i+1/2} = x_i + \Delta x/2$. These are fictitious grid points, and, as will be seen very soon, we do not need values of u at them. The result is

$$\left.\frac{\partial^2 u}{\partial x^2}\right|_{i,j} \approx \frac{f_{i+1/2,j} - f_{i-1/2,j}}{\Delta x} = \frac{(\partial u/\partial x)_{i+1/2,j} - (\partial u/\partial x)_{i-1/2,j}}{\Delta x}. \quad (4.24)$$

Derivatives $\partial u/\partial x$ can be approximated using the central difference formula (4.10) with half step $\Delta x/2$ as

$$\left.\frac{\partial u}{\partial x}\right|_{i+1/2,j} \approx \frac{u_{i+1,j} - u_{i,j}}{\Delta x}, \quad \left.\frac{\partial u}{\partial x}\right|_{i-1/2,j} \approx \frac{u_{i,j} - u_{i-1,j}}{\Delta x}. \quad (4.25)$$

Substitution into (4.24) gives the central difference formula (4.19), which, as we see now, is a result of repeated use of (4.10). Since the truncation errors of (4.24) and (4.25) are $O((\Delta x/2)^2) = O((\Delta x)^2)$ (see (4.12)), the scheme (4.19) is of the second order.

4.2.6 Mixed Derivatives

The mixed derivatives can be approximated by applying the formulas already listed, first in one direction and then in the other. For example, let us approximate the derivative $\partial^2 u/\partial x \partial y$ on the two-dimensional uniform structured grid shown in Figure 4.1a. We rewrite it as $\partial(\partial u/\partial y)/\partial x$ and approximate the x-derivative by the forward difference (4.4)

$$\left.\frac{\partial}{\partial x}\left(\frac{\partial u}{\partial y}\right)\right|_{i,j} = \frac{(\partial u/\partial y)|_{i+1,j} - (\partial u/\partial y)|_{i,j}}{\Delta x} + O(\Delta x).$$

Each y-derivative also has to be approximated. We apply the backward difference (4.9)

$$\left.\frac{\partial u}{\partial y}\right|_{i+1,j} = \frac{u_{i+1,j} - u_{i+1,j-1}}{\Delta y} + O(\Delta y), \quad \left.\frac{\partial u}{\partial y}\right|_{i,j} = \frac{u_{i,j} - u_{i,j-1}}{\Delta y} + O(\Delta y).$$

Substituting into the expression above, we obtain

$$\left.\frac{\partial^2 u}{\partial x \partial y}\right|_{i,j} = \frac{1}{\Delta x}\left(\frac{u_{i+1,j} - u_{i+1,j-1}}{\Delta y} - \frac{u_{i,j} - u_{i,j-1}}{\Delta y}\right) + O(\Delta x, \Delta y), \quad (4.26)$$

where $O(\Delta x, \Delta y)$ stands for a combination of the truncation errors of the orders of Δx and Δy.

Other combinations of finite difference formulas lead to different approximations. For example, backward differentiation in x and forward in y results in

$$\frac{\partial^2 u}{\partial x \partial y}\bigg|_{i,j} = \frac{1}{\Delta x}\left(\frac{u_{i,j+1} - u_{i,j}}{\Delta y} - \frac{u_{i-1,j+1} - u_{i-1,j}}{\Delta y}\right) + O(\Delta x, \Delta y), \quad (4.27)$$

while taking forward formula in x and central in y, we obtain

$$\frac{\partial^2 u}{\partial x \partial y}\bigg|_{i,j} = \frac{1}{\Delta x}\left(\frac{u_{i+1,j+1} - u_{i+1,j-1}}{2\Delta y} - \frac{u_{i,j+1} - u_{i,j-1}}{2\Delta y}\right) + O\left(\Delta x, (\Delta y)^2\right).$$
$$(4.28)$$

At last, the approximation of the second order in both directions is generated if we use central differences in both x and y:

$$\frac{\partial^2 u}{\partial x \partial y}\bigg|_{i,j} = \frac{1}{2\Delta x}\left(\frac{u_{i+1,j+1} - u_{i+1,j-1}}{2\Delta y} - \frac{u_{i-1,j+1} - u_{i-1,j-1}}{2\Delta y}\right)$$
$$+ O\left((\Delta x)^2, (\Delta y)^2\right). \quad (4.29)$$

Verification of these formulas requires two-dimensional Taylor series or sequential application of Taylor expansion in the x- and y-directions. The accuracy of the approximation is determined by the terms of the lowest order with respect to both Δx and Δy. The truncation error is, therefore, expressed as a function of Δx and Δy, and the order of approximation is defined as, for example, "first in x, second in y" for (4.28) or "second in x and y" for (4.29).

4.2.7   Finite Difference Approximation on Nonuniform Grids

The finite difference formulas we have introduced so far are fully valid on uniform grids, i.e. grids with constant steps Δx and Δy. A uniform grid is often not the best choice. As we have seen at the end of Section 4.2.2, achieving the same level of accuracy requires smaller grid steps in the areas of strong gradients of the approximated function than in the areas where the gradients are weak. It makes sense to use a nonuniform grid with the grid steps reduced only where it is necessary. The matter is further discussed in Chapter 12. Here we show on simple examples how the performance of the standard discretization formulas changes on nonuniform grids.

We first consider approximations of $\partial u / \partial x$ at (x_i, y_j). Let the grid be nonuniform in the x-direction, so $x_{i+1} - x_i \neq x_i - x_{i-1} \neq x_{i-1} - x_{i-2} \cdots$. The Taylor series expansions around x_i become

$$u_{i+1,j} = u_{i,j} + \left.\frac{\partial u}{\partial x}\right|_{i,j} \Delta^+ + \frac{1}{2} \left.\frac{\partial^2 u}{\partial x^2}\right|_{i,j} (\Delta^+)^2$$

$$+ \frac{1}{6} \left.\frac{\partial^3 u}{\partial x^3}\right|_{i,j} (\Delta^+)^3 + O\left((\Delta^+)^4\right), \tag{4.30}$$

$$u_{i-1,j} = u_{i,j} - \left.\frac{\partial u}{\partial x}\right|_{i,j} \Delta^- + \frac{1}{2} \left.\frac{\partial^2 u}{\partial x^2}\right|_{i,j} (\Delta^-)^2$$

$$- \frac{1}{6} \left.\frac{\partial^3 u}{\partial x^3}\right|_{i,j} (\Delta^-)^3 + O\left((\Delta^-)^4\right), \tag{4.31}$$

where we use the shorthand notation $\Delta^+ = x_{i+1} - x_i$, $\Delta^- = x_i - x_{i-1}$. In the following analysis we assume that the grid steps Δ^+ and Δ^-, while different, have the same order of magnitude, so we can introduce some typical grid step Δx, such that $\Delta^+ \sim \Delta^- \sim \Delta x$.

Rearranging the terms in (4.30) and (4.31), we find the formulas equivalent to (4.4) and (4.9) derived on a uniform grid:

$$\left.\frac{\partial u}{\partial x}\right|_{i,j} = \frac{u_{i+1,j} - u_{i,j}}{x_{i+1} - x_i} + O(\Delta x), \tag{4.32}$$

$$\left.\frac{\partial u}{\partial x}\right|_{i,j} = \frac{u_{i,j} - u_{i-1,j}}{x_i - x_{i-1}} + O(\Delta x). \tag{4.33}$$

We see that the forward and backward approximations remain consistent and of the first order. The magnitude of the lowest-order term of the truncation error remains proportional to $\partial^2 u / \partial x^2 |_{i,j}$. We conclude that switching to a nonuniform grid does not affect the performance of the two-point first-order schemes (4.4) and (4.9).

The outcome is different in the case of the central difference scheme (4.10). Subtracting (4.31) from (4.30) we obtain

$$u_{i+1,j} - u_{i-1,j} = \left.\frac{\partial u}{\partial x}\right|_{i,j} (\Delta^+ + \Delta^-) + \frac{1}{2} \left.\frac{\partial^2 u}{\partial x^2}\right|_{i,j} [(\Delta^+)^2 - (\Delta^-)^2]$$

$$+ \frac{1}{6} \left.\frac{\partial^3 u}{\partial x^3}\right|_{i,j} ((\Delta^+)^3 - (\Delta^-)^3) + O\left((\Delta x)^4\right). \tag{4.34}$$

Denoting the difference between the two neighboring grid steps as $\delta = \Delta^+ - \Delta^-$ and dividing (4.34) by $(x_{i+1} - x_{i-1}) = \Delta^+ + \Delta^-$, we obtain

$$\left.\frac{\partial u}{\partial x}\right|_{i,j} = \frac{u_{i+1,j} - u_{i-1,j}}{x_{i+1} - x_{i-1}} - \frac{1}{2}\left.\frac{\partial^2 u}{\partial x^2}\right|_{i,j}\delta + O\left((\Delta x)^2\right). \qquad (4.35)$$

Unlike the situation with a uniform grid, there is no cancelation of the terms proportional to $\partial^2 u/\partial x^2|_{i,j}$. The central difference formula becomes formally of the first order of approximation. If, however, the difference between the neighboring steps is small:

$$\delta \ll x_{i+1} - x_i \sim x_i - x_{i-1} \sim \Delta x, \qquad (4.36)$$

the first-order term in (4.35) is also small, and the scheme approaches the second order of approximation.

The situation is more complex for approximations of higher derivatives. As an example, we consider the central difference approximation (4.19) of $\partial^2 u/\partial x^2|_{i,j}$. At first glance, it is unclear what to take as the grid step Δx in the denominator. It is easy to see by direct substitution of (4.30) (4.31) that simply replacing Δx by Δ^- or Δ^+ would lead to an inconsistent scheme.

In order to derive a consistent version of (4.19) on a nonuniform grid, we repeat the procedure demonstrated in Section 4.2.5 as a derivation of (4.19) (see (4.24) and (4.25)). We pick the half-integer points so that they are located exactly in the middle between the grid points: $x_{i-1/2} = (x_i - x_{i-1})/2$, $x_{i+1/2} = (x_i - x_{i+1})/2$. The derivative $\partial u/\partial x$ at these points is approximated by central differences as

$$\left.\frac{\partial u}{\partial x}\right|_{i+1/2,j} \approx \frac{u_{i+1,j} - u_{i,j}}{\Delta^+}, \quad \left.\frac{\partial u}{\partial x}\right|_{i-1/2,j} \approx \frac{u_{i,j} - u_{i-1,j}}{\Delta^-}. \qquad (4.37)$$

The approximations are of the second order in Δ^+ and Δ^-. The second derivative is evaluated as in (4.35):

$$\left.\frac{\partial^2 u}{\partial x^2}\right|_{i,j} \approx \frac{(\partial u/\partial x)_{i+1/2,j} - (\partial u/\partial x)_{i-1/2,j}}{x_{i+1/2} - x_{i-1/2}}$$

$$= \left(\frac{u_{i+1,j} - u_{i,j}}{\Delta^+} - \frac{u_{i,j} - u_{i-1,j}}{\Delta^-}\right)\left(\frac{\Delta^+ + \Delta^-}{2}\right)^{-1}. \qquad (4.38)$$

Similar to (4.35), the scheme is of the first order of approximation but approaches the second order when δ is small.

The examples illustrate the general rule that transition to a nonuniform grid should be done with care. The accuracy of the simple schemes for the first derivative based on two neighboring points is not affected. Other schemes may become inconsistent or of reduced order of accuracy. In the latter case, the loss of accuracy can be diminished by choosing grids with low degree of nonuniformity (with the ratio between the neighboring grid steps close to 1).

In Section 4.3 and Chapter 12, we will discuss how to design the discretization schemes avoiding these issues and having the desired accuracy on nonuniform grids.

4.3 DEVELOPMENT OF FINITE DIFFERENCE SCHEMES

How do we develop finite difference schemes? The simplest and, probably, most efficient way is to look up the desired formula in a CFD book. If such a solution is satisfactory, the reader can skip this section without much harm for future understanding. If this is not the case, one of the two methods considered here can be used. There is the third method based on the principles of finite volume approximation. Recognizing its importance for modern CFD, we describe it separately and in more detail in Chapter 5.

The general task is formulated as follows: *To develop a finite difference approximation of a given order of accuracy for a partial derivative using a given set of grid points.*

4.3.1 Taylor Series Expansions

This method is a generalization of the procedure applied in Section 4.2 to develop our first finite difference formulas. We will demonstrate it on two examples using, for simplicity, a uniform grid. In the first, we are asked to develop a finite difference scheme of the second order for $\partial^2 u/\partial x^2$ based on the values of u at three points: x_i, $x_{i-1} = x_i - \Delta x$, and $x_{i+1} = x_i + \Delta x$. The general structure of such a scheme is

$$\frac{\partial^2 u}{\partial x^2}\bigg|_i = au_{i-1} + bu_i + cu_{i+1} + O\left((\Delta x)^2\right), \qquad (4.39)$$

where a, b, and c are undetermined constants, and we for brevity drop the indices other than those along the x-coordinate. We use the Taylor series expansions of u_{i-1} and u_{i+1} around x_i:

$$u_{i-1} = u_i - \left.\frac{\partial u}{\partial x}\right|_i \Delta x + \left.\frac{1}{2}\frac{\partial^2 u}{\partial x^2}\right|_i (\Delta x)^2 - \left.\frac{1}{6}\frac{\partial^3 u}{\partial x^3}\right|_i (\Delta x)^3$$
$$+ O\left((\Delta x)^4\right) \tag{4.40}$$

$$u_{i+1} = u_i + \left.\frac{\partial u}{\partial x}\right|_i \Delta x + \left.\frac{1}{2}\frac{\partial^2 u}{\partial x^2}\right|_i (\Delta x)^2 + \left.\frac{1}{6}\frac{\partial^3 u}{\partial x^3}\right|_i (\Delta x)^3$$
$$+ O\left((\Delta x)^4\right). \tag{4.41}$$

The series are truncated at the fourth-order term. In general, the truncation limit is determined experimentally, but it should be not lower than the sum of the order of approximated derivative and the desired order of the truncation error.

Substitution of the expansions into (4.39) gives

$$au_{i-1} + bu_i + cu_{i+1} = (a+b+c)u_i + (-a+c)\left.\frac{\partial u}{\partial x}\right|_i \Delta x$$
$$+ (a+c)\left.\frac{1}{2}\frac{\partial^2 u}{\partial x^2}\right|_i (\Delta x)^2 + (-a+c)\left.\frac{1}{6}\frac{\partial^3 u}{\partial x^3}\right|_i (\Delta x)^3$$
$$+ (a+c)O\left((\Delta x)^4\right). \tag{4.42}$$

To obtain the desired scheme, we require that the coefficients at u_i and $\partial u/\partial x|_i$ are 0 and the coefficient at $\partial^2 u/\partial x^2|_i$ is 1:

$$\begin{cases} a+b+c &= 0 \\ -a+c &= 0 \\ (a+c)\frac{\Delta x^2}{2} &= 1. \end{cases}$$

The solution of this system is

$$a = c = \frac{1}{\Delta x^2}, \quad b = -\frac{2}{\Delta x^2}.$$

The expression (4.42) reduces to the final answer

$$\left.\frac{\partial^2 u}{\partial x^2}\right|_i = \frac{u_{i-1} - 2u_i + u_{i+1}}{\Delta x^2} + 2\frac{O\left((\Delta x)^4\right)}{\Delta x^2}$$
$$= \frac{u_{i-1} - 2u_i + u_{i+1}}{\Delta x^2} + O\left((\Delta x)^2\right), \tag{4.43}$$

which is, of course, the central difference scheme (4.19).

As another example, we derive a one-sided formula that approximates $\partial u/\partial x$ on three grid points with the truncation error $O\left((\Delta x)^2\right)$. Such schemes are useful for Neumann boundary conditions (see Section 4.4.2). The approximation is done at x_i using u_{i-2}, u_{i-1}, and u_i. Again, we start with the general formula

$$\left.\frac{\partial u}{\partial x}\right|_i = au_{i-2} + bu_{i-1} + cu_i + O\left((\Delta x)^2\right) \qquad (4.44)$$

and use the Taylor series expansions of u_{i-2} and u_{i-1} around x_i:

$$u_{i-2} = u_i - 2\left.\frac{\partial u}{\partial x}\right|_i \Delta x + 2\left.\frac{\partial^2 u}{\partial x^2}\right|_i (\Delta x)^2 + O\left((\Delta x)^3\right)$$

$$u_{i-1} = u_i - \left.\frac{\partial u}{\partial x}\right|_i \Delta x + \frac{1}{2}\left.\frac{\partial^2 u}{\partial x^2}\right|_i (\Delta x)^2 + O\left((\Delta x)^3\right).$$

The linear combination in the right-hand side of (4.44) is

$$au_{i-2} + bu_{i-1} + cu_i = (a + b + c)u_i + (-2a - b)\left.\frac{\partial u}{\partial x}\right|_i \Delta x$$

$$+ \left(2a + \frac{b}{2}\right)\left.\frac{\partial^2 u}{\partial x^2}\right|_i (\Delta x)^2 + (a + b)O\left((\Delta x)^3\right). \quad (4.45)$$

Requiring that the coefficients at u_i and $\partial^2 u/\partial x^2|_i$ are 0 and the coefficient at $\partial u/\partial x|_i$ is 1, we obtain

$$\begin{cases} a + b + c & = 0 \\ (-2a - b)\Delta x & = 1 \\ 2a + \frac{b}{2} & = 0. \end{cases}$$

The solution is

$$a = \frac{1}{2\Delta x}, \quad b = -\frac{2}{\Delta x}, \quad c = \frac{3}{2\Delta x}.$$

Substitution into (4.45) and rearranging produces the desired formula

$$\left.\frac{\partial u}{\partial x}\right|_i = \frac{u_{i-2} - 4u_{i-1} + 3u_i}{2\Delta x} + O\left((\Delta x)^2\right). \qquad (4.46)$$

4.3.2 Polynomial Fitting

The idea of the method is to assume that, locally, the unknown function is approximated by a polynomial of a certain order. The polynomial is "fitted" to the function's behavior by determining its coefficients so that the polynomial is equal to the function exactly at the selected grid points.

As an example, we consider the approximation of the boundary condition

$$\left.\frac{\partial u}{\partial x}\right|_{x=0} = q \tag{4.47}$$

by a scheme of the fourth order. We use the same uniform grid as before and assume that the boundary is at the grid point $x_0 = 0$. The neighboring points are $x_1 = \Delta x$, $x_2 = 2\Delta x$, etc.

First, we locally represent $u(x)$ as a polynomial of third degree

$$u(x) = a + bx + cx^2 + dx^3. \tag{4.48}$$

Obviously, the condition (4.47) applied to (4.48) means

$$b = q. \tag{4.49}$$

What remains is to find a, c, and d such that the polynomial fits the values of u at the grid points adjacent to the boundary. The number of points to be taken must be equal to the number of the still unknown coefficients. Calculating (4.48) at the three nearest points, we obtain the system

$$\begin{cases} u_1 = a + q\Delta x + c(\Delta x)^2 + d(\Delta x)^3 \\ u_2 = a + 2q\Delta x + 4c(\Delta x)^2 + 8d(\Delta x)^3 \\ u_3 = a + 3q\Delta x + 9c(\Delta x)^2 + 27d(\Delta x)^3 \end{cases},$$

which can be solved for a, c, and d. Doing that and calculating the polynomial at the boundary x_0, we find the approximation of (4.47):

$$u_0 = \frac{1}{11}(18u_1 - 9u_2 + 2u_3 - 6q\Delta x) + O\left((\Delta x)^4\right). \tag{4.50}$$

Note that the order of the truncation error does not follow directly from the derivation. It has to be determined separately in a Taylor series analysis.

4.3.3   Development on Nonuniform Grids

An effective way of designing finite difference schemes for nonuniform grids is described in Chapter 12. It is based on the transformation of the coordinate system (mapping) in such a way that the grid becomes uniform in the new coordinates. One can also directly apply the undetermined coefficient techniques described in Sections 4.3.1 and 4.3.2 modifying the algebra for a variable grid step.

The transition from uniform to nonuniform grid typically results in more complex finite difference formulas. Keeping the same order of approximation may require including function values at additional grid points. This is, in particular, true for the central difference formulas, for which the cancelation of higher-order terms does not occur.

Let us, as an example, again consider the second-order approximation of $\partial^2 u/\partial x^2$. On a nonuniform grid, the Taylor expansion series (4.40) (4.41) become

$$u_{i-1} = u_i - \frac{\partial u}{\partial x}\Big|_i \Delta^- + \frac{1}{2}\frac{\partial^2 u}{\partial x^2}\Big|_i (\Delta^-)^2 - \frac{1}{6}\frac{\partial^3 u}{\partial x^3}\Big|_i (\Delta^-)^3 + O\left((\Delta^-)^4\right),$$

$$u_{i+1} = u_i + \frac{\partial u}{\partial x}\Big|_i \Delta^+ + \frac{1}{2}\frac{\partial^2 u}{\partial x^2}\Big|_i (\Delta^+)^2 + \frac{1}{6}\frac{\partial^3 u}{\partial x^3}\Big|_i (\Delta^+)^3 + O\left((\Delta^+)^4\right),$$

where the shorthand notation $\Delta^- = x_i - x_{i-1}$, $\Delta^+ = x_{i+1} - x_i$ is used. We attempt to design a scheme of the type (4.39). Substituting the expansions into $au_{i-1} + bu_i + cu_{i+1}$ and requiring that the coefficients at u_i, $du/dx|_i$, and $d^3u/dx^3|_i$ are 0, while the coefficient at $d^2u/dx^2|_i$ is 1, we obtain the system of four equations:

$$\begin{cases} a + b + c & = 0 \\ -a\Delta^- + c\Delta^+ & = 0 \\ a(\Delta^-)^2 + c(\Delta^+)^2 & = 1 \\ -a(\Delta^-)^3 + c(\Delta^+)^3 & = 0, \end{cases} \tag{4.51}$$

which does not have a solution. We must either agree to the first order of approximation (i.e. drop the last equation in (4.51)) or add another point and another undetermined coefficient, i.e. derive a scheme

$$\frac{\partial^2 u}{\partial x^2}\Big|_i = au_{i-1} + bu_i + cu_{i+1} + du_{i+2} + O\left((\Delta x)^2\right) \tag{4.52}$$

or

$$\frac{\partial^2 u}{\partial x^2}\Big|_i = du_{i-2} + au_{i-1} + bu_i + cu_{i+1} + O\left((\Delta x)^2\right). \tag{4.53}$$

4.4 FINITE DIFFERENCE APPROXIMATION OF PARTIAL DIFFERENTIAL EQUATIONS

4.4.1 Approach and Examples

A finite difference representation of an entire partial differential equation (PDE) is obtained by replacing each term of the equation by its finite difference approximation. It is imperative that *all the terms are approximated at the same grid point and the same time layer*.

Let us consider, as our first example, the modified heat equation

$$\frac{\partial u}{\partial t} = a^2 \frac{\partial^2 u}{\partial x^2} + f(x, t) \qquad (4.54)$$

and derive, in a more educated manner, the scheme introduced at the end of Chapter 3.

The scheme is developed on a uniform grid with steps Δx and Δt (see Figure 3.8). We write the approximation at the point x_i and time layer t^n. A forward difference of the first order is used for the time derivative:

$$\left. \frac{\partial u}{\partial t} \right|_i^n = \frac{u_i^{n+1} - u_i^n}{\Delta t} + O(\Delta t). \qquad (4.55)$$

The central difference of the second order is used for the x-derivative:

$$\left. \frac{\partial^2 u}{\partial x^2} \right|_i^n = \frac{u_{i+1}^n - 2u_i^n + u_{i-1}^n}{(\Delta x)^2} + O\left((\Delta x)^2\right). \qquad (4.56)$$

The nonderivative term in the right-hand side is represented by its value at the discretization point as $f_i^n = f(x_i, t^n)$.

Substitution into the PDE yields

$$\frac{u_i^{n+1} - u_i^n}{\Delta t} + O(\Delta t) = a^2 \frac{u_{i+1}^n - 2u_i^n + u_{i-1}^n}{(\Delta x)^2} + O\left((\Delta x)^2\right) + f_i^n,$$

or, after truncation,

$$\frac{u_i^{n+1} - u_i^n}{\Delta t} = a^2 \frac{u_{i+1}^n - 2u_i^n + u_{i-1}^n}{(\Delta x)^2} + f_i^n. \qquad (4.57)$$

The truncation error is a combination of the truncated terms $O(\Delta t)$ and $O\left((\Delta x)^2\right)$:

$$\text{TE} = O(\Delta t) + O\left((\Delta x)^2\right) = O\left(\Delta t, (\Delta x)^2\right). \qquad (4.58)$$

This is *the truncation error of discretization of a partial differential equation*. Its order, which is also called *the order of the finite difference scheme*, is defined by the lowest truncated terms with respect to Δt and Δx. We see in (4.58) that the forward in time central in space scheme for the heat equation is of the first order in time and second order in space.

The next example shows how to discretize a PDE with variable coefficients. We consider one-dimensional heat transfer with variable thermal conductivity $\kappa(x)$. The PDE is derived from (2.29) by setting the velocity V to 0 and assuming that the temperature field T depends only on x and t:

$$\rho C \frac{\partial T}{\partial t} = \frac{\partial}{\partial x} \left(\kappa \frac{\partial T}{\partial x} \right). \tag{4.59}$$

The new question is how to correctly approximate the spatial derivative term in the right-hand side. The finite difference formula approximating the external derivative should be written for the *entire* expression under the derivative, including the variable coefficient κ.

Let us, for example, derive a scheme of the second order in x and first order in t based on the same grid points as in (4.57). Similarly to the procedure used to derive the central difference formula (4.19), we repeatedly apply the approximation (4.10) of the first derivative (see (4.24) and (4.25)). The external derivative is approximated at (x_i, t^n) using the values at the half-integer grid points $x_{i-1/2} = x_i - \Delta x/2$ and $x_{i+1/2} = x_i + \Delta x/2$:

$$\frac{\partial}{\partial x} \left(\kappa \frac{\partial T}{\partial x} \right)\Big|_i^n = \frac{\kappa_{i+1/2}(\partial T/\partial x)_{i+1/2} - \kappa_{i-1/2}(\partial T/\partial x)_{i-1/2}}{\Delta x}, \tag{4.60}$$

where $\kappa_{i\pm1/2}$ stands for the values of $\kappa(x)$ at $x_{i\pm1/2}$. We now approximate $\partial u/\partial x$ at $(x_{i\pm1/2}, t^n)$ by the central differences to obtain

$$\frac{\partial}{\partial x} \left(\kappa \frac{\partial T}{\partial x} \right)\Big|_i^n = \frac{1}{\Delta x} \left[\kappa_{i+1/2} \frac{(T_{i+1}^n - T_i^n)}{\Delta x} - \kappa_{i-1/2} \frac{(T_i^n - T_{i-1}^n)}{\Delta x} \right]. \tag{4.61}$$

If κ is a known function of x, its values at the half-integer points can be calculated directly. If not, for example, when κ depends on the solution (e.g. the conductivity coefficient varies with temperature), only the integer point values of κ are available. In this case, we have to apply an interpolation, for example,

$$\kappa_{i-1/2} \approx \frac{\kappa_i + \kappa_{i-1}}{2}, \quad \kappa_{i+1/2} \approx \frac{\kappa_i + \kappa_{i+1}}{2}. \tag{4.62}$$

This operation is of the second order of accuracy, as discussed in Section 4.4.8.

For the time derivative, we use the same forward scheme of the first order as before. The final finite difference approximation of the PDE is

$$(\rho C)_i^n \frac{T_i^{n+1} - T_i^n}{\Delta t} = \frac{1}{\Delta x} \left[\kappa_{i+1/2} \frac{(T_{i+1}^n - T_i^n)}{\Delta x} - \kappa_{i-1/2} \frac{(T_i^n - T_{i-1}^n)}{\Delta x} \right]. \tag{4.63}$$

The truncation error is of the first order in t and second order in x.

The Burgers equation

$$\frac{\partial u}{\partial t} + u\frac{\partial u}{\partial x} = \mu\frac{\partial^2 u}{\partial x^2} \tag{4.64}$$

is principally different from the equations in the two previous examples because of the presence of the product $u\partial u/\partial x$, i.e. nonlinearity in terms of the unknown function. The profound implications of nonlinearity for CFD analysis will be discussed later in this book. At the moment, we develop a finite difference scheme based on the same grid points and achieving the same order of approximation as the two schemes above.

We apply (4.55) for $\partial u/\partial t$, (4.56) for $\partial^2 u/\partial x^2$, and the central difference

$$\left.\frac{\partial u}{\partial x}\right|_i^n = \frac{u_{i+1}^n - u_{i-1}^n}{2\Delta x} + O\left((\Delta x)^2\right). \tag{4.65}$$

For the function u in the nonlinear term, we use the obvious

$$u(x_i, t^n) = u_i^n. \tag{4.66}$$

The resulting finite difference approximation of (4.64) is

$$\frac{u_i^{n+1} - u_i^n}{\Delta t} + u_i^n\frac{u_{i+1}^n - u_{i-1}^n}{2\Delta x} = \mu\frac{u_{i+1}^n - 2u_i^n + u_{i-1}^n}{(\Delta x)^2}. \tag{4.67}$$

The truncation error is $\mathrm{TE} = O\left((\Delta x)^2, \Delta t\right)$.

Our last example is the approximation of the linear convection equation

$$\frac{\partial u}{\partial t} + c\frac{\partial u}{\partial x} = 0, \tag{4.68}$$

where $c > 0$ is a constant. We approximate the equation at the grid point (x_i, t^n) and apply the forward difference for the time derivative and backward difference for the x-derivative. The result is the so-called *upwind scheme*:

$$\frac{u_i^{n+1} - u_i^n}{\Delta t} + c\frac{u_i^n - u_{i-1}^n}{\Delta x} = 0. \tag{4.69}$$

Both finite difference approximations of the derivatives are of the first order. The truncation error of discretization of the equation is, therefore, $O(\Delta t, \Delta x)$.

4.4.2 Boundary and Initial Conditions

The finite difference approximation of the PDE is applied only at the internal grid points of the computational domain. At the grid points lying at the boundaries, the scheme should approximate the boundary and initial conditions. Usual finite difference formulas can be employed with the obvious limitation that they can only use the values of the unknown function at the points within the domain, that is, on only one side of the boundary.

There is one important requirement. The order of approximation of the boundary conditions should not be lower than the order of the scheme used to approximate the PDE. The reason is that, formally, the order of the approximation of the entire solution is equal to the lowest order of approximation of any of its parts. If a lower-order scheme is used for the boundary conditions, the larger truncation error propagates into the solution domain and compromises otherwise higher accuracy of computations.

Let us consider the example of the modified heat equation (4.54) approximated within the solution domain by the scheme (4.57). The problem is solved within the domain $0 < x < L$, $t_0 < t < T_{end}$. It has the Dirichlet and Neumann boundary conditions

$$u(0, t) = a(t), \quad \frac{\partial u}{\partial x}(L, t) = b(t), \text{ at } t_0 < t < t_{end},$$

where a and b are known functions of time and the initial conditions

$$u(x, t_0) = g(x) \quad \text{at} \quad 0 \leq x \leq L.$$

Let the grid points x_0 and x_N correspond to the boundaries $x = 0$ and $x = L$, and let the time layer t^0 correspond to the initial moment of time $t = t_0$.

Approximation of the Dirichlet boundary condition is easy. We assume that

$$u_0^n = a(t^n) = a^n, \quad n = 1, \dots, M, \tag{4.70}$$

which is an exact (no error is introduced) representation. For the Neumann condition, care must be taken to approximate the partial derivative by a formula of second order so as to retain the second order of spatial approximation of the entire scheme. For example, the backward difference

$$\frac{\partial u}{\partial x}(L, t) = \frac{u_N^n - u_{N-1}^n}{\Delta x} + O(\Delta x)$$

would be unsuitable. A correct solution is to apply one of the one-sided finite difference formulas of the second order, for example, (4.14), and write

$$\frac{3u_N^n - 4u_{N-1}^n + u_{N-2}^n}{2\Delta x} = b(t^n) = b^n, \quad n = 1, \dots, M. \tag{4.71}$$

This can be used to compute u_N^n after the interior grid point values of u_i^n at the current time layer t^n have already been found:

$$u_N^n = \frac{1}{3}(2\Delta x b^n + 4u_{N-1}^n - u_{N-2}^n).$$

At last, the initial conditions are represented exactly by

$$u_i^0 = g(x_i), \quad i = 1, \dots, N. \tag{4.72}$$

There exists a technique of imposing boundary conditions that does not require the grid points lying at the boundary. The "ghost" points located symmetrically on the outer side of the boundary are used instead. Two examples are presented in Figure 4.4. In one, shown in Figure 4.4a, the Dirichlet boundary condition $u|_{\partial\Omega} = a$ is approximated with the second order of accuracy by assigning the value of u at the ghost point x_{N+1} to $u_{N+1}^n = a - (u_N^n - a)$. The second example shown in Figure 4.4b illustrates how one can approximate the Neumann condition $\partial u/\partial x|_{\partial\Omega} = b$ by choosing u_{N+1}^n so that the slope of the line connecting u_N^n and u_{N+1}^n is equal to b.

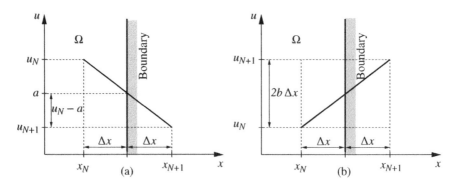

Figure 4.4 Setting boundary conditions using ghost points. (a) Dirichlet boundary condition $u|_{\partial\Omega} = a$. (b) Neumann boundary condition $\partial u/\partial x|_{\partial\Omega} = b$.

4.4.3 Difference Molecule and Difference Equation

The finite difference schemes are often illustrated by the *difference molecules*. This is particularly convenient for describing the schemes applied to the model equations, such as the one-dimensional heat equation, which have only two independent variables, for example, x and t. A difference molecule shows all the grid points that are involved into the finite difference approximation of the equation at the point (x_i, t^n). As an example, Figure 4.5a shows the difference molecule of the forward in time, central in space schemes (4.57), (4.63), and (4.67) for the heat and Burgers equations. The difference molecule for the upwind scheme (4.69) for the linear convection equation is shown in Figure 4.5b.

A finite difference equation, such as (4.57), (4.63), (4.67), (4.69), (4.70), (4.71), or (4.72), can be viewed as an algebraic equation connecting the approximate values of the solution at the neighboring grid points composing the difference molecule. Such equations are called *difference* or, in more general terms, *discretization* equations.

Identifying a specific molecule by the grid point P at which the PDE is approximated, we can write the general form of such an equation as

$$a_P u_P + \sum_{\ell} a_{\ell,P} u_P = Q_P, \qquad (4.73)$$

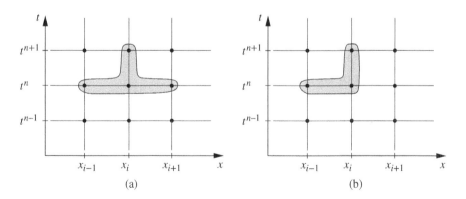

Figure 4.5 (a) Difference molecule for the scheme (4.57), (4.63), and (4.67) for heat and Burgers equations. (b) Difference molecule for the upwind scheme (4.69) for linear convection equation.

where the index ℓ runs through all the neighbors of P belonging to the molecule and Q is the source term, such as f_i^n in (4.57). In some cases, the coefficients a are functions of the solution u, which makes the difference equation nonlinear.

As examples, we consider the heat and Burgers equations (4.54) and (4.64) discretized according to (4.57) and (4.67). Taking the point (x_i, t^n) as P and the points (x_{i-1}, t^n), (x_{i+1}, t^n), and (x_i, t^{n+1}) as corresponding to, respectively, $\ell = 1, 2$, and 3, we can write the difference equation in the form (4.73) with

$$a_P = -\frac{1}{\Delta t} + \frac{2a^2}{(\Delta x)^2}, \quad a_{1,P} = a_{2,P} = -\frac{a^2}{(\Delta x)^2}, \quad a_{3,P} = \frac{1}{\Delta t}, \quad Q_P = f_i^n$$

for (4.57) and with

$$a_P = -\frac{1}{\Delta t} + \frac{2\mu}{(\Delta x)^2}, \quad a_{1,P} = -\frac{\mu}{(\Delta x)^2} - \frac{u_P}{2\Delta x}, \quad a_{2,P} = -\frac{\mu}{(\Delta x)^2} + \frac{u_P}{2\Delta x},$$

$$a_{3,P} = \frac{1}{\Delta t}$$

for (4.67).

A trivial but very important observation is illustrated in (4.73). A difference equation is, by the nature of the finite difference method, compact. Among all the unknowns of the problem (the approximations of the solution at all the grid points), it connects only the few corresponding to the points of a single finite difference molecule.

4.4.4 System of Difference Equations

The compactness of the difference equations mentioned at the end of Section 4.4.3 does not change the fact that such equations taken together for the entire computational grid are coupled to each other, i.e. form one large system of algebraic equations. The system is irreducible to smaller independent systems and should be solved as a whole. This can be a difficult and computationally intensive task. It is not uncommon for a CFD analysis to use grids consisting of millions of nodes. We should also take into account that several variables usually have to be calculated at every node, each variable requiring a separate equation. For example, computing a flow of an incompressible fluid with heat transfer would require at least five variables: three velocity components, pressure and temperature.

The need to solve huge systems of algebraic equations has always been a critical aspect, even a bottleneck of CFD analysis. The available computing power limits the size of the system that can be solved in reasonable time and, thus, the type of the flow and heat transfer processes that can be accurately analyzed.

Significant efforts have been invested over the years to develop effective methods of solving large systems of discretization equations. The most important of them are discussed in this book. The first rather simple idea appears in the next section.

4.4.5   Implicit and Explicit Methods

Let us consider a finite difference scheme applied to a *time-dependent problem*. The natural and commonly used approach to solution of discretization equations for such problems is illustrated in Figure 4.6. We start with the initial conditions at the time layer t^0 and use the discretization equations approximating the PDE and boundary conditions to find the unknowns at the time layer t^1, then at t^2, and so on, marching ahead one layer at a time. Each such time step requires solution of a system that consists of equations for all the space grid points. Albeit still large, such systems are much smaller than the cumulative system that includes the equations for all the time layers.

With respect to the marching procedure, the discretization schemes can be of one of two different types: *explicit* and *implicit*. The types are

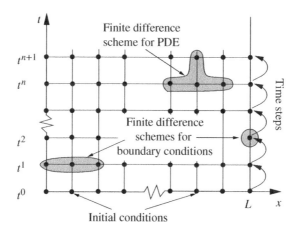

Figure 4.6   Illustration of the marching procedure.

presented here using the example of a finite difference solution of the modified heat equation (4.57). It will be clear from the discussion that the concept equally applies to other equations and other discretization methods, for example, finite element or spectral.

Explicit Schemes: Let us first consider the already introduced scheme

$$\frac{u_i^{n+1} - u_i^n}{\Delta t} = a^2 \frac{u_{i+1}^n - 2u_i^n + u_{i-1}^n}{(\Delta x)^2} + f_i^n. \tag{4.74}$$

The difference molecule is shown in Figure 4.5a. The difference equation includes only one value of u from the $(n+1)$th time layer. The transition from layer t^n to layer t^{n+1} is easy. One has to rewrite (4.74) as

$$u_i^{n+1} = u_i^n + \Delta t \left(a^2 \frac{u_{i+1}^n - 2u_i^n + u_{i-1}^n}{(\Delta x)^2} + f_i^n \right), \tag{4.75}$$

that is, to rewrite the difference equation so as to *explicitly* express the unknown quantity u_i^{n+1} at the layer t^{n+1} through the quantities u_i^n at the layer t^n already known from the previous time step. The schemes, for which this is possible, are called *explicit* schemes.

Implicit Schemes: Another scheme for the heat equation can be developed if we, for example, approximate the PDE at the point (x_i, t^{n+1}) using the backward difference of the first order for the time derivative and the central difference of the second order for the space derivative. The result is the scheme

$$\frac{u_i^{n+1} - u_i^n}{\Delta t} = a^2 \frac{u_{i+1}^{n+1} - 2u_i^{n+1} + u_{i-1}^{n+1}}{(\Delta x)^2} + f_i^{n+1} \tag{4.76}$$

with the truncation error TE $= O(\Delta t, (\Delta x)^2)$. The difference molecule is shown in Figure 4.7. If we now move all the terms needed to be calculated at the new time layer into the left-hand side, we obtain

$$u_i^{n+1} - a^2 \frac{\Delta t}{(\Delta x)^2} (u_{i+1}^{n+1} - 2u_i^{n+1} + u_{i-1}^{n+1}) = u_i^n + \Delta t f_i^{n+1}. \tag{4.77}$$

We see that a time step of the marching procedure cannot be completed as easily as before. No explicit formula exists that expresses the unknown u_i^{n+1} through the known values such as $u_i^n, u_{i-1}^n, u_{i+1}^n$. Coupled equations (4.77) for

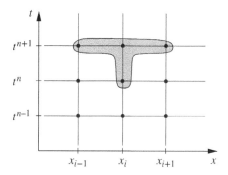

Figure 4.7 Difference molecule of the implicit scheme (4.76) for the heat equation (4.54).

$i = 1, \dots, N - 1$ together with the equations approximating the boundary conditions at $i = 0$ and $i = N$ must be solved simultaneously as a system of linear algebraic equations. Scheme (4.76) is an example of an *implicit* scheme.

One time step of an implicit scheme requires larger amount of computational work than one step of an explicit scheme, since a system of algebraic equations has to be solved. The implicit methods are, nevertheless, in wide use for the reason of their better stability properties. The issue is further discussed in Section 6.3. For now, we only mention that larger time steps can be used with an implicit scheme, so the solution can be computed in a smaller number of steps.

We also have to note that efficient methods for solution of systems of difference equations have been developed. The methods exploit the compact nature of the equations. One method, the Thomas algorithm suitable for 1D problems, is presented in Chapter 7, while the general discussion is postponed till Chapter 8.

4.4.6 Consistency of Numerical Approximation

The truncation error introduced by the finite difference approximation of a PDE problem is a combination of the errors of approximation of all derivative and nonderivative terms and boundary conditions at all grid points and time layers.

The necessary (but not sufficient) condition for an accurate finite difference solution is the generalization of the consistency condition introduced earlier for partial derivatives. The truncation errors of the finite difference approximation of the equation and boundary and initial conditions should vanish as all the grid steps approach 0:

$$\lim_{\Delta x, \Delta y, \Delta z, \Delta t \to 0} \text{TE} = 0. \tag{4.78}$$

If this is the case, the finite difference scheme is called *consistent*. Obviously, the consistency is achieved if the truncation error is at least of the first order in every coordinate and time at all points and time layers of the computational grid. *The scheme has to be at least of the first order to be consistent.*

4.4.7 Interpretation of Truncation Error: Numerical Dissipation and Dispersion

The consistency and the order of the truncation error are the main characteristics of a finite difference approximation. Further information about the truncation error and its possible effect on the solution can be obtained from the *modified equation* as is briefly discussed here using the example of the upwind scheme (4.69). A more detailed discussion can be found in the books listed at the end of this chapter.

To interpret the truncation error of the upwind scheme, we substitute the Taylor series expansions

$$u_i^{n+1} = u_i^n + \left(\frac{\partial u}{\partial t} \right)\Big|_{i,j}^n \Delta t + \left(\frac{\partial^2 u}{\partial t^2} \right)\Big|_{i,j}^n \frac{(\Delta t)^2}{2} + \cdots$$

$$u_{i-1}^n = u_i^n - \left(\frac{\partial u}{\partial x} \right)\Big|_{i,j}^n \Delta x + \left(\frac{\partial^2 u}{\partial x^2} \right)\Big|_{i,j}^n \frac{(\Delta x)^2}{2} + \cdots$$

into (4.69). After rearranging, we obtain

$$u_t + c u_x = -\frac{\Delta t}{2} u_{tt} + \frac{c \Delta x}{2} u_{xx} - \frac{(\Delta t)^2}{6} u_{ttt} - c \frac{(\Delta x)^2}{6} u_{xxx} + \cdots, \tag{4.79}$$

where we use the abbreviations u_t, u_x, etc. for the partial derivatives of u taken at the grid point (x_i, t^n). Equation (4.79) is quite remarkable. Its left-hand side is the original PDE and, thus, would be equal to zero if u were the exact solution of the original PDE. Since u is an extension of the finite difference solution, it satisfies not the original PDE but the *modified equation* (4.79). The nonzero right-hand side of (4.79) represents the truncation error.

The composition and behavior of the truncation error can be better understood if we replace the time derivatives u_{tt} and u_{ttt} by spatial derivatives. It can be done by differentiating (4.79) with respect to x and t and performing algebraic operations with the resulting equations. The final equation is

$$
u_t + cu_x = \frac{c\Delta x}{2}(1 - v)u_{xx} - c\frac{(\Delta x)^2}{6}(2v^2 - 3v + 1)u_{xxx}
$$
$$
+ O\left((\Delta x)^3, (\Delta x)^2 \Delta t, \Delta x(\Delta t)^2, (\Delta t)^3\right), \qquad (4.80)
$$

where $v = c\Delta t/\Delta x$.

Let us analyze the right-hand side of (4.80). The lowest-order term $(c\Delta x/2)(1 - v)u_{xx}$ looks familiar. The second spatial derivative multiplied by a constant can be found in the right-hand sides of heat, diffusion, or Navier–Stokes equations. On all occasions, it corresponds to parabolic behavior and represents a physical process of diffusion or dissipation, which results in smoothing of gradients of the solution. In our case, the diffusion is caused by the error of discretization, i.e. has numerical rather than physical character. The commonly used name is *numerical dissipation*. The constant coefficient at u_{xx} is called *numerical viscosity* or *scheme viscosity*.

The effect of numerical dissipation is illustrated in Figure 4.8. Let the initial condition for the linear convection equation (4.68) have the form of a sharp front as in Figure 4.8a. The exact solution of the PDE retains the

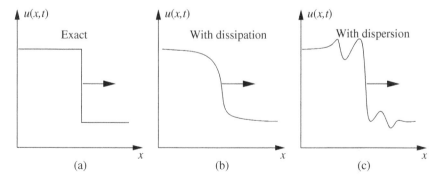

Figure 4.8 Effect of numerical dissipation and dispersion on wavelike solutions. (a) Exact solution. (b) Numerical solution with strong dissipation. (c) Numerical solution with strong dispersion.

shape and travels to the right with the constant speed c. Due to the presence of numerical dissipation, the computed solution does not behave in this manner. The front is gradually smoothed out and takes the form illustrated in Figure 4.8b.

The second term $c((\Delta x)^2/6)(2v^2 - 3v + 1)u_{xxx}$ of the truncation error in the right-hand side of (4.80) can also be identified as an artificial pseudophysical effect. A third-order spatial derivative is known to lead to dispersion, which is the phenomenon of dependence of the wave speed of a propagating wave on its wavelength. The exact solution of a PDE, for example, the solution shown in Figure 4.8a, can be represented as a superposition of Fourier harmonics with different wavelengths λ. The presence of the dispersion term in (4.80) makes them propagate with different speeds rather than with the common speed c. With time, the harmonics separate in phase and form a solution illustrated in Figure 4.8c. The dispersion, when it is caused by the truncation error, is called *numerical dispersion.*

The numerical dissipation and dispersion are common in finite difference solutions. Generally, any term with an even-order space derivative in the right-hand side of the modified equation creates numerical dissipation, while any term with an odd-order space derivative creates numerical dispersion. Of course, only the terms of the lowest order in Δx and Δt are important. For example, the truncation error of the first-order scheme (4.69) is clearly dominated by the numerical dissipation term $c\Delta x(1 - v)u_{xx}/2$. Such schemes are called *predominantly dissipative* or simply *dissipative*. We should expect the behavior illustrated in Figure 4.8b, rather than Figure 4.8c. Other schemes, some of which are discussed in the following chapters, have the lowest-order term of the truncation error proportional to an odd derivative. They are *predominantly dispersive* or *dispersive*.

We have discussed the modified equation and numerical dissipation and dispersion on the example of the hyperbolic equation (4.68). The analysis can also be applied to parabolic systems. This can sometimes produce quite interesting results (one example is given in Chapter 7 in the discussion of the simple explicit method for the heat equation). In general, however, the composition of the truncation error is less important for parabolic system than for hyperbolic ones.

The reason is the different nature of the physical phenomena represented by parabolic and hyperbolic equations. The hyperbolic equations often describe wave propagation in media with negligibly low natural dissipation and diffusion. Good examples are the supersonic aerodynamic or acoustic flows. Viscosity and heat diffusivity do not play important roles in these phenomena, except within the boundary layers and shock waves in high-speed flows. Therefore, the waves propagate in an almost ideal, nonviscous, and nondiffusive manner. For such waves, the numerical dispersion and dissipation, if left unchecked, can result in strongly distorted solutions. Obviously, evaluation and control of both dispersive and dissipative errors are not just desirable but necessary.

In the systems described by parabolic equations, strong natural dissipation is typically present as a part of the physical process. In general, schemes of second or higher orders in space are used for parabolic equations. For such schemes, assuming that the grid steps are sufficiently small to accurately resolve the space and time gradients of the solution, the physical diffusion is likely to dominate over the numerical diffusion. The dispersive errors are either small or avoided completely by the schemes used for parabolic equations.

4.4.8 Methods of Interpolation for Finite Difference Schemes

As we discuss in Chapter 5, interpolation of solution variables to points not belonging to the computational grid is a practically unavoidable component of the finite volume method. In the finite difference schemes considered here, interpolation is necessary in some situations, in particular when the PDE contains a diffusion term with variable coefficient. We have already encountered such a situation when we derived scheme (4.63) for the heat equation

$$\rho C \frac{\partial T}{\partial t} = \frac{\partial}{\partial x} \left(\kappa \frac{\partial T}{\partial x} \right). \tag{4.81}$$

In this section, the same equation is used to illustrate methods of interpolation and discuss their properties.

The most obvious and commonly utilized approach is the linear interpolation. When the interpolation point $x_{i-1/2}$ is located exactly in the middle between the two closest grid points, the averaging formula (4.62) is used. When it is not the case, e.g. in discretizations on nonuniform grids, the general formula assuming linear variation of the coefficient is used. For example, the first equation of (4.62) becomes

$$\kappa_{i-1/2} = \kappa_{i-1} + (\kappa_i - \kappa_{i-1}) \frac{x_{i-1/2} - x_{i-1}}{x_i - x_{i-1}} = \kappa_i \frac{\Delta^-}{\Delta} + \kappa_{i-1} \frac{\Delta^+}{\Delta}, \tag{4.82}$$

where we use the shorthand notation $\Delta = x_i - x_{i-1}$, $\Delta^- = x_{i-1/2} - x_{i-1}$, $\Delta^+ = x_i - x_{i-1/2}$.

The linear interpolation is an operation that introduces the approximation error of the second order. To prove this, we write the Taylor series expansions around the interpolation point:

$$\kappa_i = \kappa_{i-1/2} + \left.\frac{\partial \kappa}{\partial x}\right|_{i-1/2} \Delta^+ + \frac{1}{2}\left.\frac{\partial^2 \kappa}{\partial x^2}\right|_{i-1/2} (\Delta^+)^2 + O\left(\Delta^3\right),$$

$$\kappa_{i-1} = \kappa_{i-1/2} - \left.\frac{\partial \kappa}{\partial x}\right|_{i-1/2} \Delta^- + \frac{1}{2}\left.\frac{\partial^2 \kappa}{\partial x^2}\right|_{i-1/2} (\Delta^-)^2 + O\left(\Delta^3\right).$$

By direct substitution, we find that the right-hand side of (4.82) can be written as $\kappa_{i-1/2} + O\left((\Delta x)^2\right)$.

Interestingly, there are situations when the linear interpolation leads to decidedly inaccurate solutions, even though its error remains formally of the second order in terms of the grid step. This happens when the diffusion coefficient changes abruptly – on the distance less than one grid step – within the solution domain. We will consider a simple example of the heat conduction in composites, where the thermal conductivity may change by orders of magnitude across the thin interface between different materials. Since the interface is usually fixed in such systems, the need for interpolation can be avoided by splitting the solution domain into two parts along the interface. The example, however, provides a good illustration of the numerical technique suitable for more complex systems, where the interface is moving, for example, in multiphase fluid flows.

As an illustration, we use one-dimensional heat conduction in a thin rod, in which the conductivity increases sharply somewhere between the grid points x_{i-1} and x_i (see Figure 4.9a). The intermediate grid point $x_{i-1/2}$ is positioned at the interface.

We now recall that (4.81) is the energy conservation equation expressing the fact that the internal energy ρCT changes in the result of x-dependent heat flux $q = \kappa \partial T/\partial x$. Our foremost concern must, therefore, be that the solution reproduces the flux q correctly. At $x_{i-1/2}$, the correct value is easily evaluated using the elementary thermal resistance analysis. The resistances of the rod segments $[x_{i-1}, x_{i-1/2}]$ and $[x_{i-1/2}, x_i]$ are $R^- = \Delta^-/\kappa^-$ and $R^+ = \Delta^+/\kappa^+$ (see Figure 4.9b). The total resistance of the segment $[x_{i-1}, x_i]$ is $R = R^- + R^+$, so the heat flux between x_{i-1} and x_i is

$$q = \frac{\Delta T}{R} = \frac{T_i^n - T_{i-1}^n}{\frac{\Delta^-}{\kappa^-} + \frac{\Delta^+}{\kappa^+}}. \tag{4.83}$$

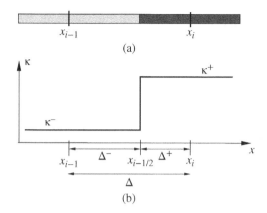

Figure 4.9 Example of the situation requiring harmonic interpolation. (a) Heat conduction in an one-dimensional rod with abrupt change of thermal conductivity. (b) Approximation using a half-integer point.

In the finite difference scheme developed in Section 4.4.1, the heat flux is represented as

$$q = \kappa_{i-1/2} \frac{T_i^n - T_{i-1}^n}{\Delta}. \tag{4.84}$$

This means that the correct approximation of the conductivity is not the linear interpolation (4.82), but the *weighted harmonic mean*, which we will also call the *harmonic interpolation*:

$$\kappa_{i-1/2} = \frac{\Delta}{\frac{\Delta^-}{\kappa^-} + \frac{\Delta^+}{\kappa^+}} = \frac{x_i - x_{i-1}}{\frac{x_{i-1/2} - x_{i-1}}{\kappa_{i-1}} + \frac{x_i - x_{i-1/2}}{\kappa_i}}. \tag{4.85}$$

To illustrate the advantage of (4.85) over (4.82), we consider the limit $\kappa_{i-1} \ll \kappa_i$. In this case, the total resistance R is well approximated by R^-, so a correct approximation of $\kappa_{i-1/2}$ in (4.84) must be

$$\kappa_{i-1/2} \approx \frac{\Delta}{\Delta^-} \kappa_{i-1}.$$

This limit is provided by (4.85). On the contrary, the linear interpolation (4.82) would grossly overpredict the conductivity and, thus, the heat flux providing

$$\kappa_{i-1/2} \approx \frac{\Delta}{\Delta^+} \kappa_i.$$

We note that the usefulness of the harmonic interpolation is limited to the solutions with interfaces on which the diffusion coefficients change abruptly. In the problems without such interfaces or with the grid steps sufficiently small to locate several grid points across the interface, the linear interpolation (4.82) is usually perfectly adequate.

BIBLIOGRAPHY

Fletcher, C.A.J. (1991). *Computational Techniques for Fluid Dynamics: Fundamental and General Techniques*, vol. **1**. Berlin: Springer-Verlag.

Hirsch, C. (2007). *Numerical Computation of Internal and External Flows*. Elsevier.

Patankar, C.H. (1980). *Numerical Heat Transfer and Fluid Flow*. CRC Press.

Tannehill, J.C., Anderson, D.A., and Pletcher, R.H. (1997). *Computational Fluid Mechanics and Heat Transfer*. Philadelphia, PA: Taylor & Francis.

PROBLEMS

1. Consider the function $u = \sin x$. Apply the forward difference (4.4), backward difference (4.9), and central difference (4.10) to evaluate the derivative du/dx at $x = 1.0$. Use the uniform grids with steps $\Delta x = 1.0, 0.5, 0.1$, and 0.01. Compare the results with the exact value of the derivative. Determine which grid steps give accurate results (e.g. with the error less than 0.01) for each scheme. Explain why one scheme is more accurate than the two others.

2. In the same way as in Problem 1, compare the performance of the central difference scheme (4.10) and the fourth-order scheme (4.15).

3. Write the schemes similar to (4.4), (4.9), and (4.10) for $\partial u/\partial y|_{i,j}$.

4. Write the schemes similar to (4.17), (4.18), and (4.19) for $\partial^2 u/\partial y^2|_{i,j}$.

5. Write the central finite difference formulas of the second order for $\partial u/\partial x$ and $\partial^2 u/\partial x^2$ of function $u(x, y, t)$ at the grid point (x_{i+1}, y_j, t^{n+1}).

6. Apply the central difference formula (4.10) to approximate the first derivatives of the functions $u_1(x) = \sin(x)$, $u_2(x) = \sin(3x)$, $u_3(x) = \sin(10x)$ at $x = 0$. Use the uniform grid with step $\Delta x = 0.1$. Compare the results with the exact values of the derivatives, and

explain why the error of approximation is different for the three functions.

7. Verify the consistency and order of approximation of the schemes (4.13) and (4.14) using Taylor series expansions.

8. Verify the consistency and order of approximation of the scheme (4.20) using Taylor series expansions.

9. Verify the statement made in Section 4.2.7 that the scheme

$$\frac{\partial^2 u}{\partial x^2}\bigg|_{i,j} = \frac{u_{i+1,j} - 2u_{i,j} + u_{i-1,j}}{(x_i - x_{i-1})^2}$$

is inconsistent if used with a nonuniform grid.

10. Using the method of Taylor series expansion or the method of polynomial fitting, develop the following finite difference schemes.
    a) (4.13)
    b) (4.17)
    c) (4.25)

11. Consider the generic transport equation $\phi_t + u\phi_x = \mu\phi_{xx}$, where u is a known function of x and t and μ is a constant coefficient. Assume that the computational grid is uniform with steps Δx and Δt. Write the finite difference schemes satisfying the following requirements.
    a) Explicit scheme of the first order in time and second order in space. Use central differences for the space derivatives.
    b) The same requirements as in (a), but the scheme is implicit.
    c) Scheme of the first order in time. Use implicit central difference approximation for the diffusion term $\mu\phi_{xx}$ and explicit backward difference approximation for the convection term $u\phi_x$.

12. The Neumann boundary condition $\partial u/\partial x = a$ at $x = 0$ has to be implemented in a finite difference scheme. The grid is uniform with step Δx. Select a finite difference approximation of the boundary condition if the scheme's order of approximation in x is:
    a) First
    b) Second.

13. Consider the PDE problem

$$\frac{\partial u}{\partial t} = a^2 \frac{\partial^2 u}{\partial x^2} + \sin(bx) \text{ at } \quad 0 < x < L, \quad 0 < t \le t_{end},$$

$$\frac{\partial u}{\partial x}(0, t) = \sin(\omega t), \quad \frac{\partial u}{\partial x}(L, t) = 0, \quad u(x, 0) = \cos\left(\frac{2\pi x}{L}\right),$$

where a, b, and ω are constants. Write the full finite difference discretization of the problem on a uniform grid. The discretization must be of the first order in time and second order in space. The answer should include the description of the grid and the finite difference schemes for the PDE and the boundary and initial conditions with specification of the grid points, at which each equation is applied.

14. A finite difference scheme was developed to solve the heat equation. Testing in comparison with a known exact solution showed that the error is of the same order of magnitude as the solution itself and does not tend to zero as the grid steps decrease. Assuming there was no coding errors in the computer program, what was the reason for such behavior?

15. The modified equations of certain finite difference schemes are given below. In each case, determine the order of approximation, and find whether the dominant error is due to numerical dissipation or numerical dispersion:

$$u_t + cu_x = \left(\frac{1}{2}c^2 \Delta t\right) u_{xx} - \left(\frac{1}{6}c(\Delta x)^2 + \frac{1}{3}c^3(\Delta t)^2\right) u_{xxx} + \cdots$$

$$u_t + cu_x = \frac{1}{6}c(\Delta x)^2(v^2 - 1)u_{xxx}$$

$$- \frac{1}{120}c(\Delta x)^4(9v^4 - 10v^2 + 1)u_{xxxxx} + \cdots .$$

$$u_t - a^2 u_{xx} = \left[\frac{1}{2}a^4 \Delta t + \frac{(a\Delta x)^2}{12}\right] u_{xxxx} + \cdots$$

$$\left[\frac{1}{3}a^6(\Delta t)^2 + \frac{1}{12}a^4 \Delta t(\Delta x)^2 + \frac{1}{360}a^2(\Delta x)^4\right] u_{xxxxxx} + \cdots .$$

Programming Exercise Calculate the finite difference solution of Equation (4.54) with $a = 0.5$ and $f(x, t) = \sin 5x$ in the domain $0 < x < \pi$, $0 < t < 50$ with the initial condition $u(x, 0) = x(\pi - x)$ and the boundary conditions $u(0, t) = 0$, $u(\pi, t) = 0$. Use the fully explicit scheme (4.74). Try two grids, one with 100 space points and 100 time layers and another with the number of time layers increased to 100,000. Compare the profile of u at $t = t^{end} = 50$ with the exact solution of the steady-state problem $u_{exact} = (25a^2)^{-1} \sin 5x$.

5

FINITE VOLUME SCHEMES

5.1 INTRODUCTION AND GENERAL FORMULATION

5.1.1 Introduction

This chapter is entirely devoted to one class of discretization schemes – the finite volume schemes. Such attention is fully warranted by the special place these schemes occupy within the field of applied CFD. Many general-purpose codes for fluid flows, including the majority of commercial CFD programs, are based on the finite volume approach. The reasons for popularity will become clear in the discussion that follows. Right now, we only mention the main two: (i) convenience of use with unstructured grids and (ii) the global conservation property.

The finite volume schemes are principally different from the classical finite difference schemes in the way they are derived. Instead of discretizing the partial differential equations, we start with the physical conservation laws in the integral form, such as those presented in Section 2.4. Discretization is applied directly to the integral equations written for small control

Essential Computational Fluid Dynamics, Second Edition. Oleg Zikanov.
© 2019 John Wiley & Sons, Inc. Published 2019 by John Wiley & Sons, Inc.
Companion Website: www.wiley.com/go/zikanov/essential

volumes. The distinctive character of the finite volume method is sometimes taken as far as claiming it to be a completely separate approach unrelated to the finite difference discretization. We adhere to a more moderate view, according to which the finite volume method is, albeit special, still a type of the general finite difference technique.

In the following discussion, the method will be illustrated using the formal conservation equation introduced in Section 2.4:

$$\frac{d}{dt}\int_\Omega \Phi \, d\Omega = -\int_S \Phi V \cdot n \, dS + \int_S \chi \nabla \Phi \cdot n \, dS + \int_\Omega Q \, d\Omega. \qquad (5.1)$$

It is written for an arbitrary control volume Ω with surface S and outward-facing unit-length normal to the surface n (see Figure 5.1) and expresses the "conservation" of the scalar field Φ within Ω. Each term in (5.1) has counterparts in the integral equations of conservation of mass, momentum, and energy:

- The term $\frac{d}{dt}\int_\Omega \Phi \, d\Omega$ in the left-hand side represents the rate of change of the total amount of Φ within Ω.
- The term $-\int_S \Phi V \cdot n \, dS$ is the integral of the *convective flux* $q_{conv} = \Phi V \cdot n$. It describes the transport of Φ across the boundary by the flow velocity V. The negative sign in front of the integral reflects the convention of always using the normal n directed outside the control volume. A flow with positive $V \cdot n$ removes Φ from Ω.
- The term $\int_S \chi \nabla \Phi \cdot n \, dS$ is the integral of the *diffusive flux* $q_{conv} = -\chi \nabla \Phi \cdot n$. It describes the cross-boundary transport of Φ by a diffusive mechanism, such as viscous friction or heat conduction. χ is the always positive coefficient of diffusion. The sign in front of

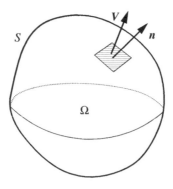

Figure 5.1   Control volume used for derivation of finite volume approximation.

the integral is, again, chosen so as to account for the conventional orientation of n.

- The volume integral $\int_\Omega Q\, d\Omega$ represents the change of the amount of Φ within Ω caused by an *internal (volumetric) source* of intensity Q.

Let us identify the control volume Ω. In the finite volume method, the computational domain is divided into small, usually nonoverlapping[1] subdomains called *cells*. A finite volume scheme is derived by applying integral balance equation, such as (5.1), to every cell. The next step described later in this chapter is to approximate the integrals using the values of the solution variables at the grid points. In the result of this operation, each integral equation is replaced by an algebraic discretization equation. Taken for all cells, the discretization equations form a system that has to be solved in the same way as for a finite difference scheme.

Note that the following discussion concerns only spatial discretization. The time discretization in the finite volume schemes is not different from the time discretization applied to the finite difference schemes or other discretization schemes in general.

5.1.2 Finite Volume Grid

Examples of finite volume grids are shown in Figure 5.2. Although they do not include a three-dimensional grid, it should be understood that the principles of the finite volume approximation discussed in this chapter are fully valid in the general three-dimensional case.

The finite volume grids applied to one-dimensional problems are the simplest. The control volumes are intervals. For example, Figure 5.2a shows a grid with cells $\Omega_i = [x_{i-1/2}, x_{i+1/2}]$. The outward-facing normal vector n is in the negative x-direction on the left-hand boundary $x_{i-1/2}$ and in the positive x-direction on the right-hand boundary $x_{i+1/2}$.

In the two-dimensional and three-dimensional cases, we have a choice between structured and unstructured grids. In the structured grids, the cells are quadrilateral (in the two-dimensional case) or hexahedral (in the three-dimensional case) and arranged in a structured pattern along the lines of a Cartesian (as in Figure 5.2b) or curvilinear (as in Figure 5.2c) coordinate system. In unstructured grids, the cells may have various shapes, such as prisms, tetrahedra, hexahedra, or pyramids, in three dimensions and

[1]Overlapping cells are rarely used but allowed under certain restrictions.

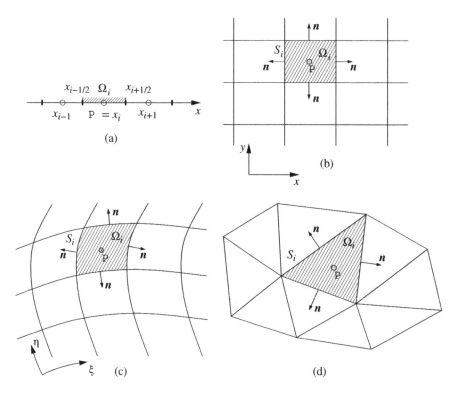

Figure 5.2 Examples of finite volume grids. (a) One-dimensional grid. (b) Two-dimensional Cartesian structured grid. (c) Two-dimensional curvilinear structured grid. (d) Two-dimensional unstructured grid.

plane figures, such as triangles, quadrilaterals, or other convex polygons, in two dimensions. As an example, Figure 5.2d shows a grid with triangular cells.

The finite volume schemes can be designed entirely in terms of the cell-related quantities, such as the cell-averaged variables $|\Omega|^{-1}\int_{\Omega}\Phi\,d\Omega$, where $|\Omega|$ is the volume of the cell. It is, however, customary and convenient to introduce a *grid point* within each cell and write the schemes in terms of the approximate values of variables at these points. One commonly used approach is based on the cell-centered arrangement illustrated in Figure 5.2. In each example, one grid point is shown and marked by the letter P. The grid point location coincides with the cell's center, so that the grid point value of a variable serves as a good approximation of the mean value of this variable in the cell.

There are other possible arrangements. In some of them, the grid points are positioned so that the faces of the control cells are located exactly midway between the two neighboring grid points. In others, the so-called cell-vertex schemes, the grid points are at the vertices of the cell boundaries. Our discussion will be limited to cell-centered arrangements. A broader and more detailed description can be found in the books focused on the subject of finite volume method – for example, in the books listed at the end of the chapter.

5.1.3 Consistency, Local, and Global Conservation Property

A finite volume scheme has to be *consistent* in the same sense as a finite difference one (see Section 4.2.2). The error of the approximation must vanish in the asymptotic limit of infinitely small cells. In other words, the scheme must be at least of the first order in terms of the cell size.

Another important property follows directly from the way we write the governing equation when we develop finite volume schemes (see (5.1)). Let us consider two neighboring cells, for example, the cells with the grid points P and F in Figure 5.3. The property concerns the convective and diffusive fluxes at the cell face separating the neighbors, S_1 in our example. Evidently, the principle of conservation of Φ expressed by (5.1) requires that the fluxes on the two sides of the boundary are exactly the same. Since the outward-facing normals n belonging to the two cells are in the opposite directions ($n_F = -n_P$), this implies the following *local conservation property*: the surface flux integrals calculated for two adjacent cells over their

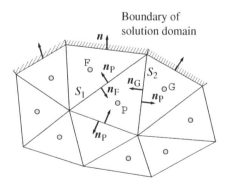

Figure 5.3 Illustration of the local and global conservation properties of finite volume methods.

common boundary must be of the same magnitude and opposite signs. In our example, this is expressed as

$$\int_{S_1} \Phi V \cdot n_P \, dS = -\int_{S_1} \Phi V \cdot n_F \, dS, \tag{5.2}$$

$$\int_{S_1} \chi \nabla \Phi \cdot n_P \, dS = -\int_{S_1} \chi \nabla \Phi \cdot n_F \, dS. \tag{5.3}$$

The expressions are mathematically trivial. Less trivially, they must also be satisfied by the discrete approximations of the integrals via the grid point values of Φ, V, and χ used by the finite volume method (see Sections 5.2 and 5.3). Violating the local conservation property by the numerical solution would result in the boundary between the cells serving as an artificial infinitely thin sink or source of the conserved quantity – normally a strongly undesirable behavior.

Remarkably, the form of the governing equations (5.1) implies that a locally conserving scheme also satisfies the *global conservation property*. The approximate finite volume solution *exactly* reproduces the physical conservation principle for the *entire* solution domain.

We will illustrate this statement using the grid in Figure 5.3. The global conservation of Φ corresponding to (5.1) is expressed as

$$\frac{d}{dt} \int_{\Psi} \Phi \, d\Omega = -\int_{\partial \Psi} (q \cdot n) dS + \int_{\Psi} Q \, d\Omega, \tag{5.4}$$

where Ψ is the solution domain, $\partial \Psi$ is its boundary, n is the unit-length outward-facing normal to $\partial \Psi$, q is the flux of Φ through the boundary imposed by the boundary conditions, and Q is the rate of generation of Φ by internal sources.

The finite volume solution satisfies the system of the algebraic equations – discrete approximations of (5.1) for individual cells. If we add these equations together, the approximations of the volume integrals in (5.1) add up to the discrete versions of the respective volume integrals in (5.4). According to (5.2) and (5.3), the approximations of the surface integrals from any two adjacent cells over a common boundary cancel each other. Only the surface integrals over the cell boundaries lying on the boundary $\partial \Omega$ of the solution domain remain uncanceled. Their sum approximates the surface integral in the right-hand side of (5.4). We see that the summation of the finite volume equations over all cells replicates (5.4) exactly, without any additional discretization-related terms.

5.2 APPROXIMATION OF INTEGRALS

The algebraic discretization equations that constitute a finite volume scheme are derived by approximating the integral equations, such as (5.1), written for individual cells. The main elements of the approximation are discussed in this section. The discussion is general and valid for all kinds of finite volume grids. As illustrations, two-dimensional structured Cartesian and unstructured grids shown in Figure 5.4 are used.

5.2.1 Volume Integrals

The simplest way to approximate a volume integral is to replace it by the product of the cell's volume (in the three-dimensional case) or area (in two dimensions) $|\Omega|$ and the mean value of the integrand $\overline{\Phi}$ approximated through the grid point values. If the cell-centered arrangement of the grid points is used, we can simply replace the mean by the value at the central grid point (see Figure 5.4):

$$\int_{\Omega} \Phi \, d\Omega = \overline{\Phi}|\Omega| \approx \Phi_{\mathrm{P}}|\Omega|, \quad \int_{\Omega} Q \, d\Omega = \overline{Q}|\Omega| \approx Q_{\mathrm{P}}|\Omega|. \qquad (5.5)$$

This approximation generates the truncation error, which has the second order of magnitude in terms of the size of the cell. For example, for the structured grid in Figure 5.4a, the dimensions of the cell are $(\Delta x, \Delta y)$, and the truncation error is $O((\Delta x)^2, (\Delta y)^2)$. If a finite volume scheme of the second order of accuracy is designed, (5.5) is sufficient and should be used. If, however, a higher-order scheme is desired, $\overline{\Phi}$ has to be replaced by a more accurate approximation that uses the values of the integrand at other points

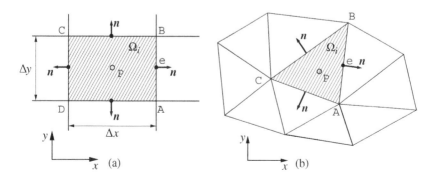

Figure 5.4 Illustration of the approximation of volume and surface integrals.

within the cell, such as the vertex points A, B, C, and D or the midpoints of the cell faces, such as e. If these points do not belong to the grid (as in the cell-centered arrangement), the values of the solution variables at them have to be interpolated from the neighboring grid points. The order of accuracy of the interpolation should, of course, not be lower than the desired order of the scheme.

5.2.2 Surface Integrals

The surface $\partial\Omega$ of the cell Ω consists of several faces, which are curves in the two-dimensional and surfaces in the three-dimensional case. Their shape and number vary with the design of the grid. For example, in the two-dimensional quadrilateral cell of the structured grid in Figure 5.4a, the faces are the four intervals AB, BC, CD, and DA, while the cell in Figure 5.4b has three faces: AB, BC, and CA. Every surface integral in the finite volume equation, such as (5.1), breaks down into the sum of integrals over the faces, which are computed separately. We will use the face AB for demonstration. The procedure can be easily generalized to other faces, to other shapes of the cell, and to the three-dimensional case.

The key component of the surface integral approximation is the midpoint rule. The integral is evaluated as a product of the area of the face and the mean value of the integrand. The mean is approximated by the value of the integrand at the midpoint of the face. In our example, this is expressed for an arbitrary integrand f as

$$\int_{AB} f \, dS = \bar{f} S_{AB} \approx f_e S_{AB}, \tag{5.6}$$

where S_{AB} is the area (length if the cell is two-dimensional) of the face and the overline stands for the mean over on the face. The approximation generates the error $\sim O((S_{AB})^2)$, which is the error of the second order in terms of the size of the cell.

Applied to the convective and diffusive flux integrals in (5.1), the approximations are

$$\int_{AB} \Phi V \cdot n \, dS = \overline{\Phi V \cdot n} S_{AB} \approx (\Phi V \cdot n)_e S_{AB} \tag{5.7}$$

and

$$\int_{AB} \chi \nabla \Phi \cdot n \, dS = \overline{\chi \nabla \Phi \cdot n} S_{AB} \approx (\chi \nabla \Phi \cdot n)_e S_{AB}, \tag{5.8}$$

where e is the midpoint of the face AB.

The surface integrals contain projections of the solution vectors, such as V and $\nabla\Phi$, on the direction of the normal vector $\boldsymbol{n}$. If the grid is structured and the cell boundaries follow the lines of a Cartesian or curvilinear coordinate system, finding the projection is simple. For example, for the face AB in Figure 5.4a, we have $\boldsymbol{n}$ in the direction of the positive x-axis, and the projections are

$$(\Phi V \cdot \boldsymbol{n})_{\mathrm{e}} = (\Phi V_x)_{\mathrm{e}}, \quad (\chi \nabla\Phi \cdot \boldsymbol{n})_{\mathrm{e}} = (\chi \partial\Phi/\partial x)_{\mathrm{e}}. \tag{5.9}$$

In many cases, however, the normal vector is not aligned with an axis of a global coordinate system. This is, in particular, true for unstructured grids as illustrated in Figure 5.4b. The evaluation of the projections requires both (or all three in three-dimensional problems) components of the solution vectors. For example, using the global Cartesian coordinate system shown in Figure 5.4b, we can write

$$(\Phi V \cdot \boldsymbol{n})_{\mathrm{e}} = \Phi_{\mathrm{e}}(V_x n_x + V_y n_y)_{\mathrm{e}}, \tag{5.10}$$

$$(\chi \nabla\Phi \cdot \boldsymbol{n})_{\mathrm{e}} = \chi_{\mathrm{e}}((\partial\Phi/\partial x)n_x + (\partial\Phi/\partial y)n_y)_{\mathrm{e}}. \tag{5.11}$$

The orientation of the normal vector $\boldsymbol{n}$ is defined by the orientation of the corresponding cell face. In the three-dimensional case, we use

$$\boldsymbol{n} = \frac{U \times V}{|U \times V|}, \tag{5.12}$$

where U and V are nonparallel vectors lying in the face, for example, the vectors connecting its corner vertices. For two-dimensional grids, such as those in Figure 5.4, we define

$$\boldsymbol{n} = \frac{1}{S_{\mathrm{AB}}}(S_y \boldsymbol{i} - S_x \boldsymbol{j}), \tag{5.13}$$

where $S = \mathrm{AB} = S_x \boldsymbol{i} + S_y \boldsymbol{j}$ is the vector connecting the vertices A and B written in the Cartesian coordinates.

This allows us to rewrite the approximations (5.7) and (5.8) in a compact form

$$\int_{\mathrm{AB}} \Phi V \cdot \boldsymbol{n} \, dS \approx \Phi_{\mathrm{e}}(V_{x\mathrm{e}} S_y - V_{y\mathrm{e}} S_x) \tag{5.14}$$

and

$$\int_{\mathrm{AB}} \chi \nabla\Phi \cdot \boldsymbol{n} \, dS \approx \chi_{\mathrm{e}}((\partial\Phi/\partial x)_{\mathrm{e}} S_y - (\partial\Phi/\partial y)_{\mathrm{e}} S_x). \tag{5.15}$$

5.3   METHODS OF INTERPOLATION

We have seen in the previous section that the approximation of surface integrals requires knowledge of solution variables and their derivatives at the
midpoints of the cell faces. These points are usually not part of the grid, and
the values have to be obtained by *interpolation* from the grid points. Among
the numerous possible schemes, we will only consider several, which are
simple and commonly used. Further information can be found in specialized
texts on finite volume methods.

For simplicity, the methods of interpolation will be introduced using the
two-dimensional Cartesian grid shown in Figure 5.5a. More specifically,
we will show how the values of Φ and $\nabla\Phi \cdot \boldsymbol{n}$ at the face midpoint e can
be approximated using the values of Φ at the grid points P, E, W, N, S,
EE, etc. of a cell-centered grid. As even simpler examples, schemes for the
one-dimensional linear convection and heat equations will be derived using
the grid shown in Figure 5.5b. The problems and principles of interpolation on nonorthogonal and unstructured grids will be briefly discussed in
Section 5.4.

5.3.1   Upwind Interpolation

The simplest method of approximation of Φ_e is to use the value at a neighboring grid point. In our example, this means approximating by either Φ_P
or Φ_E. In the *upwind interpolation*, the choice, P or E, is dictated by the
direction of the flow:

$$\Phi_e = \begin{cases} \Phi_P & \text{if} \quad (\boldsymbol{V} \cdot \boldsymbol{n})_e > 0 \\ \Phi_E & \text{if} \quad (\boldsymbol{V} \cdot \boldsymbol{n})_e < 0 \end{cases} . \tag{5.16}$$

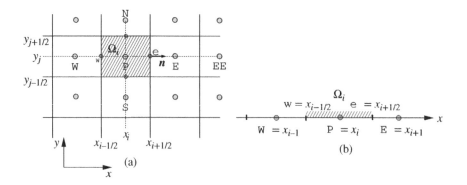

Figure 5.5   Two-dimensional Cartesian grid (a) and one-dimensional grid (b) used
to illustrate interpolation methods.

The value at the nearest upwind (upstream) grid point is taken. The choice seems natural since the upwind value is convected by the flow toward the point e. In agreement with this logic, the upwind schemes are usually considered in connection with hyperbolic problems, in which the convection velocity V determines the direction and speed of propagation of information in the solution.

When applied to hyperbolic problems, the schemes based on the upwind interpolation demonstrate valuable properties. In particular, the schemes satisfy the boundedness criterion, meaning that spurious oscillations never evolve in the solutions for propagating waves or sharp fronts. Another useful property is that such schemes tend to be numerically stable if the time step is sufficiently small. (The numerical stability is defined and discussed in Chapter 6.)

At the same time, the procedure of upwind interpolation has one very significant drawback. It is of only the first order of accuracy. This is easy to see in (5.16). The Taylor series expansion along the x-axis gives

$$\Phi_P = \Phi_e - \left.\frac{\partial \Phi}{\partial x}\right|_e |eP| + O(|eP|^2)$$

and the similar formula for Φ_E. Applying the interpolation we introduce the truncation error $\sim \partial \Phi / \partial x|_e |eP| \sim \partial \Phi / \partial x|_e |eE|$, i.e. the error of the first order in terms of the cell size.

To illustrate the use of the upwind interpolation and the effect of the truncation error, we derive the finite volume scheme for the linear convection equation

$$\frac{\partial u}{\partial t} + c\frac{\partial u}{\partial x} = 0, \quad c > 0. \tag{5.17}$$

The example seems appropriate, since the solutions of the linear convection equation are characterized by purely hyperbolic behavior. Integrating (5.17) over the cell Ω_i (see Figure 5.5b), we obtain the finite volume equation

$$\frac{d}{dt} \int_{x_{i-1/2}}^{x_{i+1/2}} u \, dx + cu_e - cu_w = 0.$$

We see that the linear convection equation can be considered as a one-dimensional version of the general conservation equation (5.1) with constant velocity c in the positive x-direction and zero diffusion flux and volume sources. The volume integral is approximated according to (5.5) as

$$\int_{x_{i-1/2}}^{x_{i+1/2}} u \, dx \approx u_P \Delta x.$$

The upwind interpolation is used for the midpoint values u_e and u_w. At the cell boundary $e = x_{i+1/2}$, the normal $\boldsymbol{n}$ is in the positive x-direction, and (5.16) gives

$$u_e \approx u_P.$$

At the boundary $w = x_{i-1/2}$, the normal is in the direction of negative x. We should take the value at the nearest grid point west of the cell:

$$u_w \approx u_W.$$

The finite volume approximation of (5.17) is, thus,

$$\frac{d}{dt}(u_P \Delta x) + c u_P - c u_W = 0.$$

The last step is to introduce the time layers $t^n = t^0 + n\Delta t$ and apply the first-order explicit time discretization. The result is

$$\frac{u_i^{n+1} - u_i^n}{\Delta t} + c\frac{u_i^n - u_{i-1}^n}{\Delta x} = 0. \tag{5.18}$$

The scheme (5.18) is identical to the finite difference scheme (4.69). This allows us to use the results of the analysis of the truncation error based on the modified equation performed in Section 4.4.7. The lowest-order term of the truncation error is $(c\Delta x/2)(1 - c\Delta t/\Delta x)u_{xx}$, which corresponds to *numerical dissipation*. The x-variations of the solution are artificially smeared out.

The strong numerical dissipation (numerical viscosity) effect is not limited to the upwind scheme for Eq. (5.17). It is a general feature of the schemes, in which the upwind interpolation (5.16) is applied to approximate the convective flux integrals. This becomes a serious issue in predominantly hyperbolic problems, where there is no natural strong diffusion, and the sharp variations easily develop and persist. Very fine grids are needed to obtain accurate solutions of such problems using the upwind schemes.

The choice of the neighboring grid point in the one-point first-order interpolation is not limited to (5.16). We could also approximate Φ_e by Φ_N or Φ_S (see Figure 5.5a) or inverse the conditions for selection of Φ_P and Φ_E (a "downwind" interpolation). According to the arguments made when we introduced the upwind formula, such interpolations would make little sense. Moreover, it is shown in Chapter 7 that such interpolations may lead to numerically unstable and, thus, useless solutions.

5.3.2 Linear Interpolation of Convective Fluxes

Another commonly used approximation is based on the linear interpolation between two neighboring grid points. It can be viewed as an approximation based on the assumption of linear profile of the variable along the line connecting these points. For our examples in Figure 5.5, this interpolation is

$$\Phi_e = \gamma \Phi_P + (1 - \gamma)\Phi_E, \tag{5.19}$$

where $\gamma = |eE|/|PE|$ is the interpolation factor. It is easy to derive using the Taylor series expansions that for the grids in Figure 5.5, the interpolation is of the second order of accuracy (a similar derivation was done in Section 4.4.8 in the context of finite difference schemes).

If the cells adjacent to the face e are of the same size, the point e is exactly in the middle between P and E, $\gamma = 0.5$, and (5.19) becomes the simple averaging formula

$$\Phi_e = \frac{\Phi_P + \Phi_E}{2}. \tag{5.20}$$

The formulas (5.19) and (5.20) are directly applicable when we approximate the convective flux integrals, such as (5.7). In the diffusive flux integrals, they can be used to evaluate variable diffusion coefficients, such as χ in (5.8) (see Section 5.3.4 for further discussion).

5.3.3 Central Difference (Linear Interpolation) Scheme for Diffusive Fluxes

The approximation based on the assumption of piecewise linear behavior of solution variables can also be applied to evaluate the gradient terms in the diffusive flux integrals, such as (5.8). In our example, the outward-facing normal n_e is in the positive x-direction, so the approximation is

$$(\chi \nabla \Phi \cdot n)_e = \chi_e \frac{\partial \Phi}{\partial x}\bigg|_e \approx \chi_e \frac{\Phi_E - \Phi_P}{|EP|} = \chi_e \frac{\Phi_E - \Phi_P}{x_E - x_P}. \tag{5.21}$$

Let us analyze the accuracy, with which this formula evaluates the real function's derivative. Using the Taylor series expansion of Φ_P and Φ_E

around the point e, we find that the truncation error is, up to the factor of χ_e,

$$
\text{TE} = \frac{(x_e - x_P)^2 - (x_e - x_E)^2}{2(x_E - x_P)} \left(\frac{\partial^2 \Phi}{\partial x^2} \right)_e
$$
$$
- \frac{(x_e - x_P)^3 + (x_E - x_e)^3}{6(x_E - x_P)} \left(\frac{\partial^3 \Phi}{\partial x^3} \right)_e + \cdots . \tag{5.22}
$$

The first term in the right-hand side is of the first order with respect to the typical grid step, which we can estimate as $|EP|$. The interpolation scheme is formally of the first order. If the cells adjacent to the face e are of the same size, so e is in the middle of the interval PE, the first term is zero, and the scheme acquires the second order of accuracy. In this case, (5.21) becomes the familiar central difference (CD) approximation of the first derivative (4.10). For this reason, the approximation (5.21) and, in general, the approximation of the derivatives in the diffusive flux integrals based on the assumption of linear profiles of solution variables are also called the *central difference* (CD) interpolation.

In the general case, when the cells are nonuniform, the first-order term in the truncation error is nonzero but becomes small when $|Pe|$ and $|eE|$ are close to each other. The CD scheme can be considered a scheme of nearly the second order on nonuniform grids if the variation of the grid size is not strong.

As an illustration, we will develop a finite volume scheme for the purely parabolic one-dimensional version of (5.1):

$$
\frac{\partial u}{\partial t} = a^2 \frac{\partial^2 u}{\partial x^2}, \tag{5.23}
$$

which is, of course, the heat equation. Integrating over the finite volume cell Ω_i of the grid in Figure 5.5b, we obtain

$$
\frac{d}{dt} \int_{x_{i-1/2}}^{x_{i+1/2}} u \, dx = a^2 \left(\frac{\partial u}{\partial x} \right)_e - a^2 \left(\frac{\partial u}{\partial x} \right)_w .
$$

The volume integral is approximated by $u_P \Delta x$, while the derivatives are approximated using the linear interpolation as

$$
\left(\frac{\partial u}{\partial x} \right)_e \approx \frac{u_E - u_P}{\Delta x}, \quad \left(\frac{\partial u}{\partial x} \right)_w \approx \frac{u_P - u_W}{\Delta x} .
$$

Applying the explicit first-order time discretization, we obtain

$$\frac{u_i^{n+1} - u_i^n}{\Delta t} \Delta x = a^2 \frac{u_{i+1}^n - u_i^n}{\Delta x} - a^2 \frac{u_i^n - u_{i-1}^n}{\Delta x}.$$

The formula can be rearranged so that it becomes identical to the familiar finite difference scheme for the heat equation (4.57):

$$\frac{u_i^{n+1} - u_i^n}{\Delta t} = a^2 \frac{u_{i+1}^n - 2u_i^n + u_{i-1}^n}{(\Delta x)^2}. \tag{5.24}$$

The fact that the finite volume schemes for the one-dimensional heat and linear convection equations coincide with the simple finite difference schemes is not at all surprising. The reason is the one-dimensionality of the problems and the obvious analogy between the finite volume grid shown in Figure 5.5b and the uniform finite difference grid used in Chapter 4. The real difference between the finite volume and finite difference schemes appears when multidimensional problems are solved using unstructured grids.

5.3.4 Interpolation of Diffusion Coefficients

The approximation of diffusive flux integrals requires evaluation of the diffusion coefficients, such as χ in (5.1), at the midpoints of the cell's faces. This is a nonissue when the coefficient is a constant or a known function of space and time. Certain steps, however, need to be taken when the coefficient is a function of the solution itself. A good example is the thermal conductivity or viscosity varying with temperature. Interpolation from grid points is required in this case.

To illustrate the possible approaches, let us assume that $\chi = \chi(\Phi)$ and evaluate χ_e in (5.8). One choice is the first-order approximation by the value of χ at one of the neighboring grid points, e.g. according to the upwind rule (5.16). Another choice is the linear interpolation

$$\chi_e = \gamma \chi_P + (1 - \gamma) \chi_E, \quad \gamma = |eE|/|PE|, \tag{5.25}$$

which has the second order of accuracy. The latter interpolation is typically sufficient and most widely used.

The exceptional situation, in which the linear interpolation leads to unacceptable loss of accuracy, has already been discussed in Section 4.4.8 for the finite difference methods. It appears when the diffusion coefficient has a discontinuity in the solution domain. Let, for example, the diffusion coefficient change abruptly at a line located somewhere between points P and E in Figure 5.5. The change is as in Figure 4.9, i.e. from a small value on the side of P to a much larger value on the side of E. The value of the diffusive flux integral across the face AB is primarily determined by the strong resistance to diffusion where χ is small. The one-dimensional resistance analysis carried out in the same way as in Section 4.4.8 leads us to the conclusion that the accurate approximation is based on the harmonic interpolation

$$\chi_e = \frac{|PE|}{\frac{|Pe|}{\chi_P} + \frac{|eE|}{\chi_E}}. \tag{5.26}$$

The linear interpolation provides in this situation less accurate or, at large difference between χ_P and χ_E, utterly incorrect results.

5.3.5 Upwind Interpolation of Higher Order

A popular higher-order scheme for interpolation of Φ_e is the *quadratic upstream interpolation for convective kinematics* or QUICK. It has the third order of accuracy and is obtained by assuming a parabolic profile of the interpolated variable and fitting the profile to its values at two grid points upstream and one point downstream of e. The resulting formulas are quite complex in the general case. In our example, however, the grid points are arranged along the coordinate lines, and the QUICK interpolation can be expressed by a simple formula. Let $(V \cdot n)_e > 0$, which means that the convective transport is in the direction of grid point E (see Figure 5.5a). Fitting a parabola through the values of Φ at the upstream points P and W and the downstream point E, we obtain

$$\Phi_e = \Phi_P + c_1(\Phi_E - \Phi_P) + c_2(\Phi_E - \Phi_W), \tag{5.27}$$

where

$$c_1 = \frac{(x_e - x_P)(x_e - x_W)}{(x_E - x_P)(x_E - x_W)}, \quad c_2 = \frac{(x_e - x_P)(x_E - x_e)}{(x_P - x_W)(x_E - x_W)}.$$

Other interpolation schemes can be used, some having the order of accuracy higher than 3. Details can be found in the specialized literature. The schemes based on the high-order interpolation often become excessively complex when used with nonuniform unstructured grids and, therefore, are rarely applied in such cases.

5.4 FINITE VOLUME METHOD ON UNSTRUCTURED GRIDS

So far, we have kept the discussion relatively simple by limiting the implementation of the finite volume method to one-dimensional or structured Cartesian grids shown in Figure 5.5. An extension to a three-dimensional Cartesian grid is straightforward and similarly simple. Such grids are, however, often insufficient. The reasons will be explained in Chapter 12. At the moment, we just say that the majority of engineering CFD simulations are performed using unstructured grids.

The basic principles of the finite volume discretization presented earlier in this chapter fully apply to unstructured grids. The integral governing equations, such as (5.1), remain the same. The volume and surface integrals are usually approximated by the same midpoint formulas (5.5)–(5.8). New challenges, however, appear in the interpolation of variables to the midpoints of cell faces. In this section, we summarize the main issues and briefly present some ways of solution. A detailed discussion is left to specialized texts on the finite volume method.

The main issues are illustrated by the two-dimensional unstructured grid in Figure 5.6. They are:

1. In general, any three neighboring grid points are not aligned on a straight line.
2. The point at which a line connecting two neighboring points crosses the cell face can be different from the midpoint of the face (see the line PE and points e and d in Figure 5.6).
3. A line between two neighboring points can be nonparallel to the normal n to the face (see the line PE and normal n in Figure 5.6).

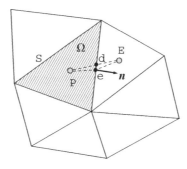

Figure 5.6 Difficulties of interpolation on unstructured grids.

Neither of these issues is significant when the first-order one-point interpolation schemes, such as the upwind interpolation (5.16), are used to evaluate convection flux integrals and variable diffusion coefficients. The interpolation formulas remain valid, retain their first order, and can be applied without modification.

The consequence of the first issue is that interpolation from three and more grid points becomes complex. In theory, two- and three-dimensional interpolations from arbitrarily located points are possible. Practically, the geometric complexity means that schemes based on more than two points are rarely used. This implies the order of approximation not higher than second.

The second issue presents a problem to the standard linear interpolation, such as (5.25) or (5.19). It is easy to see in Figure 5.6 that the linear interpolation reduces its order of approximation from second to first. The additional first-order error is proportional to $|\text{ed}|$. The error becomes negligibly small, and the scheme recovers its second order when $|\text{ed}|$ tends to zero, i.e. the angle between $e\text{P}$ and $e\text{E}$ tends to $180°$.

The third issue implies that the CD scheme (5.21) now evaluates the derivative along the line PE rather than along the normal $\boldsymbol{n}$. If we still use the scheme for evaluation of $\nabla\Phi \cdot \boldsymbol{n}|_e$, the answer contains a significant error. Evidently, the additional first-order error vanishes when the angle between PE and $\boldsymbol{n}$ tends to zero.

What can be done to recover the second order of accuracy of two-point interpolation schemes? One answer is to design a grid, in which the second and third issues are absent. In such a grid, *any line connecting the centers of two neighboring cells is perpendicular to the face separating the cells and crosses this face at its midpoint*. The grids, structured or unstructured, possessing this property are called *orthogonal grids*.

Interestingly, the cells of such grids do not have to be rectangular. As an example, Figure 5.7 shows that a two-dimensional unstructured grid with equilateral triangle cells belongs to this class.

Another popular approach is to simply ignore the issues and apply the standard two-point interpolation schemes even though the grid is not orthogonal. Obviously, it is a matter of good CFD practice and common sense to avoid grids deviating too far from orthogonality when using this approach. As an example, a near-orthogonal grid is shown in Figure 5.8a. On the contrary, the cell combinations shown in Figure 5.8b,c are far from orthogonal. If included into a grid, they are likely to compromise the accuracy of the entire solution and, therefore, must be avoided. This and other aspects of grid quality are further discussed in Chapter 12.

Various other methods, some of them quite sophisticated, are used to rectify the loss of accuracy on unstructured nonorthogonal grids. Information

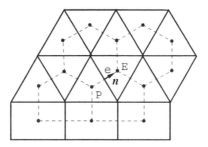

Figure 5.7   Example of a two-dimensional orthogonal unstructured grid.

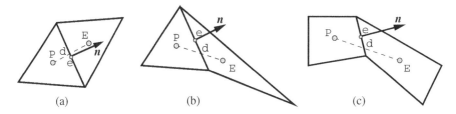

(a)   (b)   (c)

Figure 5.8   Examples of nearly orthogonal and strongly nonorthogonal (and therefore undesirable) cell combinations. In the near-orthogonal case (a), the line EP that connects grid points of two neighboring cells is nearly parallel to the normal n to the face separating the cells and crosses the face at the point d close to the midpoint e. In the strongly nonorthogonal cases (b) and (c), the angle between EP and n and the distance between d and e are large.

on them can be found in specialized texts on finite volume methods. Here we describe, as an example, an elegant way of evaluation of diffusive flux integrals.

The approach is to evaluate the gradient of the integrated variable at the cell-centered grid points and then interpolate the results to the face mid-points. The grid point gradients can be effectively estimated using the divergence theorem. For example, let us approximate $(\partial\Phi/\partial x)_{\mathrm{P}}$. We apply the second-order approximation

$$\left(\frac{\partial\Phi}{\partial x}\right)_{\mathrm{P}} \approx \frac{\int_{\Omega}(\partial\Phi/\partial x)\,d\Omega}{|\Omega|} \tag{5.28}$$

and transform the integral in the right-hand side:

$$\int_{\Omega}(\partial\Phi/\partial x)\,d\Omega = \int_{\Omega}\mathrm{div}(\Phi\boldsymbol{i})\,d\Omega = \int_{S}(\Phi\boldsymbol{i}\cdot\boldsymbol{n})dS$$
$$= \int_{S}\Phi n_x\,dS = \sum_{j}\int_{S_j}\Phi n_x\,dS,$$

where j is the index that marks the cell's faces. The surface integral over every face is replaced by the second-order approximation

$$\int_{S_j} \Phi n_x \, dS \approx \Phi_{e_j} S_{x_j},$$

where Φ_{e_j} is the value of Φ at the corresponding midpoint. The final formula is

$$\left(\frac{\partial \Phi}{\partial x}\right)_P \approx \frac{\sum_j \Phi_{e_j} S_{x_j}}{|\Omega|}. \tag{5.29}$$

Note that, typically, the midpoint values Φ_{e_j} have to be calculated anyway to approximate the convective fluxes.

5.5   IMPLEMENTATION OF BOUNDARY CONDITIONS

Near the boundary of the computational domain, the boundary conditions should be incorporated into the integral balance equations for the cells and, thus, into the finite volume discretization. The special treatment concerns only the surface integrals over the faces lying on the boundary. The cumulative (convective plus diffusive) flux should be determined on the basis of the boundary conditions.

Let us consider the two-dimensional example shown in Figure 5.9. One face lying on the boundary is AB. We have to replace the combination $-\int_{AB} \Phi V \cdot n \, dS + \int_{AB} \chi \nabla \Phi \cdot n \, dS$ by the integral that gives the flux due to the boundary conditions. For the Neumann condition, when the normal

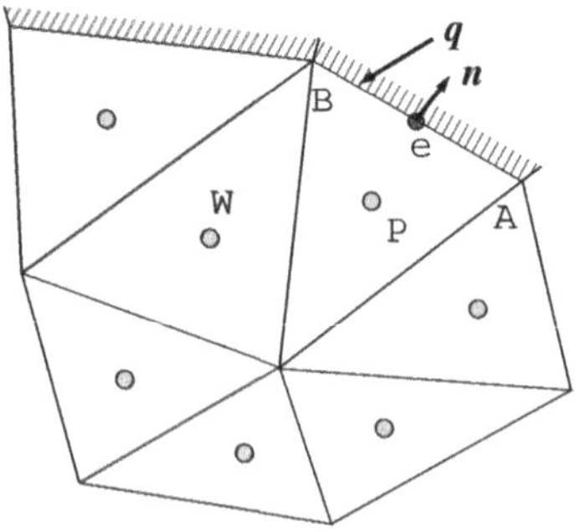

Figure 5.9   Treatment of boundary conditions in finite volume methods.

component of the boundary flux q_n is prescribed, this can be done in a straightforward manner. We simply replace the entire combination of the surface integrals by

$$-\int_{AB} q_n \, dS \approx q_{ne} S_{AB}. \tag{5.30}$$

For the Dirichlet and Robin conditions, the flux is unknown and has to be approximated using the boundary conditions and values of Φ at nearby interior points. For example, in the case of the Dirichlet condition, when Φ at the face AB is prescribed, we assume that the flux is provided by a diffusive mechanism activated by the gradient of Φ at the boundary. The boundary flux is

$$\int_{AB} (\boldsymbol{q} \cdot \boldsymbol{n}) dS = -\int_{AB} \chi \nabla \Phi \cdot \boldsymbol{n} \, dS \approx -\chi \left(\frac{\partial \Phi}{\partial n} \right)_e S_{AB}. \tag{5.31}$$

To approximate the gradient of Φ, we can use the scheme of the first order

$$\left(\frac{\partial \Phi}{\partial n} \right)_e \approx \frac{\Phi_e - \Phi_P}{|Pe|} \tag{5.32}$$

or the interpolation of higher order, which uses values of Φ at more than one interior grid points.

BIBLIOGRAPHY

Blazek, J. (2005). *Computational Fluid Dynamics: Principles and Applications.* Amsterdam: Elsevier.

Ferziger, J.H. and Perić, M. (2001). *Computational Methods for Fluid Dynamics,* 3e. New York, NY: Springer.

Versteeg, H. and Malalasekra, W. (2007). *An Introduction to Computational Fluid Dynamics: The Finite Volume Method.* Upper Saddle River, NJ: Prentice Hall.

PROBLEMS

1. Transform the following equations into the integral form similar to (5.1). For each equation, identify, if present, the rate of change term, volumetric source term, convective flux term, and diffusive flux term.

a) Linear convection equation $u_t + cu_x = 0$, where $c > 0$ is a constant.

b) One-dimensional heat equation $u_t = a^2 u_{xx} + f(x)$.

c) Three-dimensional heat equation $u_t = \nabla^2 u + g(\boldsymbol{x})$. (*Hint*: Use the identity $\nabla^2 u = \operatorname{div}(\nabla u)$ and the divergence theorem.)

d) Three-dimensional Poisson equation $\nabla^2 u = g(\boldsymbol{x})$.

e) One-dimensional Burgers equation $u_t + uu_x = \mu u_{xx}$, where $\mu > 0$ is a constant.

2. A finite volume scheme is developed using a two-dimensional structured Cartesian grid with constant steps Δx and Δy (see Figure 5.5a). Write the following approximations using the values of the function u at the grid points, such as P, E, EE, and W.

a) Approximation of the second order for the volume integral $\int_{\Omega_i} u\, d\Omega$.

b) Upwind approximation for the convective flux integral $\int_{S_e} u V \cdot \boldsymbol{n}\, dS$, where S_e is the face containing the point e, $\boldsymbol{V}$ is the constant velocity $\boldsymbol{V} = (1, 0.5)$, and $\boldsymbol{n}$ is the outward-facing unit-length normal to S_e.

c) The same as in (b) but with velocity $\boldsymbol{V} = (-0.5, 1)$.

d) Central difference approximation of the second order for the diffusive flux integral $\int_{S_e} (\partial u / \partial x)\, dS$.

3. A finite volume scheme is developed using a two-dimensional structured Cartesian grid with constant steps Δx and Δy (see Figure 5.5a). Write the following approximations using the values of the function u at the grid points, such as P, E, EE, and W.

a) Approximation based on linear interpolation of the convective flux integral $\int_{S_e} u V \cdot \boldsymbol{n}\, dS$, where S_e is the face containing the point e, $\boldsymbol{V}$ is the constant velocity $\boldsymbol{V} = (1, 0.5)$, and $\boldsymbol{n}$ is the outward-facing unit-length normal to S_e.

b) The same as in (a) but with $\boldsymbol{V} = (0.0, 2.5)$.

c) The same as in (a) but the approximation must be based on the QUICK interpolation.

d) Second-order approximation of the diffusive flux integral $\int_{S_e} \chi(u) \nabla u \cdot \boldsymbol{n}\, dS$.

4. The Burgers equation $u_t + uu_x = \mu u_{xx}$ is solved by the finite volume method on a one-dimensional grid with constant step Δx (see Figure 5.5b). Develop the schemes based on the following principles.

a) Upwind interpolation for convective flux and linear interpolation for diffusive flux.

b) Linear interpolation for convective and diffusive fluxes.

5. The conservation equation (5.1) is solved by the finite volume method on a two-dimensional structured Cartesian grid with constant steps Δx and Δy (see Figure 5.5a). Develop the schemes based on the following principles.

a) Upwind interpolation for convective flux and linear interpolation for diffusive flux.

b) Linear interpolation for convective and diffusive fluxes.

6. The two-dimensional heat equation $\partial u / \partial t = \nabla \cdot (\chi \nabla u) + f$, where χ is a function of u and f is a known function of x and t, is solved in a rectangular domain $0 \le x \le L_x$, $0 \le y \le L_y$. The boundary conditions correspond to perfect insulation on all four boundaries. Write the integral conservation equation, and approximate all integrals for an internal cell and for cells adjacent to the boundaries. Use the structured Cartesian grid with constant steps Δx and Δy (see Figure 5.5a). The resulting finite volume scheme must be of the second order of approximation.

6

NUMERICAL STABILITY FOR MARCHING PROBLEMS

6.1 INTRODUCTION AND DEFINITION OF STABILITY

Numerical stability is an essential ingredient of a successful computational solution of any marching problem. Absence of this ingredient renders the solution completely useless, as illustrated by the following example.

6.1.1 Example

We return to our first finite difference scheme introduced at the end of Chapter 3 and further discussed in Chapter 4. The partial differential equation (PDE) problem for the inhomogeneous heat equation with Dirichlet boundary conditions

$$\frac{\partial u}{\partial t} = a^2 \frac{\partial^2 u}{\partial x^2} + \sin 5x, \quad u(0, t) = 0, \quad u(L, t) = 0, \quad u(x, 0) = x(\pi - x)$$

(6.1)

is solved in the domain $0 < x < \pi$, $0 < t < 50$ with $a = 0.5$. The uniform grid $x_i = i\Delta x$, $t^n = n\Delta t$ with $i = 0, \ldots, N$, $n = 0, \ldots, M$ is used. The finite

Essential Computational Fluid Dynamics, Second Edition. Oleg Zikanov.
© 2019 John Wiley & Sons, Inc. Published 2019 by John Wiley & Sons, Inc.
Companion Website: www.wiley.com/go/zikanov/essential

difference scheme is explicit and based on the first-order forward difference in time and second-order central difference in space:

$$u_i^{n+1} = u_i^n + \Delta t \left(a^2 \frac{u_{i+1}^n - 2u_i^n + u_{i-1}^n}{(\Delta x)^2} + f_i^n \right),$$

$$u_0^n = 0, \quad u_N^n = 0, \quad u_i^0 = x_i(\pi - x_i). \tag{6.2}$$

The problem has an analytical solution, according to which $u(x, t)$ differs from the equilibrium solution by less than 10^{-5} at $t = 50$. We accept this error and consider the equilibrium state as the exact solution of the problem at $t = 50$. It can be easily found by assuming $\partial u / \partial t = 0$, which leads to

$$a^2 \frac{\partial^2 u_{exact}}{\partial x^2} = -\sin 5x,$$

so

$$u_{exact}(x, t = 50) = \frac{1}{25a^2} \sin 5x = 0.16 \sin 5x. \tag{6.3}$$

We can evaluate the accuracy of a numerical solution by calculating the absolute error

$$\epsilon_{abs}(x_i) = \left| u_{exact}(x_i, t = 50) - u_i^M \right|, \quad i = 0, \ldots, N \tag{6.4}$$

and the relative error

$$\epsilon_{rel}(x_i) = \frac{\epsilon_{abs}(x_i)}{u_{mean}}, \text{ where } u_{mean} = \left(\frac{1}{N+1} \sum_{i=0}^{N} (u_i^M)^2 \right)^{1/2}. \tag{6.5}$$

The results are presented in Figure 6.1. The computed u is shown as a function of time at $x = x_{50} = L/2$ in Figure 6.1a,c and as a function of x at a fixed time in Figure 6.1b,d.

Let us first consider the solution obtained with $\Delta t = 0.01$. The results presented in Figure 6.1a,b are obviously incorrect. Soon after the start, the solution begins to grow rapidly and reaches the upper limit set by the computer's working memory. In the state calculated at $t = 2$ (see Figure 6.1b), we see wild oscillations, which have nothing in common with the exact solution (6.3). The apparent amplitude of 10^{36} is, in fact, the limit allowed by the plotting software.

By contrast, when the solution is repeated with the same Δx but the time step reduced to $\Delta t = 10^{-4}$, the behavior is much more reasonable (see

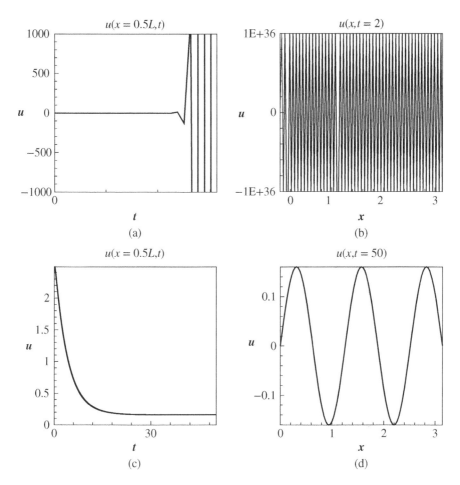

Figure 6.1 Solution of problem (6.1) using finite difference scheme (6.2). Grid step is $\Delta x = 0.01\pi$. Time step is $\Delta t = 0.01$ in (a) and (b) and $\Delta t = 10^{-4}$ in (c) and (d).

Figure 6.1c,d). The numerical solution follows closely the analytical exact solution. At $t = 50$, the relative error (6.5) is less than 10^{-4}.

Why does the size of the time step have such a dramatic impact on the solution? This question is answered in the rest of this chapter.

6.1.2 Discretization and Round-Off Error

Let us consider more carefully the errors that are generated in the process of numerical solution of a PDE problem. The following discussion is general in the sense that it applies to all kinds of discretization and all time-dependent PDE, although we will use the finite difference approximation (6.2) of the heat equation (6.1) as an example.

There are two kinds of numerical error. First, the numerical approximation of a PDE problem differs from the problem itself by the truncation error. We learned in the previous chapters that, for a given PDE problem, the amplitude of this error is determined by the discretization scheme and the size of the grid steps. For example, the finite difference scheme (6.2) represents the heat equation with the truncation error of the magnitude $O(\Delta t, (\Delta x)^2)$. We will call the error resulting from the truncation the *discretization error* and define it as the difference between the exact solution of the system of algebraic equations generated by the numerical scheme and the exact analytical solution of the PDE problem.

Another source of error is the computer arithmetics. A computer does not solve the algebraic equations, such as (6.2), exactly. There is always a *round-off error* because any computer performs arithmetic operations using a finite number of digits. The number varies depending on the computational platform and programming instructions, but it is always finite. The corresponding round-off errors are normally very small. If, however, they accumulate as the marching procedure advances in time, their effect can be quite significant.

Let us introduce the notation for the vectors of the solution values at the grid points:

ua is the exact analytical solution of the PDE problem such as (6.1).

ud is the exact (obtained by an imaginary computer with perfectly accurate arithmetics) solution of the system of discretization equations such as (6.2).

un is the actually computed solution of the system of discretization equations.

The discretization error is $\epsilon_d = ud - ua$, and the round-off error is $\epsilon = un - ud$. The actually computed solution differs from the exact solution of the PDE problem as

$$un = ua + \epsilon_d + \epsilon, \qquad (6.6)$$

while its relation to the perfectly accurate solution of the discretized problem is

$$un = ud + \epsilon. \qquad (6.7)$$

We now return to the question stated at the end of Section 6.1.1. According to (6.6), the computed solution can be incorrect because of either ϵ_d or ϵ. The first possibility can be easily rejected. The discretization error has the same order of magnitude as the truncation error of the scheme. In our example, this means $O(\Delta t, (\Delta x)^2)$. At $\Delta t = 10^{-2}$ and $\Delta x = 10^{-2}\pi$, $\epsilon_d \sim 10^{-2}$, which is far from 10^{36}. We conclude that the discretization error cannot be the reason of enormous inaccuracy illustrated in Figure 6.1a,b.

Our next step is to analyze the behavior of the round-off error.

6.1.3 Definition

Let us consider one time step of a marching procedure. It advances the solution from the time layer t^n to the time layer t^{n+1}. We disregard the details of this step for the moment and view the procedure as a black box illustrated in Figure 6.2.

The main question is what happens to the round-off error as it passes through the black box? If it is amplified, one time step after another, the growth and accumulation eventually increase ϵ to an unacceptably high level. This situation is called *instability*. If it decays or stays limited to the low level determined by the precision of the computer arithmetics, its effect can be neglected. In this case, we have *stability*.

The formal definition of stability can be done in several ways. According to one, a marching scheme is unstable if the amplitude of the round-off error ϵ increases at, at least, one grid point as the solution passes from the time layer t^n to the time layer t^{n+1}. Otherwise, the scheme is called stable. Let us use the notation ϵ_i^n for the round-off error at time layer t^n and grid point x_i. The scheme is stable if

$$\left| \frac{\epsilon_i^{n+1}}{\epsilon_i^n} \right| \leq 1 \tag{6.8}$$

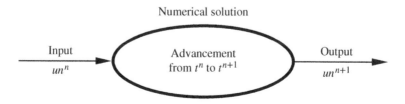

Figure 6.2   A time step of a marching scheme considered as a black box.

at every point x_i. The scheme is unstable if there is a point, where the condition is violated.

Another definition requires that the overall round-off error in the entire space domain does not grow with time. This can be expressed in terms of the vector norm of the error $\| \epsilon^n \| = \left(N^{-1} \sum_{i=1}^{N} (\epsilon_i^n)^2 \right)^{1/2}$ as

$$\frac{\| \epsilon^{n+1} \|}{\| \epsilon^n \|} \leq 1. \tag{6.9}$$

One may also require that the sign $\leq$ is replaced in (6.8) and (6.9) by strict $<$, i.e. that the round-off error is not just limited, but decreases from one time layer to the next.

It is safe to leave the differences between the definitions to mathematicians. Our only concern is that any unstable solution, in which the round-off error grows from one time layer to the next violating (6.8) or (6.9), produces incorrect results and has to be avoided at all costs in the CFD analysis.

6.2   STABILITY ANALYSIS

6.2.1   Neumann Method

Among the methods used to analyze the stability, the Neumann method based on Fourier expansions is the most widely used. The basic idea is to assume the actual numerical solution in the form (6.7), feed it as the input stream into the black box in Figure 6.2, and use the Fourier expansion to analyze the behavior of ϵ.

We will employ the scheme (6.2) for demonstration. First, note that both the actually computed (rounded-off) **un** and the exact solution of the numerical problem **ud** satisfy the finite difference equation (6.2). The only difference is that in the equation for **un**, all numbers are rounded-off to a certain fixed number of digits. Writing (6.7) at every grid point

$$un_i^n = ud_i^n + \epsilon_i^n, \tag{6.10}$$

and substituting into (6.2), we obtain

$$\frac{ud_i^{n+1} + \epsilon_i^{n+1} - ud_i^n - \epsilon_i^n}{\Delta t}$$
$$= a^2 \frac{ud_{i+1}^n + \epsilon_{i+1}^n - 2ud_i^n - 2\epsilon_i^n + ud_{i-1}^n + \epsilon_{i-1}^n}{(\Delta x)^2} + f_i^n. \tag{6.11}$$

The exact solution ud also satisfies (6.2):

$$\frac{ud_i^{n+1} - ud_i^n}{\Delta t} = a^2 \frac{ud_{i+1}^n - 2ud_i^n + ud_{i-1}^n}{(\Delta x)^2} + f_i^n. \tag{6.12}$$

Subtraction of (6.12) from (6.11) yields the equation for the round-off error

$$\frac{\epsilon_i^{n+1} - \epsilon_i^n}{\Delta t} = a^2 \frac{\epsilon_{i+1}^n - 2\epsilon_i^n + \epsilon_{i-1}^n}{(\Delta x)^2}. \tag{6.13}$$

The particular form (6.13) of the error equation is for the particular PDE (heat equation (6.1)) and finite difference scheme (simple explicit scheme (6.2)) used in our example. The principal procedure, substitution of un into the finite difference equation and subtraction of the equation for ud, is, however, universal. In general, the final equation may contain ud but contains only ϵ and known coefficients if the solved PDE problem is linear.

Note that the inhomogeneous term in the right-hand side of (6.1) does not appear in the equation for the error (6.13) and, thus, does not affect the stability. Our analysis would be the same if we considered any other inhomogeneous term or the homogeneous heat equation. In general, the inhomogeneous source terms can be neglected in the stability analysis if they are completely independent of the solution u. They, however, should be taken into account in the opposite case.

Equation (6.13) can be considered as a finite difference representation of the PDE

$$\frac{\partial \epsilon}{\partial t} = a^2 \frac{\partial^2 \epsilon}{\partial x^2}, \tag{6.14}$$

which can be solved analytically using the method of separation of variables. We will follow this approach, ignore the effect of boundary conditions, and write the general solution as

$$\epsilon(x, t) = \sum_m b_m(t) e^{ik_m x} + \overline{c.c}, \tag{6.15}$$

where $\overline{c.c}$ stands for complex conjugate of the first term in the right-hand side. Formula (6.15) allows some interpretation. The round-off error at a given time layer t^n is an irregular function of x like that shown in Figure 6.3. The series (6.15) is the decomposition of this function into Fourier harmonics. We know from the theory of Fourier series that such a decomposition

Figure 6.3 Round-off error as a function of x.

is possible for any piecewise continuous function and that the wavenumbers are

$$k_m = \frac{2\pi m}{L}, \quad m = 1, 2, 3, \ldots, \quad (6.16)$$

where L is the length of the solution domain.

One correction of the classical Fourier theory is needed. In the numerical solution we deal not with a continuous function $\epsilon(x, t)$ but, rather, with its values at discrete points x_i separated from each other by the distance Δx. As illustrated in Figure 6.4, the harmonics with the wavelength smaller than $2\Delta x$ cannot be identified on the grid. We have to limit the series (6.15) by the minimum discernible wavelength, i.e. the maximum discernible wavenumber

$$k_{max} = \frac{2\pi}{2\Delta x} = \frac{\pi}{\Delta x} = \frac{2\pi}{L}\frac{N}{2}, \quad (6.17)$$

where N is the number of the grid points. The set of the wavenumbers in (6.16) is, thus, limited by $m = 1, \ldots, N/2$. The corrected decomposition

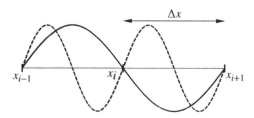

Figure 6.4 Solid curve illustrates the shortest Fourier wave discernible on a grid with step Δx. It has the wavelength $2\Delta x$. The waves with smaller wavelengths cannot be identified on the grid. For example, the presence of the harmonic with the wavelength Δx shown by the dashed curve remains unrecognized since the harmonic has the same values at the grid points as the long-wave harmonic.

formula is

$$\epsilon(x, t) = \sum_{m=1}^{N/2} b_m(t)e^{\iota k_m x} + \overline{c.c}, \quad k_m = \frac{2\pi m}{L}. \tag{6.18}$$

The growth or decay of the round-off error is determined by the behavior of the coefficients $b_m(t)$. If, for at least one Fourier mode $\epsilon_m(x, t) = b_m(t)e^{\iota k_m x}$, the amplitude grows with time, the entire error grows, and the scheme is unstable. In the case when the PDE and the numerical approximation equation are linear, the behavior of $b_m(t)$ can be determined analytically, since the error equations, such as (6.13) and (6.14), are also linear.

Substituting (6.18) into (6.13) and using the functional orthogonality of the Fourier series, we separate the PDE into $N/2$ equations, one for every Fourier mode and its conjugate. Furthermore, each of these separate equations has an exponential solution:

$$\epsilon_m(x, t) = e^{c_m t}e^{\iota k_m x}, \tag{6.19}$$

where c_m is yet unknown complex coefficient.

The stability criterion (6.8) says that the scheme is stable if there are no grid points where ϵ is amplified. We translate this as a requirement that none of the Fourier modes ϵ_m is amplified and see immediately that the coefficients c_m provide all the necessary information. The amplification of the mode ϵ_m is the same at all grid points and is given by

$$\left| \frac{\epsilon_i^{n+1}}{\epsilon_i^n} \right| = \left| \frac{e^{c_m(t^n + \Delta t)}e^{\iota k_m x_i}}{e^{c_m t^n}e^{\iota k_m x_i}} \right| = |e^{c_m \Delta t}|. \tag{6.20}$$

This quantity is called the *amplification factor*

$$G_m = |e^{c_m \Delta t}|. \tag{6.21}$$

The stability criterion can be reformulated as follows: *the scheme is stable if the condition*

$$G_m = |e^{c_m \Delta t}| \leq 1 \tag{6.22}$$

is satisfied for all $m = 1, \ldots, N/2$.

If $G_m > 1$ for at least one m, the component ϵ_m of the round-off error grows *exponentially*, so it quickly becomes unacceptably large, even though the error produced at each time step is small.

We will now return to our example and do some algebra. At the grid points, the solution (6.19) is

$$\epsilon_m(x_i, t^n) = \epsilon_i^n = e^{c_m t^n} e^{\iota k_m x_i}, \tag{6.23}$$

$$\epsilon_m(x_i, t^{n+1}) = \epsilon_i^{n+1} = e^{c_m(t^n + \Delta t)} e^{\iota k_m x_i}, \tag{6.24}$$

$$\epsilon_m(x_{i-1}, t^n) = \epsilon_{i-1}^n = e^{c_m t^n} e^{\iota k_m(x_i - \Delta x)}, \tag{6.25}$$

$$\epsilon_m(x_{i+1}, t^n) = \epsilon_{i+1}^n = e^{c_m t^n} e^{\iota k_m(x_i + \Delta x)}, \tag{6.26}$$

and the equation for the mth mode of the round-off error is

$$\frac{e^{c_m(t^n + \Delta t)} e^{\iota k_m x_i} - e^{c_m t^n} e^{\iota k_m x_i}}{\Delta t}$$
$$= a^2 \frac{e^{c_m t^n} e^{\iota k_m(x_i + \Delta x)} - 2 e^{c_m t^n} e^{\iota k_m x_i} + e^{c_m t^n} e^{\iota k_m(x_i - \Delta x)}}{(\Delta x)^2}.$$

Dividing by $e^{c_m t^n} e^{\iota k_m x_i}$ we obtain

$$\frac{e^{c_m \Delta t} - 1}{\Delta t} = a^2 \frac{e^{\iota k_m \Delta x} - 2 + e^{-\iota k_m \Delta x}}{(\Delta x)^2},$$

which can be regrouped as

$$e^{c_m \Delta t} = 1 + \frac{a^2 \Delta t}{(\Delta x)^2}(e^{\iota k_m \Delta x} - 2 + e^{-\iota k_m \Delta x}). \tag{6.27}$$

We use the identity

$$\cos(k_m \Delta x) = \frac{e^{\iota k_m \Delta x} + e^{-\iota k_m \Delta x}}{2}, \tag{6.28}$$

so (6.27) becomes

$$e^{c_m \Delta t} = 1 + \frac{2a^2 \Delta t}{(\Delta x)^2}(\cos(k_m \Delta x) - 1) = 1 - \frac{4a^2 \Delta t}{(\Delta x)^2} \sin^2\left(\frac{k_m \Delta x}{2}\right). \tag{6.29}$$

It is convenient to introduce the fictitious angle $\beta = k_m \Delta x$ and use it instead of the wavenumber index m. With the wavenumber k_m running from $2\pi/L$ to $N\pi/L$ (see (6.16) and (6.17)), and $\Delta x = L/N$, the angle is always within the limits

$$\frac{2\pi}{N} \leq \beta \leq \pi. \tag{6.30}$$

The stability criterion is formulated in terms of β as

$$G(\beta) \leq 1 \text{ for all } \beta. \tag{6.31}$$

The rest is easy. We use (6.22) and (6.29) with the abbreviation

$$r = \frac{a^2 \Delta t}{(\Delta x)^2} \tag{6.32}$$

to express the stability criterion as

$$\left| 1 - 4r \sin^2 \left(\frac{\beta}{2} \right) \right| \leq 1, \text{ or } -1 \leq 1 - 4r \sin^2 \left(\frac{\beta}{2} \right) \leq 1,$$

$$\text{or } \begin{cases} 4r \sin^2(\beta/2) \geq 0 \\ 4r \sin^2(\beta/2) \leq 2. \end{cases}$$

Since r is always positive and $0 \leq \sin^2(\beta/2) \leq 1$, the first condition is satisfied automatically. The second condition gives the *stability criterion for the simple explicit scheme* (6.2) *for the heat equation*

$$r = \frac{a^2 \Delta t}{(\Delta x)^2} \leq \frac{1}{2}. \tag{6.33}$$

We are now prepared to answer the question of what was wrong with the first of the numerical solutions shown in Figure 6.1. At $\Delta x = 0.01\pi$, $a = 0.5$, and $\Delta t = 0.01$, we have the stability parameter $r \approx 2.54$, which is far above the stability limit. What is seen in the top two plots of Figure 6.1 is the typical example of a *numerically unstable solution*. For the second simulation, we use $\Delta t = 10^{-4}$, which corresponds to $r \approx 0.0253$. The solution is stable, and accurate results are obtained as shown in the other two plots of Figure 6.1. We can evaluate the maximum time step at which the scheme is stable as

$$\Delta t_{max} = \frac{(\Delta x)^2}{2a^2} \approx 1.97 \times 10^{-3}.$$

It is not always possible to find an analytical estimate of the upper bound of $G(\beta)$, as we did in (6.33). Numerical evaluation may be needed. Furthermore, the information on the amplification of the round-off error at different wavelengths – that is, the information on the behavior of the function $G(\beta)$ – can be useful for understanding the properties of a finite difference scheme. It is, therefore, convenient and customary to plot the entire $G(\beta)$

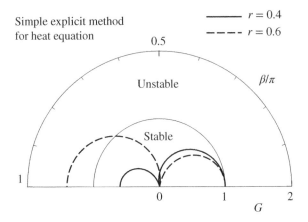

Figure 6.5 Amplification factor G of the simple explicit scheme (6.2) for the heat equation (6.1).

in Cartesian or, as illustrated in Figure 6.5, polar coordinates. In the latter case, β and G serve as the polar angle and radius, respectively. A scheme is deemed stable if the entire curve $G(\beta)$ at $0 < \beta < \pi$ lies within the unit radius circle $G = 1$. For example, the curves in Figure 6.5 illustrate the fact that the simple explicit scheme (6.2) applied to the one-dimensional heat equation is stable at $r = 0.4$ and unstable at $r = 0.6$.

The specific stability results, such as (6.29) and (6.33), are only valid for the particular example when the scheme (6.2) is applied to the heat equation, but the procedure we used to arrive at these results is universal and can be used for other schemes and other equations. Let us reiterate the main steps:

- The equation for the round-off error is derived by substituting the expression (6.10) into the finite difference scheme.
- The Fourier decomposition (6.15) is assumed for the round-off error.
- The error equation is solved for separate Fourier modes to find the amplification factor $G(\beta)$.
- The stability criterion, for example, (6.33), is derived.

The comment is in order concerning whether we can apply the methods and results of the Neumann stability analysis developed for simple model problems, such as (6.1), to significantly more complex equations such as the full Navier–Stokes system. The mathematically rigorous answer is no. Most importantly, the model equations are linear with constant coefficients. Only for such systems the exponential solution (6.19) exists. On the contrary, the realistic equations of fluid flows and convective heat transfer are almost always nonlinear and, often, include variable coefficients. We will return to this question in the following chapters and show that the stability of time integration of complex equation can be analyzed in approximate sense using the criteria derived for model systems.

Let us consider another example – fully implicit scheme for the heat equation (6.1). The scheme was introduced in Chapter 4:

$$\frac{u_i^{n+1} - u_i^n}{\Delta t} = a^2 \frac{u_{i+1}^{n+1} - 2u_i^{n+1} + u_{i-1}^{n+1}}{(\Delta x)^2} + f_i^{n+1}. \tag{6.34}$$

Following the procedure of the Neumann stability analysis (see (6.10)–(6.13)), we derive the equation for the round-off error

$$\frac{\epsilon_i^{n+1} - \epsilon_i^n}{\Delta t} = a^2 \frac{\epsilon_{i+1}^{n+1} - 2\epsilon_i^{n+1} + \epsilon_{i-1}^{n+1}}{(\Delta x)^2}.$$

Substituting the expressions analogous to (6.23)–(6.26) and dividing by $e^{c_m t^n} e^{\imath k_m x_i}$, we obtain

$$\frac{e^{c_m \Delta t} - 1}{\Delta t} = a^2 e^{c_m \Delta t} \frac{e^{\imath k_m \Delta x} - 2 + e^{-\imath k_m \Delta x}}{(\Delta x)^2},$$

which can be rewritten as

$$e^{c_m \Delta t} \left[1 - r(e^{\imath k_m \Delta x} - 2 + e^{-\imath k_m \Delta x}) \right] = 1$$

or, with (6.28),

$$e^{c_m \Delta t} [1 + 2r - 2r \cos \beta] = 1.$$

Using the trigonometric identity we find the amplification factor as

$$G(\beta) = \frac{1}{1 + 4r \sin^2(\beta/2)}. \tag{6.35}$$

The plot of $G(\beta)$ is shown in Figure 6.6. It illustrates the fact that *the stability condition $G(\beta) \leq 1$ is satisfied for any $r \geq 0$.* The scheme is stable for any choice of the time and space grid steps. This situation is typical for implicit schemes. Many of them (but not all!) are *unconditionally stable.*

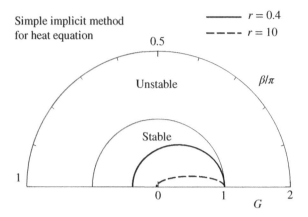

Figure 6.6   Amplification factor G of the simple implicit scheme (6.34) applied to the heat equation (6.1).

6.2.2   Matrix Method

The matrix method is another approach to the stability analysis. It can be used for two-layer schemes, i.e. for the schemes with equations containing values of u at only two consecutive time layers. The equation for the round-off error, for example, (6.13), can be expressed in the matrix form

$$\epsilon^{n+1} = \mathbf{A} \cdot \epsilon^n \text{ or } \epsilon^{n+1} = \underbrace{\mathbf{A} \cdot \mathbf{A} \cdots \mathbf{A}}_{n+1} \cdot \epsilon^0 = \mathbf{A}^{n+1} \cdot \epsilon^0, \tag{6.36}$$

where $\mathbf{A}$ is a square matrix $N \times N$. In our example of the simple explicit scheme (6.2) applied to the heat equation, the matrix has zero elements except for the three main diagonals:

$$\mathbf{A} = \begin{pmatrix} (1-2r) & r & 0 & \cdots & \cdots & 0 \\ r & (1-2r) & r & 0 & \cdots & 0 \\ 0 & r & (1-2r) & r & 0 & \cdots \\ \cdots & \cdots & \cdots & \cdots & \cdots & \cdots \\ \cdots & \cdots & 0 & r & (1-2r) & r \\ 0 & \cdots & \cdots & 0 & r & (1-2r) \end{pmatrix},$$

where $r = a^2 \Delta t / (\Delta x)^2$. In the cases when the boundary conditions affect the stability, their approximation must be included into the matrix.

It can be shown that the vector norm $\|\epsilon^{n+1}\|$ of the round-off error evolving according to (6.36) remains bounded if the eigenvalues of A are all different and have absolute values less than or equal to 1:

$$|\lambda_m| \leq 1, \forall \quad m = 1, \ldots, N. \tag{6.37}$$

In our example, the tridiagonal form of the matrix simplifies the task of finding the eigenvalues. There is a direct formula:

$$\lambda_m = 1 - 4r \sin^2 \left(\frac{m\pi}{2(N+1)} \right).$$

In more general cases, when the analytical formulas do not exist, the eigenvalues can be found using one of the approximate numerical methods. The stability criterion (6.37) leads to

$$-1 \leq 1 - 4r \sin^2 \left(\frac{m\pi}{2(N+1)} \right) \leq 1.$$

The right-hand side condition is always satisfied, while the left-hand side condition gives

$$r \sin^2 \left(\frac{m\pi}{2(N+1)} \right) \leq \frac{1}{2}.$$

The inequality is true for any m if

$$r \leq \frac{1}{2}. \tag{6.38}$$

The Neumann and matrix methods deal with the same subject – the round-off error – and have the same purpose, namely, to establish the conditions under which the error remains bounded. There are some differences that are briefly discussed here.

The stability criteria produced by the two methods can, in general, be different. This is not a serious concern for us because they coincide in almost all practically important cases.

There is a difference in the range of applicability. The matrix method just described can only be applied to two-layer schemes. The limitation is quite strong since, as we will see in the next chapters, there are many powerful and popular multilevel schemes, in which the finite difference equations connect values of u at three or more time layers. On the contrary, the Neumann method can be applied, at least theoretically, to schemes with any number of layers.

The matrix method has an advantage that it can include the boundary conditions of Dirichlet and Neumann type into the stability analysis. The Neumann method disregards the boundary conditions.

6.3 IMPLICIT VERSUS EXPLICIT SCHEMES – STABILITY AND EFFICIENCY CONSIDERATIONS

We saw on the example of the two schemes for the heat equation that the implicit and explicit methods have very different stability characteristics. For the explicit schemes, small Δt and, thus, large number of time steps can be required to avoid numerical instability. For example, a solution by the simple explicit method (6.2) with $\Delta x = 0.01\pi$ is stable only if $\Delta t < 1.97 \times 10^{-3}$, which means that about 2.5×10^4 time steps are needed to cover the interval $0 < t < 50$. On the contrary, many implicit methods, such as the simple implicit method (6.34), are unconditionally stable. The solution can be completed in much fewer time steps, say, 10^3 or even 10^2.

At first glance, the implicit methods are much more efficient and have to be invariably used. A more careful consideration, however, shows that the situation is complex and case specific.

Two important factors have to be considered. The first is that, although we can choose the time step of an implicit scheme as large as desired, caution must be exercised. If Δt is too large, the numerical solution, albeit stable, suffers from large truncation errors. For example, the simple implicit method (6.34) has $\text{TE} = O\left(\Delta t, (\Delta x)^2\right)$, and selecting, say, $\Delta t = 1$ would not be a good idea if one is looking for an accurate representation of the time evolution of u.

The situation is different when we know that there is a unique final equilibrium state of the solution and are interested only in this state. In this case, using an implicit method with large Δt is justified. As illustrated in Figure 6.7, the numerical solution obtained with a large time step does not accurately represent the evolution of the system. The final state, however, is a solution of a steady-state equation. The accuracy of numerical approximation depends on Δx and order of spatial approximation, but not on Δt.

The second factor is the computational cost. As opposite to the explicit schemes, where the time advancement is a relatively simple and computationally inexpensive task (see, e.g. (6.2)), the implicit approach requires that we solve a system of coupled linear algebraic equations at every time

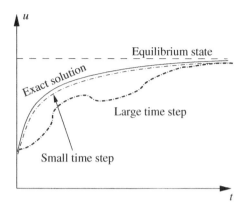

Figure 6.7 Effect of time step on solution by an implicit unconditionally stable scheme. Small Δt guarantees accurate representation of the entire solution, including the transient part. Accuracy is low for the transient part if Δt is large, but the final equilibrium state is reached faster.

Table 6.1 Comparison between implicit and explicit approaches to solution of marching problems.

	Explicit schemes	Implicit schemes
Advantages	• Small amount of computations needed for one time step	• No or weak stability constraints. Large Δt can be used to achieve an equilibrium state in shorter time
	• Accurate solution is generated if stability criteria are satisfied, and sufficiently small Δx, Δt are used	• Accurate solution is generated if sufficiently small Δx, Δt are used
	• Easy to program	
Disadvantages	• Stability constraints. Often, very small Δt must be taken, which may lead to large total amount of computations	• Larger amount of computations is needed for one time step, which may lead to large total amount of computations
		• More difficult to program

layer. The total number of equations is equal to the number of unknowns multiplied by the number of the space grid points, i.e. can be quite large.

Table 6.1 summarizes the advantages and disadvantages of the implicit and explicit approaches. They were illustrated in this section on the example of just two schemes for the heat equation but apply to the solution of all marching problems in general.

BIBLIOGRAPHY

Fletcher, C.A.J. (1991). *Computational Techniques for Fluid Dynamics: Fundamental and General Techniques*, vol. **1**. Berlin: Springer-Verlag.

Tannehill, J.C., Anderson, D.A., and Pletcher, R.H. (1997). *Computational Fluid Mechanics and Heat Transfer*. Philadelphia, PA: Taylor & Francis.

PROBLEMS

1. Can the numerical instability be avoided by simply using higher precision (larger number of decimal digits) in the computations, thus reducing the round-off error? If not, what would be the effect of higher precision?

2. Verify your answers to problem 1 in a simple computational experiment. Use the simple explicit scheme (6.2) to compute the solution of the example problem discussed in the beginning of the chapter. Take $r = 0.6$, and run computations with different levels of precision, for example, with simple and double precision in FORTRAN.

3. Consider the heat equation (6.1) with $a = 0.1$ solved in the interval $0 < x < 0.1$ using the simple explicit method (6.2). The number of grid points in the x-direction is $N = 101$ including the two points at the boundaries of the interval. What is the maximum time step Δt that allows us to avoid instability?

4. Answer the same question as in problem 3 but for the simple implicit method.

5. The heat equation problem

$$\frac{\partial u}{\partial t} = 100\frac{\partial^2 u}{\partial x^2} \text{ in } 0 \le x \le 1,$$

$$u(0, t) = 0, \ u(1, t) = 1, \ u(x, 0) = \sin(6\pi x)$$

is solved using the explicit scheme (6.2). Find the maximum grid steps Δx and Δt at which the expected error of the solution is less than $\sim 10^{-4}$.

6. The question is the same as in problem 5, but now the simple implicit scheme (6.34) is used. Consider two situations.
   a) The expected error must be less than $\sim 10^{-4}$ for the entire solution.
   b) We are only interested in the accurate reproduction of the equilibrium temperature distribution at $t \to \infty$.

7. Use the Neumann analysis to determine stability properties of the scheme

$$\frac{u_i^{n+1} - u_i^n}{\Delta t} + c\frac{u_i^n - u_{i-1}^n}{\Delta x} = 0$$

applied to the linear convection equation $u_t + cu_x = 0$, where c is a positive constant.

8. Answer the same question as in the previous problem but for the scheme

$$\frac{u_i^{n+1} - u_i^n}{\Delta t} + c\frac{u_{i+1}^n - u_{i-1}^n}{2\Delta x} = 0.$$

Part II

METHODS

7

APPLICATION TO MODEL EQUATIONS

In this chapter we present a few selected schemes for one-dimensional unsteady model equations. Linear convection and heat equations are solved as the simplest examples of parabolic and hyperbolic partial differential equations (PDEs). We also solve the Burgers equation to demonstrate the difficulties brought by nonlinearity and the possible approach to resolving them. Our goal is not to provide an extensive review of the numerous schemes developed over the last 60 or so years. On the contrary, the variety will be limited to the minimum needed to illustrate the distinctive features, pitfalls, and successive strategies typical for each type of PDE. The one-dimensional model equations are particularly helpful in this task, since they provide the advantages of simplicity and existence of exact analytical solutions.

Since the equations are one-dimensional, there is no or little difference between the finite difference and finite volume approaches. The schemes discussed in this chapter can be developed following either of these spatial discretization techniques.

Essential Computational Fluid Dynamics, Second Edition. Oleg Zikanov.
© 2019 John Wiley & Sons, Inc. Published 2019 by John Wiley & Sons, Inc.
Companion Website: www.wiley.com/go/zikanov/essential

7.1   LINEAR CONVECTION EQUATION

The one-dimensional wave equation, a representative of the hyperbolic equations of second order, is

$$\frac{\partial^2 u}{\partial t^2} = c^2 \frac{\partial^2 u}{\partial x^2}. \tag{7.1}$$

The solution can be described as a combination of two waves propagating in the opposite directions:

$$u(x, t) = F(x + ct) + G(x - ct), \tag{7.2}$$

where F and G are functions defined by initial and boundary conditions. Although (7.1) is simple already, the principal features of the finite difference schemes for hyperbolic equations can be investigated on the example of the even simpler linear convection equation:

$$\frac{\partial u}{\partial t} + c \frac{\partial u}{\partial x} = 0, \quad c > 0. \tag{7.3}$$

Our interest in this equation is further justified by the direct similarity between its structure and the structure of the material derivative $Du/Dt \equiv \partial u/\partial t + (V \cdot \nabla)u$ present in the governing equations of fluid mechanics and heat transfer.

A solution of the linear convection equation has the form of a single wave

$$u(x, t) = F(x - ct) \tag{7.4}$$

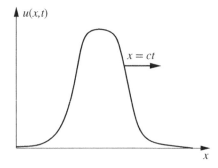

Figure 7.1   A solution of the linear convection equation (7.3).

propagating in the positive x-direction with the constant speed c (see Figure 7.1). The solution can be easily found if we know the initial condition

$$u(x, 0) = F(x) \tag{7.5}$$

and the boundary condition at the low-x boundary of the computational domain. Note that since the equation is of the first order, only one boundary condition is needed. It has to be at the low-x boundary in agreement with the wave propagation to the right.

7.1.1 Simple Explicit Schemes

We start with simple schemes. Equation (7.3) is approximated at the grid point (x_i, t^n) using the first-order forward difference for the time derivative and the first-order forward, backward, or second-order central formula for the space derivative. This results in the three schemes:

$$\frac{u_i^{n+1} - u_i^n}{\Delta t} + c \frac{u_{i+1}^n - u_i^n}{\Delta x} = 0, \quad \text{TE} = O(\Delta t, \Delta x), \tag{7.6}$$

$$\frac{u_i^{n+1} - u_i^n}{\Delta t} + c \frac{u_i^n - u_{i-1}^n}{\Delta x} = 0, \quad \text{TE} = O(\Delta t, \Delta x), \tag{7.7}$$

$$\frac{u_i^{n+1} - u_i^n}{\Delta t} + c \frac{u_{i+1}^n - u_{i-1}^n}{2\Delta x} = 0, \quad \text{TE} = O(\Delta t, (\Delta x)^2). \tag{7.8}$$

The difference molecules are shown in Figure 7.2.

The same schemes can be developed with the finite volume approach. The upwind scheme (7.7) was, in fact, derived in Section 5.3.1 using the upwind interpolation (5.16) of the convective fluxes $u_{i+1/2}$ and $u_{i-1/2}$ at the boundaries of the cell $\Omega_i = [x_{i-1/2}, x_{i+1/2}]$. We would obtain the scheme (7.8) if the convective fluxes were evaluated using the linear interpolation (5.20). In order to obtain (7.6), we would have to use the "downwind" interpolation $u_{i+1/2} \approx u_{i+1}, u_{i-1/2} \approx u_i$.

Let us analyze the stability of the derived schemes using the Neumann method (see Section 6.2.1). Since the PDE under consideration is linear, the same difference equations (7.6)–(7.8) hold for the round-off error ϵ. Dividing the equations by $\epsilon_i^n = \epsilon(x_i, t^n)$, we obtain, for the ratio $e^{a\Delta t} = \epsilon_i^{n+1}/\epsilon_i^n$,

$$\frac{e^{a\Delta t} - 1}{\Delta t} + c \frac{e^{t\beta} - 1}{\Delta x} = 0, \tag{7.9}$$

$$\frac{e^{a\Delta t} - 1}{\Delta t} + c\frac{1 - e^{-\iota\beta}}{\Delta x} = 0, \tag{7.10}$$

$$\frac{e^{a\Delta t} - 1}{\Delta t} + c\frac{e^{\iota\beta} - e^{-\iota\beta}}{\Delta x} = 0, \tag{7.11}$$

where $2\pi/N \le \beta \le \pi$ stands for $k_m\Delta x$ and k_m is the wavenumber of the Fourier harmonics constituting the error.

The amplification factor $G = |e^{a\Delta t}|$ is, respectively,

$$G = |1 - v(e^{\iota\beta} - 1)| = |1 + v - ve^{\iota\beta}| \text{ for scheme (7.6)}, \tag{7.12}$$

$$G = |1 - v(1 - e^{-\iota\beta})| = |1 - v + ve^{-\iota\beta}| \text{ for scheme (7.7)}, \tag{7.13}$$

$$G = |1 - v(e^{\iota\beta} - e^{-\iota\beta})| = |1 - 2v\iota\sin\beta| \text{ for scheme (7.8)}, \tag{7.14}$$

where we use the *Courant coefficient*

$$v = c\frac{\Delta t}{\Delta x}. \tag{7.15}$$

The Courant coefficient plays a special role in the stability of schemes for hyperbolic equations, similarly to the coefficient r for parabolic equations (see (6.32)).

The amplification factors (7.12)–(7.14) are plotted in Figure 7.2 as functions of β. As we know from Chapter 6, numerical stability of a scheme requires that $G \le 1$ for all β. We see that the forward and central schemes (7.6) and (7.8) are unconditionally unstable, which makes them worthless. The scheme (7.7) is stable if v satisfies the Courant–Friedrichs–Lewy (often abbreviated to CFL) stability condition:

$$0 \le v = c\frac{\Delta t}{\Delta x} \le 1. \tag{7.16}$$

The scheme (7.7) is called the first-order *upstream* or *upwind* method.

The modified equation for the upwind scheme is

$$u_t + cu_x = \frac{c\Delta x}{2}(1 - v)u_{xx} - \frac{c(\Delta x)^2}{6}(2v^2 - 3v + 1)u_{xxx}$$
$$+ O((\Delta x)^3, (\Delta x)^2\Delta t, \Delta x(\Delta t)^2, (\Delta t)^3). \tag{7.17}$$

The leading term of the truncation error is a numerical dissipation term. The fact that this term is of the first order means that the numerical dissipation

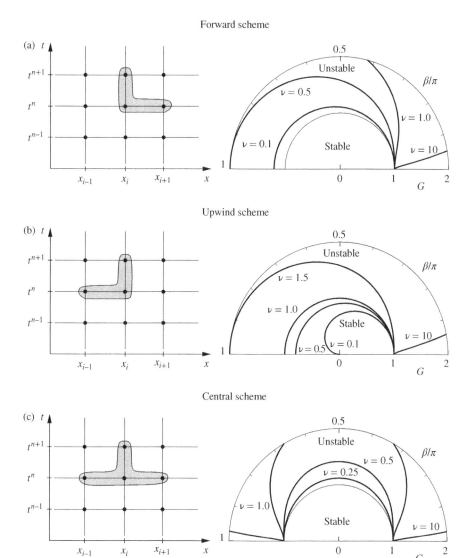

Figure 7.2 Difference molecules and amplification factors of the simple explicit schemes applied to the linear convection equation (7.3). (a) Forward scheme (7.6). (b) Upwind scheme (7.7) (c) Central scheme (7.8).

is strong, unless a very fine grid is used. This is a serious problem, not only for the particular scheme (7.7) but also, in general, for the schemes based on the upwind interpolation.

The last comment concerns a remarkable simplification of the upwind scheme, which is achieved if $v = 1$ is chosen. As one can see in (7.17), coefficients at dominating dissipation and dispersion errors become zero. In fact, the entire truncation error can be shown to vanish, and the scheme reduces to the so-called shift method:

$$u_i^{n+1} = u_{i-1}^n, \tag{7.18}$$

which is an exact solution obtained by the method of characteristics.

7.1.2 Simple Implicit Scheme

The simple implicit scheme for (7.3) is obtained by approximating the equation at (x_i, t^{n+1}) using backward difference for the time derivative and central difference for the space derivative:

$$\frac{u_i^{n+1} - u_i^n}{\Delta t} + c\frac{u_{i+1}^{n+1} - u_{i-1}^{n+1}}{2\Delta x} = 0. \tag{7.19}$$

The difference molecule is shown in Figure 7.3a.

The method is unconditionally stable and has the truncation error $O(\Delta t, (\Delta x)^2)$. The modified equation

$$u_t + cu_x = \left(\frac{c^2\Delta t}{2}\right)u_{xx} - \left(\frac{c(\Delta x)^2}{6} + \frac{c^3(\Delta t)^2}{3}\right)u_{xxx} + \cdots \tag{7.20}$$

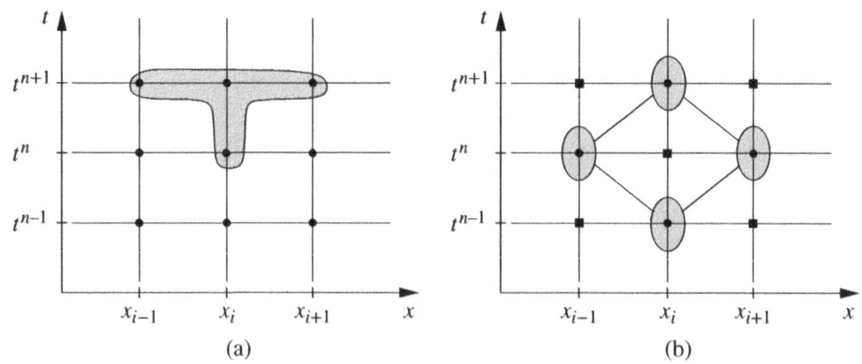

Figure 7.3 Difference molecules of schemes applied to the linear convection equation (7.3). (a) Simple implicit scheme (7.19). (b) Leapfrog scheme (7.21).

shows that, as typical for the first-order methods, the truncation error is dominated by numerical dissipation. The dissipation is particularly strong if one relies on the unconditional stability and uses large time steps Δt.

7.1.3 Leapfrog Scheme

The methods just discussed are of the first order in time. Their accuracy is low, and the solutions are distorted by strong numerical dissipation, unless very small grid steps are used. As the first example of a second-order scheme, we present the leapfrog method. This method is also the first scheme we consider that involves values of u on more than two time layers (see Figure 7.3b). For the linear convection equation (7.3), the leapfrog scheme is

$$\frac{u_i^{n+1} - u_i^{n-1}}{2\Delta t} + c\frac{u_{i+1}^n - u_{i-1}^n}{2\Delta x} = 0. \tag{7.21}$$

The formula can be designed by approximating the space and time derivatives at (x_i, t^n) by central differences. The method is explicit with the truncation error $O((\Delta t)^2, (\Delta x)^2)$. It is stable at $v \leq 1$. The modified equation is

$$u_t + cu_x = \frac{c(\Delta x)^2}{6}(v^2 - 1)u_{xxx} - \frac{c(\Delta x)^4}{120}(9v^4 - 10v^2 + 1)u_{xxxxx} + \cdots. \tag{7.22}$$

We see that the numerical dispersion dominates the truncation error. This is common for the methods of the second order in time. A remarkable feature of the leapfrog scheme is that the right-hand side of the modified equation does not contain even-order derivatives of u at all. The method generates no numerical dissipation! This is both good and bad news. Some amount of numerical dissipation is needed to suppress the round-off errors. The leapfrog method is a nice illustration of this statement. The amplification factor is

$$G = |\pm(1 - v^2\sin^2\beta)^{1/2} - iv \sin\beta|. \tag{7.23}$$

It is larger than 1 if $v > 1$. Interestingly, $G \equiv 1$ for all $v \leq 1$. The method is *neutrally stable*. The round-off error introduced at every time step neither grows nor, in the absence of numerical dissipation, decays as the solution advances.

Being, historically, one of the first second-order methods, the leapfrog scheme was quite popular during the early years of CFD (1960s and 1970s). It has lost popularity to other methods, partially because of one serious drawback associated with the leapfrog nature of the solution. It can be seen in Eq. (7.21) and Figure 7.3b that the scheme connects u_i^{n+1} with u_{i-1}^n, u_{i+1}^n, and u_i^{n-1}, but not with u_i^n, u_{i-1}^{n-1}, or u_{i+1}^{n-1}. These other values are connected with each other and with u_i^{n-2} by the finite difference equation

$$\frac{u_i^n - u_i^{n-2}}{2\Delta t} + c\frac{u_{i+1}^{n-1} - u_{i-1}^{n-1}}{2\Delta x} = 0,$$

which is the approximation of the PDE at the grid point (x_i, t^{n-1}). In fact, the entire system of discretization equations consists of two uncoupled subsystems: one with equations for the "square" grid points in Figure 7.3b and another with equations for the "circle" points. If no extra precaution is taken, the solutions to the two subsystems can deviate from each other (split) with time.

7.1.4 Lax–Wendroff Scheme

This explicit method is given by the formula

$$u_i^{n+1} = u_i^n - \frac{v}{2}(u_{i+1}^n - u_{i-1}^n) + \frac{v^2}{2}(u_{i+1}^n - 2u_i^n + u_{i-1}^n), \qquad (7.24)$$

where $v = c\Delta t/\Delta x$ is the Courant coefficient. At first glance, the scheme does not look at all as an approximation of Equation (7.3). This is, however, a consistent approximation with TE $= O((\Delta x)^2, (\Delta t)^2)$. This can be verified by the Taylor expansion method. Except for the extra term in the right-hand side, the scheme is identical to the unstable simple central difference scheme (7.8). This extra term is designed so as to cancel the dominating numerical dissipation term in the modified equation for the scheme (7.8) and, as it happens, to stabilize the scheme. The modified equation for the Lax–Wendroff scheme is

$$u_t + cu_x = -\frac{c(\Delta x)^2}{6}(1 - v^2)u_{xxx} - \frac{c(\Delta x)^3}{8}v(1 - v^2)u_{xxxx} + \cdots . \quad (7.25)$$

The truncation error is predominantly dispersive. The scheme is stable if $|v| \leq 1$. It reduces to the shift scheme (7.18) when $v = 1$.

7.1.5   MacCormack Scheme

This method, developed by R.W. MacCormack in 1969, is important for us as an example of the two-step predictor–corrector methods. For the wave equation (7.3), the scheme is

$$\text{Predictor:} \quad u_i^* = u_i^n - v(u_{i+1}^n - u_i^n).$$

$$\text{Corrector:} \quad u_i^{n+1} = \frac{1}{2}[u_i^n + u_i^* - v(u_i^* - u_{i-1}^*)]. \tag{7.26}$$

Although the particular form of the method can be different for different equations, the principal approach is always the same: we calculate an intermediate "predicted" solution and then "correct" it to achieve desired accuracy and stability. For the MacCormack method, the truncation error is $O((\Delta x)^2, (\Delta t)^2)$, and the stability criterion is $v \leq 1$.

7.2   ONE-DIMENSIONAL HEAT EQUATION

As before, we employ the one-dimensional heat equation

$$\frac{\partial u}{\partial t} = a^2 \frac{\partial^2 u}{\partial x^2} \tag{7.27}$$

as the simplest representative of the parabolic PDE family. An inhomogeneous part $f(x, t)$ can be added to the right-hand side without loss of generality. We have already solved this equation in Sections 4.4.5 and 6.2.1 using the simple explicit and implicit methods. In this section, we extend the analysis by adding another scheme and investigating the modified equations.

7.2.1   Simple Explicit Scheme

The simple explicit scheme

$$\frac{u_i^{n+1} - u_i^n}{\Delta t} = a^2 \frac{u_{i+1}^n - 2u_i^n + u_{i-1}^n}{(\Delta x)^2} \tag{7.28}$$

has the truncation error $O(\Delta t, (\Delta x)^2)$ and is stable when the stability parameter satisfies

$$r \equiv \frac{a^2 \Delta t}{(\Delta x)^2} \leq \frac{1}{2}. \tag{7.29}$$

Additional information can be extracted from the modified equation

$$u_t - a^2 u_{xx} = \left[-\frac{a^4 \Delta t}{2} + \frac{(a \Delta x)^2}{12} \right] u_{xxxx}$$

$$+ \left[\frac{a^6 (\Delta t)^2}{3} - \frac{a^4 \Delta t (\Delta x)^2}{12} + \frac{a^2 (\Delta x)^4}{360} \right] u_{xxxxxx} + \cdots . \tag{7.30}$$

There are no odd-derivative terms in (7.30). The truncation error has no dispersive part. This feature is common for the finite difference schemes of second order in space applied to parabolic equations.

Another interesting feature of the modified equation (7.30) is that we can greatly reduce the dissipation error and improve the accuracy of the scheme by choosing Δt and Δx so that $r = 1/6$. The first term of the truncation error vanishes, and the error becomes $O((\Delta t)^2, (\Delta x)^4)$. To illustrate the effect, we solve the heat equation for

$$0 < x < \pi, \ a = 2, \ u(0, t) = u(\pi, t) = 0, \ u(x, 0) = \sin(5x). \tag{7.31}$$

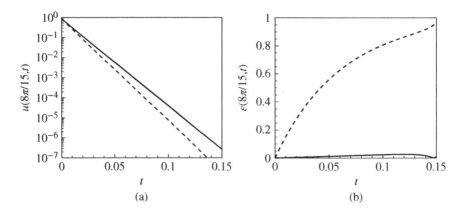

(a)   (b)

Figure 7.4 Solution of the heat equation problem (7.27), (7.31) by the simple explicit scheme. Space resolution is $\Delta x = \pi/15$. The time step is such that $r = 1/6$ (solid lines) or $r = 0.4$ (dashed lines). (a) Solution at $x = 8\pi/15$ as a function of time. The curve for the exact solution practically coincides with the curve for $r = 1/6$. (b) Relative error $\epsilon = |(u_{calculated} - u_{exact})/u_{exact}|$ at $x = 8\pi/15$.

The solution can be compared with the exact solution $u_{exact} = \sin(5x)$ $\exp(-(5a)^2 t)$. The calculations are performed at deliberately low space resolution with just $N = 15$ grid points. Two time steps are used, one corresponding to $r = 0.4$ and another to $r = 1/6$. It is clearly visible in Figure 7.4 that the solution with $r = 1/6$ has much higher accuracy.

7.2.2 Simple Implicit Scheme

The simple implicit method is

$$\frac{u_i^{n+1} - u_i^n}{\Delta t} = a^2 \frac{u_{i+1}^{n+1} - 2u_i^{n+1} + u_{i-1}^{n+1}}{(\Delta x)^2}). \tag{7.32}$$

The method is unconditionally stable and has the truncation error $O(\Delta t, (\Delta x)^2)$. Analysis of the modified equation

$$u_t - a^2 u_{xx} = \left[\frac{a^4 \Delta t}{2} + \frac{(a\Delta x)^2}{12} \right] u_{xxxx}$$

$$+ \left[\frac{a^6 (\Delta t)^2}{3} + \frac{a^4 \Delta t (\Delta x)^2}{12} + \frac{a^2 (\Delta x)^4}{360} \right] u_{xxxxxx} + \cdots \tag{7.33}$$

shows that the truncation error is purely dissipative.

Let us compare (7.33) with the modified equation (7.30) for the simple explicit method. The amplitude of the dominating part of the truncation error is determined by the coefficient at u_{xxxx}. In the explicit scheme, the terms in this coefficient are of different signs, so they cancel each other, either completely at $r = 1/6$ or partially at other values of r. No such cancelation occurs in the implicit scheme. This means that at the same Δx and Δt, the explicit method is more accurate than the implicit one.

7.2.3 Crank–Nicolson Scheme

Implicit and explicit approaches can be mixed together in one scheme, sometimes providing remarkably good results. An important example is the Crank–Nicolson scheme:

$$\frac{u_i^{n+1} - u_i^n}{\Delta t} = a^2 \frac{1}{2} \frac{(u_{i+1}^n - 2u_i^n + u_{i-1}^n) + (u_{i+1}^{n+1} - 2u_i^{n+1} + u_{i-1}^{n+1})}{(\Delta x)^2}. \tag{7.34}$$

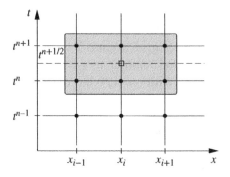

Figure 7.5 Difference molecule for the Crank–Nicolson scheme (7.34).

The scheme is illustrated by the difference molecule in Figure 7.5. We obtain this scheme by approximating the PDE at the point $(x_i, t^{n+1/2})$ located at the imaginary half-integer time layer $t^{n+1/2} = (t^n + t^{n+1})/2$. The central difference formula (4.10) with the grid step $\Delta t/2$ is applied to the time derivative $\partial u/\partial t$ at $t^{n+1/2}$:

$$\left.\frac{\partial u}{\partial t}\right|_i^{n+1/2} = \frac{u_i^{n+1} - u_i^n}{\Delta t} + O((\Delta t)^2). \tag{7.35}$$

For the space derivative, we use the second-order central difference (4.19) applied to the result of the linear interpolation $u_i^{n+1/2} = (u_i^n + u_i^{n+1})/2$:

$$\left.\frac{\partial^2 u}{\partial x^2}\right|_i^{n+1/2} = \frac{1}{2}\left(\frac{u_{i+1}^n - 2u_i^n + u_{i-1}^n}{(\Delta x)^2} + \frac{u_{i+1}^{n+1} - 2u_i^{n+1} + u_{i-1}^{n+1}}{(\Delta x)^2}\right).$$

As discussed in Section 4.4.8, the interpolation procedure generates the error of the order of $O\left((\Delta t)^2\right)$. The cumulative truncation error of (7.34) is T. E. $= O\left((\Delta t)^2, (\Delta x)^2\right)$. The time discretization by the Crank–Nicolson scheme (7.34) is more accurate than the discretization by the fully explicit and implicit schemes (7.28) and (7.32).

Applying the Neumann stability analysis, we find the amplification factor

$$G = \left|\frac{1 - r(1 - \cos\beta)}{1 + r(1 - \cos\beta)}\right|. \tag{7.36}$$

Since $r > 0$, the scheme is unconditionally stable.

7.3 BURGERS AND GENERIC TRANSPORT EQUATIONS

We consider the Burgers equation

$$\frac{\partial u}{\partial t} + u\frac{\partial u}{\partial x} = \mu\frac{\partial^2 u}{\partial x^2} \tag{7.37}$$

and the generic transport equation

$$\frac{\partial \phi}{\partial t} + u\frac{\partial \phi}{\partial x} = \mu\frac{\partial^2 \phi}{\partial x^2} \tag{7.38}$$

introduced in Section 3.1.1. Each of them combines three important features of the Navier–Stokes equations: unsteadiness, nonlinear convection, and dissipation (diffusion). The equations are hyperbolic at $\mu = 0$. They are formally parabolic at $\mu > 0$ although they may be viewed as of mixed hyperbolic–parabolic type. In the latter view, the hyperbolic behavior is produced by the combination of the first and second terms, while the parabolic behavior is the result of the combination of the first and third terms.

We will consider one scheme for the Burgers equation (7.37). It can be easily modified for the generic transport equation. The scheme uses central differences of the second order for the space derivatives and forward difference of the first order for the time derivative:

$$\frac{u_i^{n+1} - u_i^n}{\Delta t} + u_i^n \frac{u_{i+1}^n - u_{i-1}^n}{2\Delta x} = \mu \frac{u_{i+1}^n - 2u_i^n + u_{i-1}^n}{(\Delta x)^2}. \tag{7.39}$$

The scheme is explicit with the truncation error TE $= O((\Delta x)^2, \Delta t)$.

The stability analysis of (7.39) is nontrivial, primarily because the equation is nonlinear. The standard Neumann stability analysis cannot, strictly speaking, be applied because it assumes existence of exponential solutions of the equations for the round-off error, which is true only for analysis of linear PDE with constant coefficients. The common approach in the case of nonlinearity is to "freeze" the coefficients, such as u in (7.37) and (7.38), that is, to act as if it were a constant. For the scheme (7.39), this means conducting the stability analysis for the linearized discretization equation with u_i^n replaced by a constant c. Such analysis leads to two stability constraints:

$$c^2\Delta t \le 2\mu, \text{ which is equivalent to } v^2 \le 2r, \text{ and } r \le 1/2, \tag{7.40}$$

where, as before, $v = c\Delta t/\Delta x$ and $r = \mu\Delta t/(\Delta x)^2$.

The constraints must be satisfied for *all* values of $c = u_i^n$ in the solution. Since the values of u_i^n are unknown, we must find a way to estimate their upper bound:

$$u_{max} = \max_{\Omega} |u|.$$

The stability criteria are then rewritten as

$$(u_{max})^2 \Delta t \leq 2\mu \text{ and } r \leq 1/2. \tag{7.41}$$

The same criteria should be satisfied by the scheme (7.39) applied to the generic transport equation.

Another aspect of the stability analysis of (7.39) and other more complex equations, such as the Navier–Stokes system, concerns the possibility of using the stability criteria derived for the model linear convection and heat equations (7.3) and (7.27). One may be tempted to consider (7.39) as a combination of the central difference scheme (7.8) and the simple explicit scheme (7.28) applied, respectively, to the convective and diffusive parts of the Burgers equation. It seems logical that the stability criteria for (7.39) should be a combination of the criteria developed for (7.8) and (7.28). Quite often, the logic works. We can assume that a scheme is stable if the criteria derived for the schemes used for the convective and diffusive parts are satisfied and, then, verify the assumption in test calculations.

Sometimes, however, the approach results in excessively restrictive conditions. For example, applying it to (7.39) would lead to the conclusion that the scheme is unconditionally unstable since the scheme (7.8) used to approximate the convective part is unconditionally unstable (see Section 7.1.1). A more accurate stability analysis of (7.39) shows that the scheme is stabilized by the presence of the diffusion term, provided the diffusion coefficient is large enough, as specified by the first condition in (7.40).

7.4 METHOD OF LINES

The schemes considered so far were based on simultaneous discretization of time and space derivatives. This approach has the advantage that the scheme is formulated in its final form from the beginning. Its order of approximation and stability properties can be fully analyzed. Furthermore, some schemes can be optimized in such a way that certain terms of the truncation error cancel each other so that the accuracy is improved (the improved accuracy of the simple explicit scheme for the heat equation at $r = 1/6$ is a good example).

There is an alternative approach called *method of lines*, according to which the operations of spatial and time discretization are separated. The method of lines can be applied to any PDE or a system of PDEs, which can be represented in the form that includes one time derivative of the first order in each equation. The spatial discretization is performed first, and the resulting system of equations is written down as a system of ordinary differential equations:

$$u_t = R[u], \tag{7.42}$$

where $u(t)$ is the vector of spatially discretized variables (grid point values in the case of finite difference or finite volume schemes or amplitudes of trial functions in the spectral methods) and $R[u]$ is an operator that includes approximations of spatial derivatives of u and other terms.

The system (7.42) is solved using one of the time-integration methods developed for the ordinary differential equations. We have already used this approach when we illustrated the spectral methods in Section 3.3.1.

The numerical stability to round-off errors is determined by the type of the time-integration scheme and by the properties of the operator $R[u]$. Obtaining the detailed information including the amplification factor and exact stability criteria typically requires analysis of a complete discretized system, although some tendencies can often be predicted on the basis of the time-integration scheme alone.

Many methods of time integration of the general equation (7.42) are available. We will consider two families of multilevel schemes: Adams and Runge–Kutta methods. They can be effectively applied, in the way just outlined, to PDE of both hyperbolic and parabolic types.

7.4.1 Adams Methods

The Adams methods can be explicit (Adams–Bashforth) or implicit (Adams–Moulton). The general formulas are

$$u^{n+1} = u^n + \Delta t \sum_{j=0}^{q} A_j R^{n-j} \quad \text{Adams–Bashforth,} \tag{7.43}$$

$$u^{n+1} = u^n + \Delta t \sum_{j=-1}^{q-1} A_j R^{n-j} \quad \text{Adams–Moulton,} \tag{7.44}$$

where u^n is the spatially discretized vector of solution at the time layer t^n and $R^{n-j} = R[u^{n-j}]$. The scheme coefficients A_j are determined by assuming polynomial behavior of $u(t)$ and fitting it to several consecutive time layers. The number q of terms in the right-hand side determines the order of accuracy. For example, at $q = 0$, we obtain the well-known explicit and implicit Euler methods with TE $\sim O(\Delta t)$:

$$u^{n+1} = u^n + \Delta t R^n, \tag{7.45}$$

$$u^{n+1} = u^n + \Delta t R^{n+1}. \tag{7.46}$$

Formulas with $q = 1$ give schemes of the second order,

$$u^{n+1} = u^n + \frac{\Delta t}{2}(3R^n - R^{n-1}) \tag{7.47}$$

$$u^{n+1} = u^n + \frac{\Delta t}{2}(R^{n+1} + R^n), \tag{7.48}$$

while the schemes of the third order obtained at and $q = 2$ are

$$u^{n+1} = u^n + \frac{\Delta t}{12}(23R^n - 16R^{n-1} + 5R^{n-2}), \tag{7.49}$$

$$u^{n+1} = u^n + \frac{\Delta t}{12}(5R^{n+1} + 8R^n - R^{n-1}). \tag{7.50}$$

7.4.2 Runge–Kutta Methods

Generally, there exists an infinite number of the multistep Runge–Kutta schemes, which can be formulated with an arbitrary order of accuracy. For example, one method of the second order is

$$\text{Step 1:} \quad u^{(1)} = u^n + \frac{\Delta t}{2}R^n$$

$$\text{Step 2:} \quad u^{n+1} = u^n + \Delta t R^{(1)}, \tag{7.51}$$

where $R^n = R[u^n]$ and $R^{(1)} = R[u^{(1)}]$ are the spatial operators calculated for the solution at previous time layer u^n and for the intermediate solution $u^{(1)}$. A popular method of the fourth order is

$$\text{Step 1:} \quad u^{(1)} = u^n + \frac{\Delta t}{2}R^n$$

$$\text{Step 2:} \quad u^{(2)} = u^n + \frac{\Delta t}{2}R^{(1)}$$

Step 3: $u^{(3)} = u^n + \Delta t R^{(2)}$

Step 4: $u^{n+1} = u^n + \dfrac{\Delta t}{6}(R^n + 2R^{(1)} + 2R^{(2)} + R^{(3)}).$ (7.52)

As an example, we can solve the linear convection equation (7.3) using the Runge–Kutta scheme of the second order (7.51) and applying the central differences for spatial discretization:

$$R^n = -c\frac{u^n_{i+1} - u^n_{i-1}}{2\Delta x},$$

$$R^{(1)} = -c\frac{u^{(1)}_{i+1} - u^{(1)}_{i-1}}{2\Delta x};$$

the resulting scheme has TE=$O\left((\Delta x)^2, (\Delta t)^2\right)$.

The main advantages of the Runge–Kutta methods are their flexibility (a scheme of any order in time can be developed) and good stability properties. They are not unconditionally stable, but the largest allowed time step is typically large in comparison with the steps required by other explicit methods. The most serious disadvantage is the necessity to calculate the operator $R[u]$ several times on every time step.

7.5 SOLUTION OF TRIDIAGONAL SYSTEMS BY THOMAS ALGORITHM

Any implicit scheme requires solution of a system of linear algebraic equations at every time step. Typically, this is a computationally challenging task, since the system is very large. Its size in a three-dimensional CFD analysis can be anywhere between 10^5 and 10^9 or even larger. In solution of one- or two-dimensional problems, the systems are smaller, but still large enough to seriously consider the amount of computations involved.

There exists a variety of effective methods for solving linear algebraic systems. Some of them are discussed in Chapter 8. Here, we consider the special case when the solution of a one-dimensional PDE is calculated. We assume that the time discretization involves only two subsequent time layers and the space discretization is done using a finite difference or finite volume scheme based on not more than three neighboring grid points. The resulting system of linear equations has a *tridiagonal* matrix and can be solved using the very efficient form of the Gauss elimination called *double-sweep* or *Thomas* algorithm. The technique is credited to L.H.

Thomas, who published it in 1949, although the idea is simple and occurred independently to other specialists dealing with the same problem.

Let us illustrate the method on the example of the simple implicit scheme

$$\frac{u_i^{n+1} - u_i^n}{\Delta t} = a^2 \frac{u_{i+1}^{n+1} - 2u_i^{n+1} + u_{i-1}^{n+1}}{(\Delta x)^2} + f_i^{n+1}$$

written for the modified heat equation

$$\frac{\partial u}{\partial t} = a^2 \frac{\partial^2 u}{\partial x^2} + f(x, t).$$

We assume that the solution is sought in the interval $0 \le x \le L$ on a uniform grid $x_i = i\Delta x$, $i = 0, \ldots, N$, with $x_0 = 0$ and $x_N = L$. To demonstrate some variations of the method, we impose the Dirichlet boundary condition $u(0, t) = g_0$ at $x = 0$ and the Neumann boundary condition $(\partial u / \partial x)(L, t) = g_1$ at $x = L$.

The finite difference formula for the interior points can be rewritten as

$$a_i u_{i+1}^{n+1} + d_i u_i^{n+1} + b_i u_{i-1}^{n+1} = c_i, \quad i = 1, 2, \ldots, N - 1, \tag{7.53}$$

where

$$a_i = -a^2 \Delta t / \Delta x^2, \ b_i = -a^2 \Delta t / \Delta x^2, \ d_i = 1 + 2a^2 \Delta t / \Delta x^2,$$
$$c_i = u_i^n + \Delta t f_i^{n+1}.$$

The boundary condition at $x = 0$ can be expressed as

$$d_0 u_0^{n+1} = c_0, \ \text{with} \ d_0 = 1 \ \text{and} \ c_0 = g_0. \tag{7.54}$$

The boundary condition at $x = L$ must be approximated by a one-sided finite difference formula. It is important to maintain the second order of accuracy, so as not to compromise the accuracy of the entire solution. We employ the formula (4.14):

$$\frac{3}{2} u_N^{n+1} - 2u_{N-1}^{n+1} + \frac{1}{2} u_{N-2}^{n+1} = \Delta x g_1.$$

For future use, we have to combine this equation with Eq. (7.53) written for the point x_{N-1}

$$a_{N-1} u_N^{n+1} + d_{N-1} u_{N-1}^{n+1} + b_{N-1} u_{N-2}^{n+1} = c_{N-1}$$

and use the system of two equations to exclude the unknown u_{N-2}^{n+1} and obtain a linear relation between u_{N-1}^{n+1} and u_N^{n+1}:

$$b_N u_{N-1}^{n+1} + d_N u_N^{n+1} = c_N, \tag{7.55}$$

where $b_N = d_{N-1} + 4b_{N-1}$, $d_N = a_{N-1} - 3b_{N-1}$, and $c_N = c_{N-1} - 2\Delta x b_{N-1} g_1$.

The entire system of equations can be written as

$$
\begin{pmatrix}
d_0 & a_0 & 0 & \cdots & \cdots & \cdots & 0 \\
b_1 & d_1 & a_1 & 0 & \cdots & \cdots & 0 \\
0 & b_2 & d_2 & a_2 & 0 & \cdots & 0 \\
0 & 0 & * & * & * & 0 & 0 \\
0 & 0 & 0 & * & * & * & 0 \\
0 & \cdots & \cdots & 0 & b_{N-1} & d_{N-1} & a_{N-1} \\
0 & \cdots & \cdots & \cdots & 0 & b_N & d_N
\end{pmatrix}
\begin{pmatrix}
u_0 \\ u_1 \\ u_2 \\ \vdots \\ \vdots \\ u_{N-1} \\ u_N
\end{pmatrix}
=
\begin{pmatrix}
c_0 \\ c_1 \\ c_2 \\ \vdots \\ \vdots \\ c_{N-1} \\ c_N
\end{pmatrix}, \tag{7.56}
$$

where u_i is substituted for u_i^{n+1} for brevity.

We see that the coefficient matrix in the left-hand side is, indeed, tridiagonal. Its elements are zeros except for the main diagonal and adjacent sub- and super-diagonals.

The algorithm includes two stages, or "sweeps." During the first stage, the *forward sweep*, the elementary row operations are applied to transform the coefficient matrix into upper triangular form. As known from linear algebra, an elementary row operation applied simultaneously to the coefficient matrix and the right-hand side does not affect the solution of a matrix equation. First, we multiply the row 0 by $-b_1/d_0$ and add to the row 1 in order to remove b_1:

$$d_1' = d_1 - \frac{b_1}{d_0} a_0$$

$$b_1' = b_1 - \frac{b_1}{d_0} d_0 = 0 \tag{7.57}$$

$$c_1' = c_1 - \frac{b_1}{d_0} c_0.$$

A similar operation with the transformed row 1 is then used to remove b_2 in the row 2. The procedure continues with the rows 3, 4, etc. The general formula is

$$d_i' = d_i - \frac{b_i}{d_{i-1}'} a_{i-1}$$

$$b'_i = b_i - \frac{b_i}{d'_{i-1}} d'_{i-1} = 0 \qquad (7.58)$$

$$c'_i = c_i - \frac{b_i}{d'_{i-1}} c'_{i-1}, \quad i = 1, 2, \dots, N.$$

At the end of the forward sweep, the matrix equation becomes

$$
\begin{pmatrix}
d_0 & a_0 & 0 & \dots & \dots & \dots & 0 \\
0 & d'_1 & a_1 & 0 & \dots & \dots & 0 \\
0 & 0 & d'_2 & a_2 & 0 & \dots & 0 \\
0 & 0 & 0 & * & * & 0 & 0 \\
0 & 0 & 0 & 0 & * & * & 0 \\
0 & \dots & \dots & 0 & 0 & d'_{N-1} & a_{N-1} \\
0 & \dots & \dots & \dots & 0 & 0 & d'_N
\end{pmatrix}
\begin{pmatrix}
u_0 \\ u_1 \\ u_2 \\ \vdots \\ \vdots \\ u_{N-1} \\ u_N
\end{pmatrix}
=
\begin{pmatrix}
c_0 \\ c'_1 \\ c'_2 \\ \vdots \\ \vdots \\ c'_{N-1} \\ c'_N
\end{pmatrix}. \qquad (7.59)
$$

It now has an upper triangular coefficient matrix in the left-hand side.

In the *backward sweep*, the transformed system is used to find u_i. First, we determine

$$u_N = c'_N / d'_N. \qquad (7.60)$$

Then, the value of u_N and $(N-1)$st equation are used to find u_{N-1}

$$u_{N-1} = (c'_{N-1} - a_{N-1} u_N)/d'_{N-1}.$$

The procedure continues backward to find the entire solution according to

$$u_i = (c'_i - a_i u_{i+1})/d'_i, \quad i = N-1, N-2, \dots, 1,$$
$$u_0 = (c_0 - a_0 u_1)/d_0. \qquad (7.61)$$

The Thomas algorithm effectively utilizes the tridiagonal structure of the matrix. It avoids the unnecessary operations with zero matrix elements and completes the solution of the linear system (7.56) in only $O(N)$ arithmetic operations. This is in sharp contrast with the standard Gauss elimination procedure that would require $O(N^3)$ operations. Obviously, the Thomas algorithm should be used wherever possible.

BIBLIOGRAPHY

Fletcher, C.A.J. (1991). *Computational Techniques for Fluid Dynamics: Fundamental and General Techniques*, vol. **1**. Berlin: Springer-Verlag.

MacCormack, R. (2003). The effect of viscosity in hypervelocity impact cratering. *J. Spacecr. Rockets* **40** (5): 757–763 (reprinted from AIAA paper 69–354 (1969)).

Press, W.P., Teukolsky, S.A., Vettering, W.T., and Flannery, B.P. (2001). *Numerical Recipes: The Art of Scientific Computing*. Cambridge: Cambridge University Press.

Tannehill, J.C., Anderson, D.A., and Pletcher, R.H. (1997). *Computational Fluid Mechanics and Heat Transfer*. Philadelphia, PA: Taylor & Francis.

PROBLEMS

1. Compare the schemes introduced in Section 7.1 for the linear convection equation.
   a) Simple explicit schemes (7.6), (7.7), and (7.8)
   b) Simple implicit scheme (7.19)
   c) Leapfrog scheme (7.21)
   d) Lax–Wendroff scheme (7.24)
   e) MacCormack scheme (7.26)

   For every scheme, describe its stability properties (write the stability criterion, where appropriate) and order of approximation. Determine, where data are available, whether the numerical dissipation or numerical dispersion dominates the truncation error. Summarize the discussion stating, for every scheme, how appropriate it is in your opinion for solution of the linear convection equation. Does the scheme accurately reproduce the traveling wave solution shown in Figure 7.1? What are the main advantages and disadvantages of the scheme?

2. Repeat Problem 7.1 for the following schemes applied to the one-dimensional heat equation.
   a) Simple explicit scheme (7.28)
   b) Simple implicit scheme (7.32)
   c) Crank–Nicolson scheme (7.34)

3. Modify the scheme (7.39) for the Burgers equation by changing the central difference approximation of the convective term to the upwind approximation. Assume that u is always positive. How does this modification change the order of approximation and the character of the truncation error?

4. One-dimensional heat equation is solved using the method of lines scheme based on the central difference for space derivative and the Adams–Bashforth scheme (7.43) of the second order ($q = 1$) for time discretization. Write the full set of discretization equations solved on one time step.

5. Repeat Problem 7.4 for the scheme based on the Runge–Kutta method of the second order (7.51).

6. Consider the application of the Thomas algorithm to the simple implicit scheme used to solve the one-dimensional heat equation (see Section 7.5). Modify the algorithm for other sets of boundary conditions.

    a) Dirichlet conditions on both ends: $u(0, t) = g_0$, $u(L, t) = g_1$
    b) Neumann conditions on both ends: $(\partial u/\partial x)(0, t) = g_0$, $(\partial u/\partial x)(L, t) = g_1$

7. The one-dimensional heat equation is solved on a grid consisting of N spatial grid points and M time layers. The boundary conditions are of Dirichlet type. Calculate, in terms of N and M, the total number of the arithmetic operations required to complete the solution. Do this for:

    a) Simple explicit scheme (7.28).
    b) Simple implicit scheme (7.32). Here, assume that the system of linear algebraic equations is solved at each time step using the Thomas algorithm.
    c) Runge–Kutta method of the fourth order (7.52). Here, assume that the spatial derivative is discretized by the central difference of the second order.

8. Apply the Taylor expansion method to prove the order of approximation of two schemes for the linear convection equation.

    a) Leapfrog scheme (7.21)
    b) Lax–Wendroff scheme (7.24)

Programming Exercises: Develop the following algorithms. Test in comparison with exact solutions, where available.

1. Thomas algorithm for solution of an arbitrary tridiagonal linear system.

2. Upwind and leapfrog solution of the linear convection equation $\partial u/\partial t + c\partial u/\partial x = 0$ at $0 \le x \le 10$ and $0 \le t \le 2$, with $c = 1$, the boundary condition $u(0, t) = 0$, and the initial condition $u(x, 0) = \sin(x)$ if $x \le \pi$ and $u(x, 0) = 0$ if $x > \pi$. Use the exact solution of the form (7.4) for verification.

3. Simple explicit, simple implicit, and Crank–Nicolson solutions of the one-dimensional heat equation $\partial u/\partial t = 4(\partial^2 u/\partial x^2)$ at $0 \le x \le 1$ and $0 \le t \le 0.1$ with boundary conditions $u(0, t) = u(1, t) = 0$ and initial condition $u(x, 0) = \sin(5\pi x)$. Use the exact solution $u(x, t) = \sin(5\pi x)\exp((-10\pi)^2 t)$ for verification.

STEADY-STATE PROBLEMS

In this chapter, we consider methods that can be used to solve a system of linear algebraic equations (a matrix equation)

$$\mathbf{A} \cdot \boldsymbol{v} = \boldsymbol{c} \qquad (8.1)$$

or a system of nonlinear algebraic equations

$$\boldsymbol{F}(\boldsymbol{v}) = 0. \qquad (8.2)$$

8.1 PROBLEMS REDUCIBLE TO MATRIX EQUATIONS

The matrix equation (8.1) is very common, practically unavoidable in CFD analysis. In the case of nonlinear partial differential equations (PDEs), solution of a matrix equation appears as a part of the iteration procedure that involves linearization. This is discussed in Section 8.4. If PDEs are linear, (8.1) follows directly from discretization. Several examples are considered in this section.

Essential Computational Fluid Dynamics, Second Edition. Oleg Zikanov.
© 2019 John Wiley & Sons, Inc. Published 2019 by John Wiley & Sons, Inc.
Companion Website: www.wiley.com/go/zikanov/essential

8.1.1 Elliptic PDE

As we discussed earlier, a typical elliptic equation describes steady-state physical processes or processes with zero adjustment time, such as the final equilibrium temperature distribution in a heat transfer problem or pressure distribution in a flow of an incompressible fluid. The simplest model equations representing the class of elliptic PDE are the Laplace equation

$$\nabla^2 u = 0 \tag{8.3}$$

and the Poisson equation

$$\nabla^2 u = f. \tag{8.4}$$

We will often use the two-dimensional Cartesian versions:

$$\frac{\partial^2 u}{\partial x^2} + \frac{\partial^2 u}{\partial y^2} = 0 \tag{8.5}$$

and

$$\frac{\partial^2 u}{\partial x^2} + \frac{\partial^2 u}{\partial y^2} = f(x, y). \tag{8.6}$$

From the computational viewpoint, we deal with a steady-state equation that has to be solved in a spatial domain Ω subject to certain boundary conditions at the boundary S (see Figure 8.1). The boundary conditions are typically of Dirichlet, Neumann, or Robin types. Due to the elliptic nature of the problem, the system of discretization equations always connects together the values of u at all grid points within Ω and at the boundary.

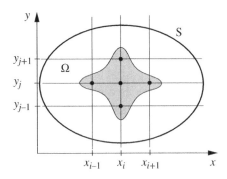

Figure 8.1 Features of the finite difference solution of two-dimensional Laplace and Poisson equations: computational domain Ω, its boundary S, rectangular grid, and difference molecule for the five-point formula.

Five-Point Formula: Let us consider the two-dimensional equations (8.5) and (8.6) discretized on a uniform rectangular grid $x_i = i\Delta x$, $y_j = j\Delta y$ with constant grid steps Δx and Δy. The common approach is to approximate the partial derivatives in (8.5) and (8.6) at internal points (x_i, y_j) using central differences of second order. The finite difference molecule consists of five grid points (x_{i-1}, y_j), (x_{i+1}, y_j), (x_i, y_j), (x_i, y_{j-1}), and (x_i, y_{j+1}) (see Figure 8.1). For example, the discretization of (8.6) is

$$\frac{u_{i+1,j} - 2u_{i,j} + u_{i-1,j}}{(\Delta x)^2} + \frac{u_{i,j+1} - 2u_{i,j} + u_{i,j-1}}{(\Delta y)^2} = f_{i,j}. \qquad (8.7)$$

To complete the system of equations, we have to add finite difference approximations of boundary conditions at all grid points lying at the boundary S.

Let us rewrite the system of the finite difference equations in the compact matrix form (8.1). For that we have to reorder the unknowns $u_{i,j}$ and the right-hand sides $f_{i,j}$ into one-dimensional arrays $\boldsymbol{v}$ and $\boldsymbol{c}$. The reordering can be done in different ways. For example, in a rectangular domain with $i = 1, \ldots, N_x$, $j = 1, \ldots, N_y$, we can choose

$$v_1 = u_{1,1}, v_2 = u_{2,1}, \ldots, v_{N_x} = u_{N_x,1}, v_{N_x+1} = u_{1,2}, \ldots,$$

$$c_1 = f_{1,1}, c_2 = f_{2,1}, \ldots, c_{N_x} = f_{N_x,1}, c_{N_x+1} = f_{1,2}, \ldots,$$

which is expressed by the formulas

$$v_\ell = u_{i,j}, \; c_\ell = f_{i,j}, \; i = 1, \ldots, N_x, \; j = 1, \ldots, N_y, \qquad (8.8)$$

where $\ell = (j - 1)N_x + i$ is the throughout index that identifies the variables corresponding to the point (x_i, y_j).

The finite difference equation (8.7) written for the point (x_i, y_j) becomes the ℓth equation of the system (8.1):

$$a_{\ell,\ell-N_x} v_{\ell-N_x} + a_{\ell,\ell-1} v_{\ell-1} + a_{\ell,\ell} v_\ell + a_{\ell,\ell+1} v_{\ell+1} + a_{\ell,\ell+N_x} v_{\ell+N_x} = c_\ell, \qquad (8.9)$$

with the nonzero coefficients of matrix $\mathbf{A}$ being

$$a_{\ell,\ell-N_x} = a_{\ell,\ell+N_x} = \frac{1}{(\Delta y)^2},$$

$$a_{\ell,\ell-1} = a_{\ell,\ell+1} = \frac{1}{(\Delta x)^2},$$

$$a_{\ell,\ell} = \left(-\frac{2}{(\Delta x)^2} - \frac{2}{(\Delta y)^2} \right). \qquad (8.10)$$

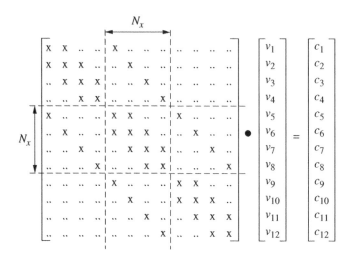

Figure 8.2 Structure of the matrix equation for the five-point discretization (8.7) of two-dimensional Poisson equation. For simplicity, we show the equation for an unnaturally crude grid with $N_x = 4$ and $N_y = 3$. In the coefficient matrix, only the elements marked by x are nonzero.

The structure of the entire matrix equation is illustrated in Figure 8.2. We see that the matrix **A** consists mostly of zeros. The nonzero elements are $A_{\ell,\ell}$ on the main diagonal, $A_{\ell,\ell+1}$ and $A_{\ell,\ell+N_x}$ on the sub- and super-diagonals, and $A_{\ell,\ell-N_x}$ and $A_{\ell,\ell-1}$ on the diagonals distanced by N_x from the main.

The five-point discretization can be applied to other two-dimensional elliptic equations of second order. For example, let us approximate the PDE

$$a(x,y)\frac{\partial^2 u}{\partial x^2} + c(x,y)\frac{\partial^2 u}{\partial y^2} + d(x,y)\frac{\partial u}{\partial x} + f(x,y)u = g(x,y), \qquad (8.11)$$

where a, c, d, f, and g are known variable coefficients and the ellipticity condition $ac > 0$ is satisfied throughout the domain Ω. Using central differences for derivatives, we obtain

$$a_{i,j}\frac{u_{i+1,j} - 2u_{i,j} + u_{i-1,j}}{\Delta x^2} + c_{i,j}\frac{u_{i,j+1} - 2u_{i,j} + u_{i,j-1}}{\Delta y^2}$$

$$+ d_{i,j}\frac{u_{i+1,j} - u_{i-1,j}}{2\Delta x} + f_{i,j}u_{i,j} = g_{i,j}. \qquad (8.12)$$

The difference molecule and the structure of the matrix equation evidently remain the same, as in Figures 8.1 and 8.2. The only change is in the expressions for the nonzero elements of matrix **A**.

Three-Dimensional Case: The approach based on the approximation of derivatives by central differences can be easily extended to the three-dimensional case (or to the cases of higher dimensions, should it become necessary). As an example, we develop the three-dimensional version of the central difference formula (8.7) for the Poisson equation

$$\frac{\partial^2 u}{\partial x^2} + \frac{\partial^2 u}{\partial y^2} + \frac{\partial^2 u}{\partial z^2} = f(x, y, z). \tag{8.13}$$

Again, we assume for simplicity that the computational grid is uniform and rectangular. In the z-direction, the grid step is Δz, and the grid points are $z_k = k\Delta z$. The approximation of u at (x_i, y_j, z_k) is denoted as $u_{i,j,k}$. The finite difference scheme of the second order in all three coordinates is

$$\frac{u_{i+1,j,k} - 2u_{i,j,k} + u_{i-1,j,k}}{\Delta x^2} + \frac{u_{i,j+1,k} - 2u_{i,j,k} + u_{i,j-1,k}}{\Delta y^2}$$
$$+ \frac{u_{i,j,k+1} - 2u_{i,j,k} + u_{i,j,k-1}}{\Delta z^2} = f_{i,j,k}. \tag{8.14}$$

Obviously, we can order the variables $u_{i,j,k}$ and the right-hand sides into one-dimensional vectors and express (8.14) as the matrix equation (8.1).

Finite Volume Discretization: Discretization of an elliptic PDE by a finite volume scheme also leads to a system of linear algebraic equations, which can be represented as (8.1). In this case, each equation connects the values of the solution variables at the grid points belonging to neighboring cells. When the finite volume grid is structured, reordering analogous to (8.8) and expressions similar to (8.9) can be written. In the more practically interesting case of an unstructured grid, reordering is not described by a single formula, but the principle remains the same. We assign throughout numbering to all cells of the grid. Since each equation of the system (8.1) includes the values of the solution variable at only several neighboring cells, the majority of the elements of the matrix **A** remain 0.

8.1.2 Marching Problems Solved by Implicit Schemes

We have seen in Section 7.5 that a matrix equation has to be solved at every time step if an implicit scheme is used for solution of an unsteady one-dimensional PDE. The situation remains the same in the case of multidimensional problems. For example, let us consider two-dimensional heat conduction described by the heat equation

$$\frac{\partial u}{\partial t} = \kappa \left(\frac{\partial^2 u}{\partial x^2} + \frac{\partial^2 u}{\partial y^2} \right) + \dot{Q} \tag{8.15}$$

and apply the Crank–Nicolson method with central differences for spatial derivatives:

$$\frac{u_{i,j}^{n+1} - u_{i,j}^n}{\Delta t} = \kappa \left(\frac{u_{i+1,j}^n - 2u_{i,j}^n + u_{i-1,j}^n}{(\Delta x)^2} + \frac{u_{i,j+1}^n - 2u_{i,j}^n + u_{i,j-1}^n}{(\Delta y)^2} \right.$$

$$\left. + \frac{u_{i+1,j}^{n+1} - 2u_{i,j}^{n+1} + u_{i-1,j}^{n+1}}{(\Delta x)^2} + \frac{u_{i,j+1}^{n+1} - 2u_{i,j}^{n+1} + u_{i,j-1}^{n+1}}{(\Delta y)^2} \right) + \dot{Q}_{i,j}. \tag{8.16}$$

This can be rewritten as

$$u_{i-1,j}^{n+1} \left[-\frac{\kappa}{(\Delta x)^2} \right] + u_{i+1,j}^{n+1} \left[-\frac{\kappa}{(\Delta x)^2} \right] + u_{i,j-1}^{n+1} \left[-\frac{\kappa}{(\Delta y)^2} \right]$$

$$+ u_{i,j+1}^{n+1} \left[-\frac{\kappa}{(\Delta y)^2} \right] + u_{i,j}^{n+1} \left[\frac{1}{\Delta t} + \frac{2\kappa}{(\Delta x)^2} + \frac{2\kappa}{(\Delta y)^2} \right] = c_{i,j}, \tag{8.17}$$

where indices i and j cover all internal points of the computational grid and

$$c_{i,j} = u_{i-1,j}^n \left[\frac{\kappa}{(\Delta x)^2} \right] + u_{i+1,j}^n \left[\frac{\kappa}{(\Delta x)^2} \right] + u_{i,j-1}^n \left[\frac{\kappa}{(\Delta y)^2} \right]$$

$$+ u_{i,j+1}^n \left[\frac{\kappa}{(\Delta y)^2} \right] + u_{i,j}^n \left[\frac{1}{\Delta t} - \frac{2\kappa}{(\Delta x)^2} - \frac{2\kappa}{(\Delta y)^2} \right] + \dot{Q}_{i,j} \tag{8.18}$$

is the coefficient known from the previous time step. Equations for boundary conditions have to be added to the system, but we ignore them for simplicity. Ordering the variables $u_{i,j}^{n+1}$ and the right-hand sides into one-dimensional arrays, for example, as in (8.8), we obtain the matrix equation (8.1) for the unknowns $u_{i,j}^{n+1}$.

An important example appears in the solution of the Navier–Stokes equations for unsteady flows of incompressible fluids. The commonly used projection algorithms, which we discuss in Chapter 10, require solution of a Poisson equation for pressure at every time step. The equation is mathematically equivalent to (8.13), and its finite difference approximation can be expressed as (8.1).

8.1.3 Structure of Matrices

We see in Figure 8.2 that even in the simple case of a two-dimensional Poisson equation discretized on a uniform structured Cartesian grid, the matrix **A** is not tridiagonal. This is true, in general, for all two-dimensional and three-dimensional PDEs and means that the efficient Thomas algorithm described in Section 7.5 cannot be used.

Nevertheless, the matrices in CFD analysis usually have one or another special structure, which can be utilized to accelerate solution of the matrix equation. One feature is common for the finite difference and finite volume discretizations. The matrices are *sparse*, i.e. the majority of their elements are zeros.

The matrices resulting from the finite difference and finite volume discretizations on structured grids have another useful property, namely, the *block structure*. The nonzero elements are contained in square submatrices (blocks) of the size corresponding to the size of the computational grid in one direction.

The pattern of blocks varies depending on the PDE solved (e.g. two-dimensional or three-dimensional), discretization schemes, and boundary conditions. For example, when a two-dimensional PDE is solved using central differences on the five-point molecule shown in Figure 8.1 and the grid points are reordered according to (8.8), the matrix has the structure illustrated in Figure 8.2. The nonzero elements are within the $N_x \times N_x$ blocks located on the main diagonal, just above and just below it. The entire matrix **A** can be viewed as a tridiagonal matrix with elements being not numbers but $N_x \times N_x$ blocks. Such a matrix is called block tridiagonal.

In our specific example of the two-dimensional Poisson equation (8.6) discretized as (8.7), the matrix is further simplified. Multiplying (8.7) by $(\Delta y)^2$ we obtain the matrix

$$\mathbf{A} = \begin{pmatrix} M & I & 0 & \ldots & \ldots & \ldots & 0 \\ I & M & I & 0 & \ldots & \ldots & 0 \\ 0 & I & M & I & 0 & \ldots & 0 \\ ** & * & * & * & * & * & * \\ 0 & \ldots & \ldots & 0 & I & M & I \\ 0 & \ldots & \ldots & \ldots & 0 & I & M \end{pmatrix} \tag{8.19}$$

with blocks

$$I = \begin{pmatrix} 1 & 0 & \dots & \dots & 0 \\ 0 & 1 & 0 & \dots & 0 \\ * & * & * & * & * \\ 0 & \dots & 0 & 1 & 0 \\ 0 & \dots & \dots & 0 & 1 \end{pmatrix}, M = \begin{pmatrix} a & b & 0 & \dots & \dots & 0 \\ b & a & b & 0 & \dots & 0 \\ * & * & * & * & * & * \\ 0 & \dots & 0 & b & a & b \\ 0 & \dots & \dots & 0 & b & a \end{pmatrix} \quad (8.20)$$

where $a = -2(\Delta y/dx)^2 - 2$, $b = (\Delta y/dx)^2$, and the right-hand side vector has elements

$$c_\ell = f_{i,j}(\Delta y)^2. \quad (8.21)$$

In deriving these formulas, we for simplicity ignore the equations for discretized boundary conditions. If the boundary conditions are of the Dirichlet type, the structure (8.19) and (8.20) of the matrix remains the same, while some modifications may be required for the other types.

The matrices appearing in CFD solutions are not necessarily block structured and sparse. One notable exception is the spectral method, for which matrix equations with *dense* (with most or all elements being nonzero) have to be solved. Even more important is the finite volume discretization on unstructured grids. Here, each discretization equation connects values of solution variables at few neighboring grid points. The matrices are, therefore, sparse. At the same time, the matrices usually do not have a clear and useful block structure.

8.2   DIRECT METHODS

The direct methods solve the matrix equation (8.1) exactly, i.e. with the maximum accuracy afforded by the computer's arithmetics. From elementary linear algebra, we know two such methods: the Cramer's rule and the standard Gauss elimination. Unfortunately, these methods, while working quite well for small matrices, are practically useless for large matrices found in CFD. Let the order of matrix A – that is, the total number of unknowns in the problem – be N. This can be a fairly large number. For example, in a three-dimensional problem with 100 grid points in each direction, N is 10^6 multiplied by the number of variables – for example, four (three velocity components and pressure) in the case of simple incompressible flows. The situation with the Cramer's rule solution is particularly hopeless, since it would require calculation of N determinants at the cost of $\sim (N+1)!$ arithmetic operations. The standard Gauss elimination procedure is much more efficient but still requires about $2N^3/3$ multiplications and additions.

The amount of calculations can be reduced dramatically if one of the methods utilizing the special structure of CFD matrices is used. Two examples of this approach are presented below.

8.2.1 Cyclic Reduction Algorithm

We will show the original (and simplest) version of the algorithm valid when the two-dimensional Poisson equation (8.6) is solved in a rectangular domain, say, $0 \leq x \leq L_x$, $0 \leq y \leq L_y$ using the discretization (8.7) on a uniform structured grid $x_0 = 0, \dots, x_{N_x} = L_x$, $y_0 = 0, \dots, y_{N_y} = L_y$. The boundary conditions at two opposite boundaries, for example, at $y = 0$ and $y = L_y$, are of the Dirichlet type:

$$u_{i,0} = g_i, \ u_{i,N_y} = h_i. \tag{8.22}$$

The conditions at the other two boundaries can be of any type. We also assume that $N_y = 2^{r+1}$, where $r > 0$ is an integer number. Extensions of the method to general separable elliptic PDEs, other combinations of boundary conditions, and other values of N_y can be found in specialized literature, including the texts listed at the end of this chapter.

For convenience of explanation, we break the vectors of solution variables and right-hand sides into $N_y + 1$ vectors so that each of them has the length $N_x + 1$ and includes all values of $u_{i,j}$ and $f_{i,j}$ at a certain y_j:

$$V_j = \begin{pmatrix} u_{0,j} \\ \vdots \\ u_{N_x,j} \end{pmatrix}, \quad j = 0, \dots, N_y, \tag{8.23}$$

$$C_j = \begin{pmatrix} f_{0,j} \\ \vdots \\ f_{N_x,j} \end{pmatrix}, \quad j = 1, \dots, N_y - 1, \tag{8.24}$$

$$C_0 = \begin{pmatrix} g_0 \\ \vdots \\ g_{N_x} \end{pmatrix}, \quad C_{N_y} = \begin{pmatrix} h_0 \\ \vdots \\ h_{N_x} \end{pmatrix}. \tag{8.25}$$

The matrix equation can be written in block-structured form as

$$
\begin{pmatrix}
I & 0 & 0 & \dots & \dots & \dots & 0 \\
I & M & I & 0 & \dots & \dots & 0 \\
0 & I & M & I & 0 & \dots & 0 \\
* & * & * & * & * & * & * \\
0 & \dots & 0 & I & M & I & 0 \\
0 & \dots & \dots & 0 & I & M & I \\
0 & \dots & \dots & \dots & 0 & 0 & I
\end{pmatrix}
\cdot
\begin{pmatrix}
V_0 \\
V_1 \\
\vdots \\
\vdots \\
V_{N_y-1} \\
V_{N_y}
\end{pmatrix}
=
\begin{pmatrix}
C_0 \\
C_1 \\
\vdots \\
\vdots \\
C_{N_y-1} \\
C_{N_y},
\end{pmatrix}
\tag{8.26}
$$

where the elements of the blocks I and M and the right-hand side vectors are as in (8.20) and (8.21). The first and last lines of the matrix in the left-hand side that do not appear in (8.19) are due to the Dirichlet boundary conditions.

Equation (8.26) can be viewed as a system of $N_y + 1$ coupled matrix equations of size $N_x + 1$. We will look at three consecutive equations for internal points:

$$
V_{j-2} + M \cdot V_{j-1} + V_j = C_{j-1}, \tag{8.27}
$$

$$
V_{j-1} + M \cdot V_j + V_{j+1} = C_j, \tag{8.28}
$$

$$
V_j + M \cdot V_{j+1} + V_{j+2} = C_{j+1}. \tag{8.29}
$$

The matrix M is tridiagonal (see (8.20)) and the same for all three equations. Left-multiplying (8.28) by $-M$ and adding the result and the equations (8.27) and (8.29) together, we obtain

$$
V_{j-2} + M^{(1)} \cdot V_j + V_{j+2} = C_j^{(1)}, \tag{8.30}
$$

where

$$
M^{(1)} = 2I - M^2, \quad C_j^{(1)} = C_{j-1} - M \cdot C_j + C_{j+1}. \tag{8.31}
$$

We write $(N_y/2) - 1$ such equations for $j = 2, 4, \ldots, N_y - 2$ and observe that they all look as the original block equations (8.27)–(8.29), except that M and C_j are replaced by $M^{(1)}$ and $C_j^{(1)}$. The same operation as above performed with any three consecutive new equations leads to

$$
V_{j-4} + M^{(2)} \cdot V_j + V_{j+4} = C_j^{(2)}, \tag{8.32}
$$

with

$$
M^{(2)} = 2I - (M^{(1)})^2, \quad C_j^{(2)} = C_{j-1}^{(1)} - (M^{(1)}) \cdot C_j^{(1)} + C_{j+1}^{(1)}. \tag{8.33}
$$

We have now reduced the system to $(N_y/2^2) - 1$ equations for $j = 4, 8, \ldots ,$ $N_y - 4$. Note that the equations written for $j = 4$ and $j = N_y - 4$ include, respectively, vectors V_0 and V_{N_y} known from the Dirichlet boundary conditions.

Since $N_y = 2^{r+1}$, the reduction procedure can be carried out r times, after which we are left with just one equation:

$$V_0 + M^{(r)} \cdot V_{N_y/2} + V_{N_y} = C_{N_y/2}^{(r)}, \tag{8.34}$$

with $M^{(r)}$ and $C_{N_y/2}^{(r)}$ calculated recurrently as shown above. The equation can be rearranged as

$$M^{(r)} \cdot V_{N_y/2} = C_{N_y/2}^{(r)} - V_0 - V_{N_y} \tag{8.35}$$

so that the right-hand side only contains the right-hand sides of the original matrix equation (8.26) and the boundary conditions and, therefore, is known.

We now note that the product of any two tridiagonal matrices is a tridiagonal matrix. Adding the diagonal matrix $2I$ leaves the matrix tridiagonal. Therefore, the matrices $M^{(1)}, \ldots , M^{(r)}$ are all tridiagonal. Equation (8.35) can be easily and efficiently solved by the Thomas algorithm (see Section 7.5) to find $V_{N_y/2}$. The two equations at the previous reduction level

$$V_0 + M^{(r-1)} \cdot V_{N_y/4} + V_{N_y/2} = C_{N_y/4}^{(r-1)}, \tag{8.36}$$

$$V_{N_y/2} + M^{(r-1)} \cdot V_{3N_y/4} + V_{N_y} = C_{3N_y/4}^{(r-1)} \tag{8.37}$$

can now be rearranged and solved by the Thomas algorithm for $V_{N_y/4}$ and $V_{3N_y/4}$. The process is then continued to find $V_{N_y/8}, \ldots , V_{7N_y/8}$ and so on, until we find all V_j solving $N_y - 1$ tridiagonal systems on the way.

The total number of the arithmetic operations needed for the solution scales as $3N^2 \log_2 N$ at large $N \sim N_x \sim N_y$. The procedure, as it is shown above, needs some modifications to avoid numerical instabilities. The matrices $M^{(k)}$ can be precomputed and stored in a memory-saving manner. A reader interested in these and other practical aspects of the method is referred to specialized literature. We note that reliable and efficient algorithms implementing the method are available in the linear algebra libraries discussed in Section 8.5.

8.2.2   Thomas Algorithm for Block-Tridiagonal Matrices

The Thomas algorithm developed in Section 7.5 can be generalized to the case of a block-tridiagonal matrix. We have already seen in (8.19) that such a matrix appears in the five-point discretization of the Poisson equation. The method described below is valid in the more general case, when the matrix equation (8.19) has the form

$$
\begin{pmatrix}
D_0 & A_0 & 0 & \dots & \dots & \dots & 0 \\
B_1 & D_1 & A_1 & 0 & \dots & \dots & 0 \\
0 & B_2 & D_2 & A_2 & 0 & \dots & 0 \\
0 & 0 & * & * & * & 0 & 0 \\
0 & 0 & 0 & * & * & * & 0 \\
0 & \dots & \dots & 0 & B_{N-1} & D_{N-1} & A_{N-1} \\
0 & \dots & \dots & \dots & 0 & B_N & D_N
\end{pmatrix}
\cdot
\begin{pmatrix}
V_0 \\ V_1 \\ V_2 \\ \vdots \\ \vdots \\ V_{N-1} \\ V_N
\end{pmatrix}
=
\begin{pmatrix}
C_0 \\ C_1 \\ C_2 \\ \vdots \\ \vdots \\ C_{N-1} \\ C_N
\end{pmatrix}.
\tag{8.38}
$$

Here A_j, B_j, and D_j are square blocks of size M containing the nonzero elements of the matrix. The vectors of unknowns and right-hand sides are divided into subvectors of length M. The integer constant $M > 1$ can, in principle, be an arbitrary factor of the matrix size N. The typical situation leading to (8.38) is when a two-dimensional elliptic equation with Dirichlet or Neumann boundary conditions is discretized in a rectangular domain using central differences. In that case, $N = N_x N_y$ and $M = N_x$ or N_y.

Each step of the forward sweep of the generalized Thomas algorithm consists of inverting a diagonal submatrix and performing matrix operations on two subsequent block rows to eliminate the subdiagonal blocks. Each such matrix operation is equivalent to a series of elementary row operations and, therefore, does not affect the solution. The first step is (compare with (7.57))

$$
\begin{aligned}
D'_1 &= D_1 - B_1 \cdot D_0^{-1} \cdot A_0 \\
B'_1 &= B_1 - B_1 \cdot D_0^{-1} \cdot D_0 = 0 \\
C'_1 &= C_1 - B_1 \cdot D_0^{-1} \cdot C_0.
\end{aligned}
\tag{8.39}
$$

The following steps are given by the general formula

$$
\begin{aligned}
D'_i &= D_i - B_i \cdot (D'_{i-1})^{-1} \cdot A_{i-1} \\
B'_i &= B_i - B_i \cdot (D'_{i-1})^{-1} \cdot D_{i-1} = 0 \\
C'_i &= C_i - B_i \cdot (D'_{i-1})^{-1} \cdot C_{i-1}.
\end{aligned}
\tag{8.40}
$$

At the end of the forward sweep, the matrix equation transforms into the one with block upper-triangular coefficient matrix:

$$
\begin{pmatrix}
D_0 & A_0 & 0 & \dots & \dots & \dots & 0 \\
0 & D'_1 & A_1 & 0 & \dots & \dots & 0 \\
0 & 0 & D'_2 & A_2 & 0 & \dots & 0 \\
0 & 0 & 0 & * & * & 0 & 0 \\
0 & 0 & 0 & 0 & * & * & 0 \\
0 & \dots & \dots & 0 & 0 & D'_{N-1} & A_{N-1} \\
0 & \dots & \dots & \dots & 0 & 0 & D'_N
\end{pmatrix}
\cdot
\begin{pmatrix}
V_0 \\ V_1 \\ V_2 \\ \vdots \\ \vdots \\ V_{N-1} \\ V_N
\end{pmatrix}
=
\begin{pmatrix}
C_0 \\ C'_1 \\ C'_2 \\ \vdots \\ \vdots \\ C'_{N-1} \\ C'_N
\end{pmatrix}.
\tag{8.41}
$$

The backward sweep is trivial. We invert the diagonal submatrices of the transformed system D'_i and calculate first

$$
V_N = (D'_N)^{-1} \cdot C'_N
\tag{8.42}
$$

and then the rest of the solution according to

$$
V_i = (D'_i)^{-1} \cdot (C'_i - A_i \cdot V_{i+1}), \quad i = N - 1, N - 2, \dots, 0.
\tag{8.43}
$$

The inversion of the matrices D' is computationally expensive, which makes the entire algorithm not particularly effective. Its use can be justified when the equation has to be solved many times with the same matrix but different right-hand sides (for example, in an implicit solution of a marching problem). In this case, the necessary inverted block matrices can be precomputed and stored for multiple use.

8.2.3 LU Decomposition

Other variations of the Gauss elimination found their use in the CFD applications, where the matrix $\mathbf{A}$ is not sparse, for example, in the spectral methods. They are often based on factorization (decomposition) of $\mathbf{A}$ into a product of a lower-triangular matrix $\mathbf{L}$ and upper-triangular matrix $\mathbf{U}$. The procedures of factorization are described in the books on numerical linear algebra. Here, we only discuss the consequences. Rewriting (8.1) as

$$
\mathbf{L} \cdot \mathbf{U} \cdot v = c
\tag{8.44}
$$

and introducing a new vector $w = \mathbf{U} \cdot v$, we can split Equation (8.1) into two:

$$
\mathbf{L} \cdot w = c
\tag{8.45}
$$

and

$$\mathbf{U} \cdot \boldsymbol{v} = \boldsymbol{w}. \tag{8.46}$$

The equations are solved sequentially, first (8.45) and then (8.46). Since the matrices $\mathbf{L}$ and $\mathbf{U}$ are triangular, the solution can be easily obtained by backward substitution (operation count $\sim N^2$). A nice feature of the method is that the only computationally expensive part, the decomposition of $\mathbf{A}$, does not use the right-hand side vector $\boldsymbol{c}$. This becomes useful if (8.1) has to be solved many times with the same matrix $\mathbf{A}$ but different right-hand sides. One can perform the decomposition once, store the matrices L and U, and apply (8.45) and (8.46) as many times as required.

8.3 ITERATIVE METHODS

Unlike direct methods, the iterative methods are not designed to find an exact (up to the computer round-off error) solution of the matrix equation (8.1). Instead, the methods rely on successive iterations to obtain a sufficiently accurate approximation of $\boldsymbol{v}$. Two important considerations justify this approach in CFD:

- The matrix equation is generated by discretization, so its solution inevitably contains some discretization error. We can allow ourselves an additional error of an iteration procedure without significant drop of the overall accuracy provided this error is much smaller than the discretization one.
- The iterative methods can be arranged so that they utilize the sparse character of matrix $\mathbf{A}$ to reduce the computational cost of the solution. At the same time, the methods are applicable to a much wider class of matrices than the effective direct methods. This makes the iterative methods especially attractive for the problem resulting from the finite volume discretizations on unstructured grids.

The main characteristics of an iterative method are the ability to converge (to achieve an approximation accurate within a given tolerance) and the computational cost of the procedure. The cost is determined by the number of iterations needed for convergence (speed of convergence) and the amount of computations required to complete a single iteration. Both should be small or, at least, not very large for the iteration procedure to be effective.

Numerous iterative methods have been developed over the last decades. We will consider some algorithms that, albeit simple, possess the principal features of the commonly used techniques.

8.3.1 General Methodology

The general iteration procedure of solution of the matrix equation (8.1) is as follows:

1. Guess an initial approximation $v_i^{(0)}$ of the solution.
2. Use $\mathbf{A}$, c, and the already achieved approximation to find the next, more accurate approximation $v^{(k)}$.
3. Repeat Step 2 iteratively until the convergence criterion is satisfied.

A few comments are in order concerning the definition and realization of the convergence criterion. The definition of the error is

$$\epsilon^{(k)} = v - v^{(k)}, \tag{8.47}$$

where v is the exact solution of (8.1) and $v^{(k)}$ is the approximation obtained after the kth iteration. Since v is typically unavailable during the solution, the error cannot be found directly from (8.47). Instead, we can find the difference between successive approximations and evaluate its norm as

$$\delta^{(k)} = \| v^{(k+1)} - v^{(k)} \| = \max_i |v_i^{(k+1)} - v_i^{(k)}| \tag{8.48}$$

or

$$\delta^{(k)} = \| v^{(k+1)} - v^{(k)} \| = \left[\sum_{i=1}^{N} (v_i^{(k+1)} - v_i^{(k)})^2 \right]^{1/2}. \tag{8.49}$$

If the iterative procedure converges, $v^{(k)}$ and $v^{(k+1)}$ both tend to the exact solution v, so the norm of the difference between them tends to 0. The magnitude of $\delta^{(k)}$ can therefore be used as a convergence criterion. The iterations are stopped when $\delta^{(k)}$ becomes smaller than the desired tolerance ϵ_0. Note that determining ϵ_0 that guarantees the desired accuracy of the solution is, by itself, a nontrivial problem. We discuss this issue in Section 13.2.1.

Another popular and effective approach is to monitor the norm of the residual vector

$$r^{(k)} = c - \mathbf{A} \cdot v^{(k)}. \tag{8.50}$$

It is easy to see that

$$\mathbf{A} \cdot \epsilon^{(k)} = \mathbf{A} \cdot (v - v^{(k)}) = c - \mathbf{A} \cdot v^{(k)} = r^{(k)} \tag{8.51}$$

so the residual vanishes with $e^{(k)}$, as $v^{(k)}$ converges to the exact solution v. The criterion of convergence is $\| r^{(k)} \| < \epsilon_0$.

The initial guess $v^{(0)}$ can, in general, be an arbitrary vector. For the sake of faster convergence, however, $v^{(0)}$ should be chosen as close to the exact solution as possible.

8.3.2 Jacobi Iterations

We start with the simplest method, the Jacobi iteration algorithm. The method is rather inefficient and is presented here solely as the simplest illustration of the basic approach. First, we rewrite the equations of the system (8.1) so that the first equation is an expression of v_1 through v_2, v_3, ..., the second equation is an expression of v_2 through v_1, v_3, ..., and so on. In general, the ℓth equation becomes

$$v_\ell = \frac{1}{a_{\ell\ell}} \left(c_\ell - \sum_{j=1}^{\ell-1} a_{\ell j} v_j - \sum_{j=\ell+1}^{N} a_{\ell j} v_j \right), \quad \ell = 1, \ldots, N. \quad (8.52)$$

The exact solution v satisfies this equation exactly. In the Jacobi method, we use the results of the previous iteration to compute the right-hand side and assign the left-hand side as the new approximation:

$$v_\ell^{(k+1)} = \frac{1}{a_{\ell\ell}} \left(c_\ell - \sum_{j=1}^{\ell-1} a_{\ell j} v_j^{(k)} - \sum_{j=\ell+1}^{N} a_{\ell j} v_j^{(k)} \right), \quad \ell = 1, \ldots, N. \quad (8.53)$$

This formula shows why the iteration methods are particularly effective if the matrix A is sparse. Only few terms of the sums in the right-hand side of (8.53) are nonzero. For example, let us write the Jacobi iteration formula for the five-point discretization of the Poisson equation (8.9)

$$v_\ell^{(k+1)} = \frac{1}{a_{\ell,\ell}} (c_\ell - a_{\ell,\ell-N_x} v_{\ell-N_x}^{(k)} - a_{\ell,\ell-1} v_{\ell-1}^{(k)} - a_{\ell,\ell+1} v_{\ell+1}^{(k)} - a_{\ell,\ell+N_x} v_{\ell+N_x}^{(k)})$$

$$(8.54)$$

or, in terms of the original structured grid,

$$u_{i,j}^{(k+1)} = \frac{1}{(2/(\Delta x)^2 + 2/(\Delta y)^2)} \left(-f_{i,j} \right.$$

$$\left. + \frac{1}{(\Delta y)^2} u_{i,j-1}^{(k)} + \frac{1}{(\Delta x)^2} u_{i-1,j}^{(k)} + \frac{1}{(\Delta y)^2} u_{i,j+1}^{(k)} + \frac{1}{(\Delta x)^2} u_{i+1,j}^{(k)} \right). \quad (8.55)$$

There are only five nonzero terms in the right-hand side. Computing only them and explicitly omitting the others reduces the computational cost of each iteration from $\sim N^2$ to $\sim N$ operations.

8.3.3 Gauss–Seidel Algorithm

The Gauss–Seidel method is an improvement of the Jacobi algorithm, which results in faster convergence. The right-hand side of (8.53) is constantly updated using already found $v_j^{(k+1)}$. If a new approximation is calculated sequentially from $\ell = 1$ to $\ell = N$, the iterations are

$$v_\ell^{(k+1)} = \frac{1}{a_{\ell\ell}} \left(c_\ell - \sum_{j=1}^{\ell-1} a_{\ell j} v_j^{(k+1)} - \sum_{j=\ell+1}^{N} a_{\ell j} v_j^{(k)} \right), \quad \ell = 1, \dots, N.$$

(8.56)

For the Laplace and Poisson equations (8.5) and (8.6), the solution by the Gauss–Seidel algorithm requires approximately two times fewer iterations than the solution by the Jacobi algorithm.

As an example, let us solve the two-dimensional Poisson equation using the five-point scheme. Each iteration can be arranged so that we begin at the lower-left corner of the computational domain and move to the right, covering one row after another (see Figure 8.3). The Gauss–Seidel iteration (8.56) is

$$v_\ell^{(k+1)} = \frac{1}{a_{\ell,\ell}} \left(c_\ell - a_{\ell,\ell-N_x} v_{\ell-N_x}^{(k+1)} - a_{\ell,\ell-1} v_{\ell-1}^{(k+1)} \right.$$

$$\left. - a_{\ell,\ell+1} v_{\ell+1}^{(k)} - a_{\ell,\ell+N_x} v_{\ell+N_x}^{(k)} \right)$$

(8.57)

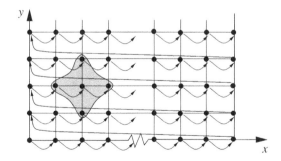

Figure 8.3 A Gauss–Seidel iteration for the five-point scheme (8.9) applied to a two-dimensional elliptic equation. Arrows illustrate the order, in which the grid point values are updated. A difference molecule is also shown.

or, in terms of the grid point indices,

$$u_{i,j}^{(k+1)} = \frac{1}{(2/(\Delta x)^2 + 2/(\Delta y)^2)} \left(-f_{i,j} \right.$$

$$+ \frac{1}{(\Delta y)^2} u_{i,j-1}^{(k+1)} + \frac{1}{(\Delta x)^2} u_{i-1,j}^{(k+1)} + \frac{1}{(\Delta y)^2} u_{i,j+1}^{(k)} + \left. \frac{1}{(\Delta x)^2} u_{i+1,j}^{(k)} \right). \quad (8.58)$$

8.3.4 Successive Over- and Underrelaxation

Successive *over- and underrelaxations* are the techniques that can be applied to accelerate convergence or avoid divergence of the iterative methods such as the Gauss–Seidel algorithm. The principal idea is to determine the direction in which the solution evolves from one iteration to the next and accelerate or decelerate this evolution. The correction is achieved by performing the simple operation

$$\boldsymbol{v}^* = \boldsymbol{v}^{(k)} + \omega(\boldsymbol{v}^{(k+1)} - \boldsymbol{v}^{(k)}) \quad (8.59)$$

and using $\boldsymbol{v}^*$ as a more accurate new approximation instead of $\boldsymbol{v}^{(k+1)}$. The technique is schematically illustrated in Figure 8.4.

The difference between the over- and underrelaxation is in the value of the coefficient ω. For the problems, where, as illustrated in Figure 8.4a, the approximations converge to the exact solution *monotonically*, the convergence can be accelerated by using $1 < \omega < 2$, which enhances the step in the right direction. This method is called *overrelaxation*. When,

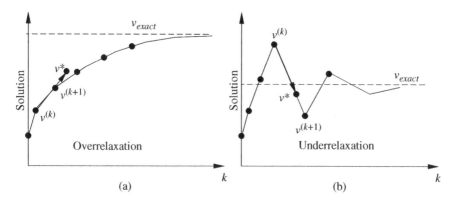

Figure 8.4 Over- and underrelaxation improvements of an iteration procedure.

as shown in Figure 8.4b, the iterations lead to nonmonotonic behavior, the overrelaxation would compromise the convergence. Acceleration can, however, be achieved if a value $0 < \omega < 1$ is used thus reducing overshoot. This method is called *underrelaxation*. The underrelaxation is also often used to achieve convergence of otherwise diverging iteration procedures.

Optimal values of ω that provide the fastest convergence have been found theoretically for simple model problems, such as the Laplace or Poisson equations in rectangular domains. For complex CFD problems, there are typically no rigorous theoretical results, but the estimates based on experience and extrapolation of theoretical data are usually available.

8.3.5 Convergence of Iterative Procedures

The question left out so far is that of convergence. Can we be sure, when we start the computations, that the iterations will converge to a final equilibrium point? What is the speed of convergence?

The answer to the first question for the Gauss–Seidel procedure is given by the following sufficient condition. The Gauss–Seidel procedure applied to a system of linear equations converges if the system is irreducible (cannot be separated into smaller decoupled systems – in other words, rearranged in such a way that some unknowns can be found without solving the entire system) and if the following is true for the coefficient matrix:

$$|a_{\ell\ell}| \geq \sum_{j=1,j\neq\ell}^{N} |a_{\ell j}| \text{ for all } \ell \text{ and } |a_{\ell\ell}| > \sum_{j=1,j\neq\ell}^{N} |a_{\ell j}| \text{ for some } \ell. \quad (8.60)$$

The condition (8.60) is referred to as the *diagonal dominance* of matrix $\mathbf{A}$.

If the system of finite difference equations does not satisfy the convergence condition (8.60), it can often be rearranged by changing the order of unknowns so that the largest coefficients lie on the diagonal of $\mathbf{A}$. The new system may satisfy the condition. The diagonal dominance is not a necessary condition. Quite often, the Gauss–Seidel iterations converge even though (8.60) is not satisfied.

As an example, let us again consider the five-point scheme for the Poisson equation, rewritten as (8.9). We see that no rearranging is required. The diagonal coefficient $a_{\ell,\ell} = [(2/\Delta x)^2 + 2/(\Delta y)^2)]$ is the largest element in each row. It is exactly equal to the sum of all the other coefficients. The first condition in (8.60) is, thus, satisfied. The second condition that, at least in one row, $|a_{\ell\ell}|$ is greater than the sum of the other coefficients is normally satisfied by the rows of **A** corresponding to the boundary points of the computational domain. For example, if the Dirichlet boundary condition $u = g$ is imposed at the point (x_0, y_j) of the grid, the equation is

$$u_{0,j} = g_j.$$

The corresponding row of matrix **A** has only one nonzero element, and this element is on the main diagonal.

The rate of convergence varies with the type of the problem, method, and the grid size. There is no general theory that would be valid for all CFD computations. The conclusions are often made based on experience. There are several well-established trends, though.

As an illustration, we solve the two-dimensional Poisson equation

$$\frac{\partial^2 u}{\partial x^2} + \frac{\partial^2 u}{\partial y^2} = \sin(\pi x)\sin(\pi y)$$

in the square domain $[0, 1] \times [0, 1]$ with homogeneous boundary conditions $u = 0$ at all boundaries. The equation is discretized on a uniform grid using the five-point scheme (8.7). The matrix equation is solved by three simple iterative methods: Jacobi (8.54), Gauss–Seidel (8.57), and Gauss–Seidel with successive overrelaxation with $\omega = 1.5$. To illustrate the effect of numerical resolution on convergence, the solution is computed using three grids: with $N_x = N_y = 10$, $N_x = N_y = 30$, and $N_x = N_y = 100$. The initial guess $u_{i,j}^{(0)} = 0$ is used for all cases. The convergence is monitored by the norm $||r^k|| = \left[\sum_{i,j}(r_{i,j}^{(k)})^2\right]^{1/2}$ of the residual vector (8.50). The calculations continue until the norm reduces below $10^{-6}||r^0||$.

The results are shown in Table 8.1 and Figure 8.5. They illustrate the commonly observed (although not absolute) trends:

1. Relaxation methods (with properly chosen ω) are faster than Gauss–Seidel, which, in turn, is faster than Jacobi.
2. Finer grid (larger N_x and N_y) requires a larger (sometimes much larger) number of iterations to achieve the same reduction of residual.

Considering the second trend, we should also remember that the computational cost of each iteration increases proportionally to the total number of grid points. We see that using a finer grid requires larger, perhaps much larger, amount of computational operations. In our example, the

Table 8.1   Example of performance of iteration methods.

Method	$N_x = N_y = 10$	$N_x = N_y = 30$	$N_x = N_y = 100$
Jacobi	276	2,516	27,992
Gauss–Seidel	139	1,259	13,997
SOR	90	837	9,329

The Poisson equation is solved in a rectangular domain using the finite difference scheme of the second order on three different grids. The table shows the number of iterations needed to reduce the norm of residual $||r^{(k)}||$ below $10^{-6}||r^{(0)}||$

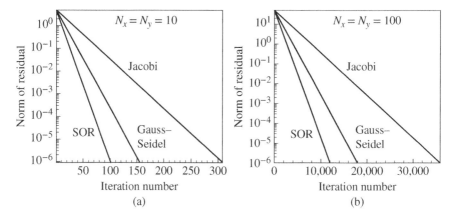

Figure 8.5 Example of performance of Jacobi, Gauss–Seidel, and SOR iteration methods (see Table 8.1 and text). Norm of residual $||r^{(k)}||$ (see (8.50)) is shown as a function of the iteration number k. Note that the abscissa scales in the two graphs are different by about 2 orders of magnitude.

computational cost of the solution with $N_x = N_y = 100$ is about 4 orders of magnitude higher than the cost of the solution with $N_x = N_y = 10$. Solutions on fine grids are nevertheless often unavoidable, since solutions computed on crude grids have unacceptably large discretization errors.

We see that there is a need for a compromise between the small discretization error achievable on fine grids and faster convergence provided by crude grids. Even better would be a strategy that combines the advantages of fine and crude grids. One simple approach is to start computations on a crude grid, quickly reach convergence, and then interpolate the computed variables to a fine grid and use it as a good initial approximation $u^{(0)}$ for a fine grid solution. A more advanced, efficient, and widely used strategy based on the multigrid method is discussed in the next section.

8.3.6 Multigrid Methods

An important observation has been made regarding the error $\epsilon^{(k)} = v^{(k)} - v$ of an iteration solution of a matrix equation. The components of the error that vary significantly on small distances comparable to the grid steps vanish quickly, after just few iterations. These components will be referred to as *small scale* in the following discussion. On the contrary, the large-scale components, for which significant variation occurs on distances much larger than the grid step, decrease slowly, require a larger number of iterations, and are, in general, responsible for the slow convergence on finer grids.

We can now conjecture the main idea of the multigrid method. The large-scale error can be accurately represented on a coarser grid. Iterations on such a grid can be performed with lower computational cost. Furthermore, the components of the error, which are large scale on a fine grid, are small scale on a coarser grid. Their removal occurs faster if iterations are performed on a coarser grid. Why not iterate on several grids of different degrees of coarseness interpolating solutions between them and allowing each grid to take care of the error on the respective length scales?

We will give a sketch of a simple version of the strategy. As an example, we will solve a two-dimensional elliptic problem expressed in the general matrix form (8.1). Two uniform structured grids shown in Figure 8.6 will be used. One is the fine grid with steps Δx and Δy, in which the actual solution is to be found. We also introduce a coarse grid that includes only every second grid point in each direction. Its steps are $2\Delta x$ and $2\Delta y$.

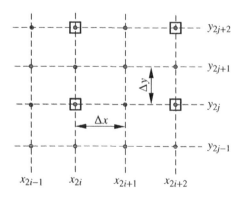

Figure 8.6 Example of a multigrid approach. Fine (circles) and coarse (squares) grids are used to solve a two-dimensional elliptic problem.

The multigrid iteration procedure is as follows:

Step 1: Conduct k iterations on the fine grid. The result is the kth approximation $v^{(k)}$. Calculate the residual $r^{(k)} = c - A \cdot v^{(k)}$.

Step 2: The error of approximation $\epsilon^{(k)} = v - v^{(k)}$ cannot be calculated directly, but it satisfies the so-called correction equation (see (8.51))

$$A \cdot \epsilon^{(k)} = r^{(k)}. \tag{8.61}$$

We can also assume that, after the k iterations of Step 1 performed on the fine grid, the error $\epsilon^{(k)}$ primarily varies on the length scales larger than Δx and Δy. Let us estimate the error applying efficient coarse grid iterations to the correction equation (8.61). For this purpose, we restrict (8.61) (both the coefficients of matrix A and the right-hand side $r^{(k)}$) to the points of the coarse grid. After several iterations, we obtain a coarse grid approximation of the error, which we denote as $\tilde{\epsilon}^{(k)}$.

Step 3: Interpolate $\tilde{\epsilon}^{(k)}$ onto the fine grid, and update the solution as $\tilde{v}^{(k)} = v^{(k)} + \tilde{\epsilon}^{(k)}$. If $\tilde{\epsilon}^{(k)}$ is an accurate approximation of $\epsilon^{(k)}$, the updated field $\tilde{v}^{(k)}$ is much closer to the exact solution than the original $v^{(k)}$.

Step 4: Check convergence. If not achieved, repeat starting with step 1 and using $\tilde{v}^{(k)}$ as a new initial guess.

It is important to stress that the procedure outlined here is only a simple example. Actual multigrid solvers are quite complex and diverse. They are routinely applied to unstructured and nonuniform grids. Moreover, the advanced multigrid solvers typically use more than two embedded grids of various degrees of coarseness.

At least a brief discussion is needed of the methods used to transfer the information between the fine and coarse grids. We will limit the discussion to the example illustrated in Figure 8.6. The variables computed on the fine grid will be denoted as $u_{i,j}^{\Delta}$, and their counterparts on the coarse grid as $u_{i,j}^{2\Delta}$. Let the points used for the coarse grid be those with even indices (x_{2i}, y_{2j}) (see Figure 8.6). The restriction of a fine grid solution onto the coarse grid can be achieved if we directly transfer the values at the common grid points and discard the others:

$$u_{2i,2j}^{2\Delta} = u_{2i,2j}^{\Delta}, \quad \text{discard } u_{i,j}^{\Delta} \text{ if } i \text{ or } j \text{ is odd.} \tag{8.62}$$

A better method is the *full weighting*, in which we evaluate a coarse grid value as an average of the fine grid solution over the surrounding area.

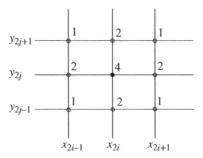

Figure 8.7 Full weighting restriction on a two-dimensional uniform grid. The coarse grid value $u^{2\Delta}_{2i,2j}$ is evaluated as a combination (8.63) of fine grid values of a 9-point stencil with weight coefficients as indicated in the figure.

Numerical approximations of integrals are used to compute the averages. The full weighting operation has several advantages over the direct transfer. It can be applied to unstructured grids, in which the points of fine and coarse grids do not necessarily coincide. It clearly makes better use of the information available from the fine grid solution. At last, it commutes with the linear interpolation procedure discussed next. In our example of two-dimensional structured uniform grids shown in Figure 8.6, the full weighting formula is (see Figure 8.7)

$$u^{2\Delta}_{2i,2j} = \frac{1}{16}[u^{\Delta}_{2i-1,2j-1} + u^{\Delta}_{2i-1,2j+1} + u^{\Delta}_{2i+1,2j-1} + u^{\Delta}_{2i+1,2j+1}$$
$$+ 2(u^{\Delta}_{2i,2j-1} + u^{\Delta}_{2i,2j+1} + u^{\Delta}_{2i-1,2j} + u^{\Delta}_{2i+1,2j}) + 4u^{\Delta}_{2i,2j}]. \qquad (8.63)$$

The inverse operation of transformation from a coarse grid to a fine grid is done by one of the interpolation methods (see Section 4.4.8). The most common is the linear interpolation in our example written as

$$u^{\Delta}_{2i,2j} = u^{2\Delta}_{2i,2j},$$
$$u^{\Delta}_{2i+1,2j} = (u^{2\Delta}_{2i,2j} + u^{2\Delta}_{2i+2,2j})/2,$$
$$u^{\Delta}_{2i,2j+1} = (u^{2\Delta}_{2i,2j} + u^{2\Delta}_{2i,2j+2})/2, \qquad (8.64)$$
$$u^{\Delta}_{2i+1,2j+1} = (u^{2\Delta}_{2i,2j} + u^{2\Delta}_{2i,2j+2} + u^{2\Delta}_{2i+2,2j} + u^{2\Delta}_{2i+2,2j+2})/4.$$

8.3.7   Pseudo-transient Approach

Another approach to solution of a steady-state problem is to introduce fictitious "time" by adding "time derivative" to the equation. For example, the
elliptic equations (8.6) and (8.5) are replaced by the parabolic equations

$$\frac{\partial u}{\partial \tau} = \frac{\partial^2 u}{\partial x^2} + \frac{\partial^2 u}{\partial y^2} - f \qquad (8.65)$$

and

$$\frac{\partial u}{\partial \tau} = \frac{\partial^2 u}{\partial x^2} + \frac{\partial^2 u}{\partial y^2}. \qquad (8.66)$$

Solutions of parabolic systems converge to steady equilibrium states, unless
they are forced to do otherwise by time-dependent boundary conditions
or source terms. In the pseudo-transient approach, the equilibrium state
corresponds to the solution of the original elliptic equation. For example,
the steady-state solutions of (8.65) and (8.66) are the solutions of (8.6)
and (8.5).

   The parameter τ does not have a physical meaning. Its only purpose is to
facilitate the convergence to the solution of a steady-state problem. Nevertheless, all the methods developed for marching problems can be applied.
For example, (8.65) and (8.66) can be solved by the simple explicit method
with central differences for $\partial^2 u / \partial x^2$ and $\partial^2 u / \partial y^2$ or by the alternating directions implicit (ADI) method described in Section 9.3.3.

8.4   SYSTEMS OF NONLINEAR EQUATIONS

In this section, we review the methods used to solve nonlinear equations. At
first glance, this discussion should have started the chapter. The equations
describing convection heat transfer and fluid flows are nonlinear, after all.
Nonlinearity is always present in the convective flux terms of the momentum, mass, and energy conservation equations. Even conduction heat transfer may require nonlinear equations – for example, if variation of physical
properties with temperature is nonnegligible. CFD analysis of any such situation involves solution of a system of nonlinear algebraic equations resulting
from finite difference, finite volume, or other discretization.

In fact, straightforward solution of nonlinear algebraic equations is avoided by virtually all practical CFD methods. The nonlinear systems are difficult and computationally expensive to solve. The common approach is to rely on linearization and/or multistep algorithms to reduce the task to the solution of linear matrix equations.

This section provides a brief review of the three methods for nonlinear systems: the Newton method included for the sake of completeness and the iteration procedures based on linearization and sequential iterations.

8.4.1   Newton's Algorithm

We solve a general system of nonlinear equations:

$$f_j(v_1, \ldots, v_n) = 0, \quad j = 1, \ldots, n, \tag{8.67}$$

which is expressed in vector form as

$$\boldsymbol{F}(\boldsymbol{v}) = 0. \tag{8.68}$$

The functions f_j have all the necessary differentiability and smoothness properties.

Many methods used to solve such systems are based on the Newton's algorithm, also called the Newton–Raphson algorithm. The basis formula is obtained by taking the multidimensional Taylor series expansion of (8.68) and truncating all but the first-order terms. Let us assume that we already have an approximation $\boldsymbol{v}^{(k)}$ of the solution. Truncated expansion around it gives

$$f_j(v_1, \ldots, v_n) = f_j(v_1^{(k)}, \ldots, v_n^{(k)})$$
$$+ \sum_{i=1}^{n} \frac{\partial f_j}{\partial v_i}(v_1^{(k)}, \ldots, v_n^{(k)})(v_i - v_i^{(k)}), j = 1, \ldots, n. \tag{8.69}$$

If $\boldsymbol{v}$ is a solution of (8.68), the left-hand sides in (8.69) are 0, and we obtain a system of linear equations for $\boldsymbol{v}$:

$$\sum_{i=1}^{n} \frac{\partial f_j}{\partial v_i}(v_1^{(k)}, \ldots, v_n^{(k)})(v_i - v_i^{(k)}) = -f_j(v_1^{(k)}, \ldots, v_n^{(k)}), \quad j = 1, \ldots, n \tag{8.70}$$

or, in matrix form,

$$\mathbf{J}^{(k)} \cdot (\boldsymbol{v} - \boldsymbol{v}^{(k)}) = -\boldsymbol{F}(\boldsymbol{v}^{(k)}), \tag{8.71}$$

where $\mathbf{J} = \{\partial f_j / \partial v_i\}$ is the Jacobian matrix of $\boldsymbol{F}(\boldsymbol{v})$. Now we recall that (8.69) is just a first-order approximation of (8.67), so $\boldsymbol{v}$ in (8.71) is not a solution but rather the next, better approximation. This leads to the Newton's iteration formula

$$\mathbf{J}^{(k)} \cdot (\boldsymbol{v}^{(k+1)} - \boldsymbol{v}^{(k)}) = -\boldsymbol{F}(\boldsymbol{v}^{(k)}) \tag{8.72}$$

and

$$\mathbf{J}^{(k)} \cdot \boldsymbol{v}^{(k+1)} = \mathbf{J}^{(k)} \cdot \boldsymbol{v}^{(k)} - \boldsymbol{F}(\boldsymbol{v}^{(k)}) \equiv \boldsymbol{c}^{(k)}, \tag{8.73}$$

which is the familiar matrix equation.

The formulas (8.72) and (8.73) may look attractive on paper but are, in fact, practically never applied in CFD problems because of the significant additional computational cost involved into the calculation of the Jacobian $\mathbf{J}$, which has to be done anew at every iteration. Furthermore, the nature of functions f_i may be such that their differentiation needed to evaluate $\mathbf{J}$ is either impossible or very difficult. A variety of more efficient methods based on modification of the Newton's algorithm have been developed. They, too, found little application in CFD.

8.4.2 Iteration Methods Using Linearization

The iteration methods discussed in this chapter for the linear systems can be utilized to solve systems of nonlinear equations. The commonly used approach is to linearize the equations replacing the unknown coefficients by their estimates taken from the previous iteration. For example, the quadratic term vw (here, v and w are the velocity components) in the momentum equation can be linearized at the $(k + 1)$st iteration as

$$v^{k+1} w^{k+1} \approx v^k w^{k+1}. \tag{8.74}$$

We will illustrate the procedure on the example of the steady-state conduction heat transfer in the situation, when the variation of temperature is large, so the temperature dependence of physical properties cannot be ignored. The heat conduction equation is

$$\nabla(\kappa \nabla u) = \dot{Q}, \tag{8.75}$$

where u is temperature, $\dot{Q}$ is the known intensity of internal heat sources, and the conductivity is $\kappa = \kappa(u(\mathbf{x}))$.

Let us apply the central difference discretization and the linear interpolation of κ as we have already done in (4.62) and (4.63) for the one-dimensional heat equation. For the two-dimensional Cartesian version of (8.75), we obtain

$$\frac{1}{\Delta x}\left(\frac{\kappa_{i+1,j} + \kappa_{i,j}}{2}\frac{u_{i+1,j} - u_{i,j}}{\Delta x} - \frac{\kappa_{i,j} + \kappa_{i-1,j}}{2}\frac{u_{i,j} - u_{i-1,j}}{\Delta x}\right)$$

$$+ \frac{1}{\Delta y}\left(\frac{\kappa_{i,j+1} + \kappa_{i,j}}{2}\frac{u_{i,j+1} - u_{i,j}}{\Delta y} - \frac{\kappa_{i,j} + \kappa_{i,j-1}}{2}\frac{u_{i,j} - u_{i,j-1}}{\Delta y}\right) = \dot{Q}_{i,j}.$$

$$(8.76)$$

The entire system of equations can be symbolically written as

$$\mathbf{N}(\boldsymbol{\kappa}, \mathbf{u}) = \mathbf{Q}, \qquad (8.77)$$

where $\mathbf{k}$, $\mathbf{u}$, and $\mathbf{Q}$ are the vectors, whose elements are the values of, respectively, κ, u, and $\dot{Q}$ at the grid points, and $\mathbf{N}$ is the vector of the left-hand sides of (8.76).

The system is linearized by approximating κ by its estimate known from the previous iteration. The system of linear equations for the unknown $\mathbf{u}^{k+1}$

$$\mathbf{N}(\boldsymbol{\kappa}^{(k)}, \mathbf{u}^{(k+1)}) = \mathbf{Q} \qquad (8.78)$$

is solved on the $(k + 1)$st iteration. Returning to the finite difference formula, we can write the linearized equation as

$$\frac{1}{\Delta x}\left(\frac{\kappa_{i+1,j}^{(k)} + \kappa_{i,j}^{(k)}}{2}\frac{u_{i+1,j}^{(k+1)} - u_{i,j}^{(k+1)}}{\Delta x} - \frac{\kappa_{i,j}^{(k)} + \kappa_{i-1,j}^{(k)}}{2}\frac{u_{i,j}^{(k+1)} - u_{i-1,j}^{(k+1)}}{\Delta x}\right)$$

$$+ \frac{1}{\Delta y}\left(\frac{\kappa_{i,j+1}^{(k)} + \kappa_{i,j}^{(k)}}{2}\frac{u_{i,j+1}^{(k+1)} - u_{i,j}^{(k+1)}}{\Delta y} - \frac{\kappa_{i,j}^{(k)} + \kappa_{i,j-1}^{(k)}}{2}\frac{u_{i,j}^{(k+1)} - u_{i,j-1}^{(k+1)}}{\Delta y}\right) = \dot{Q}_{i,j}.$$

$$(8.79)$$

The system of linearized equations can be solved by one of the direct or iterative methods discussed earlier in this chapter. The solution is used to evaluate $\kappa^{(k+1)} = \kappa(u^{(k+1)})$, which is employed at the next iteration. The procedure continues until convergence to a sufficiently accurate approximation of the solution of the nonlinear problem (8.77) is achieved.

The nonlinearity complicates the issue of convergence, which typically cannot be guaranteed a priori and should be determined experimentally. Another consequence of the nonlinearity is that the convergence of the iteration procedures may require strong underrelaxation. The efficiency can be further improved by applying multigrid techniques.

8.4.3 Sequential Solution

The linearized iteration procedure can be applied to solve steady-state problems in fluid dynamics and convection heat transfer but becomes cumbersome and computationally expensive for multidimensional flows. More efficient methods have been developed that utilize specific properties of the PDE systems describing the flows.

One such property is that a dominant variable can be assigned to every evolutionary equation in the sense that the equation describes transport and conservation of this particular variable. For example, the momentum component ρu_i is the dominant variable for the ith component of the momentum equation, while the specific internal energy e is the dominant variable of the energy conservation equation (see Chapter 2).

The iteration approach is modified so that each PDE is solved separately for its dominant variable as an unknown. The other variables in the equation are replaced by the best currently available estimates and treated as known. The procedure is conducted for the entire system but sequentially, for one PDE at a time. Of course, the result is not a correct solution, since the coupling between the equations is not satisfied. For this reason, iterations are necessary. On every cycle, we solve the equations with the results found at the previous cycle serving as the best available estimates of nondominant variables.

The algorithm, which has the name of *sequential iteration method*, can be summarized as follows:

Step 1: Solve each PDE for the dominant variable with other variables replaced by the best available estimates. The solution can be done by a direct or an iterative method for the matrix equation. One consideration explains the much more common choice of the latter. Since the equations themselves are inaccurate, it is unnecessary to invest computational resources to achieve very high accuracy at this step. An iterative procedure with relatively few cycles is typically sufficient. These cycles are called inner iterations.

Step 2:   Verify convergence of the new approximation $v^{(k)}$ substituting it into the nonlinear equations and computing the norm of residuals. If the convergence is not achieved, repeat Step 1 using $v^{(k)}$ for the new estimate of the nondominant variables. The cycles form the so-called outer iterations.

8.5   COMPUTATIONAL PERFORMANCE

CFD analysis typically requires massive calculations of linear algebra type, primarily matrix and vector multiplications. The clock time required for the analysis is the product of two variables: the number of required arithmetic operations and the speed with which the operations are performed by the computer. The former is determined by the choice of numerical model (the discretization scheme, the grid size, and the method of solving the discretization equations). The latter is a function of hardware and software.

This book is not a place to discuss computer architecture, so we will just touch the main aspects. We will argue that the computer performance is an essential factor affecting the choice of the CFD model.

The starting point is the observation that no serious CFD analysis is done nowadays using a single computer processor. Tens or hundreds or thousands of CPUs are ran in parallel. The data accessed by the CPUs are also typically distributed over many operational storage units. This means that a CFD algorithm must be *parallelizable*. It has to be possible to break the substantial parts of the computational process into many subprocesses such that they can be performed *simultaneously*, without waiting for each other's results. There is an additional requirement that the data communications between the CPUs performing concurrent tasks are kept low to minimize the clock time associated with them.

Nor all CFD methods satisfy these requirements. One example is the classical Thomas algorithm for a tridiagonal matrix equation (see Section 7.5). Each step of the forward and backward sweeps uses the results of the previous step, so it cannot be started before the previous step is completed. The classical version of the cyclic reduction method described in Section 8.2.1 has the same property. Both algorithms are *sequential*, i.e. cannot be parallelized. The situation is not hopeless, since there exist versions of the algorithms allowing some parallelization. Furthermore, in many cases, the algorithms are used as parts of a larger parallelizable procedure, so that many of them can be performed concurrently, on one CPU each.

In contrast, the iterative methods tend to have good parallelization properties. Let us consider the Jacobi method of Section 8.3.2 as an example. According to (8.53), each iteration is simply a multiplication of a matrix by a vector plus addition of a vector. This can be parallelized by dividing the vectors v and c into segments, dividing the columns of matrix $\mathbf{A}$ accordingly, and letting each CPU do the multiplication and addition for a single segment. The data communication between the CPUs is limited to the values of v_ℓ at the boundaries between the segments. The parallelization is efficient provided the size of the segments is much larger than 1.

Another important aspect concerns the speed of linear algebra operations by a single CPU or a group of CPUs working in parallel. For the same numerical method and parallelization scheme, the speed can be increased substantially, sometimes manifold, by optimizing the computational hardware and software. This is an area of active research described in specialized literature. Quite often, significant improvements can be achieved without going too deeply into the subject by simply using one of the libraries of optimized linear algebra routines. Some of such libraries are freely downloadable in the Internet, while others are developed and distributed commercially. A good example of the first group is the set of libraries available through the Netlib repository. It includes subroutines for basic linear algebra operations, as well as for higher-level procedures with matrices and vectors.

BIBLIOGRAPHY

Netlib Repository. A collection of optimized linear algebra routines and other useful material on numerical linear algebra. http://www.netlib.org.

Press, W.H., Teukolsky, S.A., Vetterling, W.T., and Flannery, B.P. (2007). *Numerical Recipes 3rd Edition: The Art of Scientific Computing*. Cambridge University Press.

Trefethen, L.N. and Bau, D. III. (1997). *Numerical Linear Algebra*. Cambridge: Cambridge University Press.

PROBLEMS

1. If you have access to a CFD software, study the manual to determine which methods are available for solution of linear and nonlinear systems of algebraic equations. Does the code use any of the algorithms

discussed in this chapter – for example, iterative methods with over- and underrelaxation, linearization, sequential solution of coupled equations, or multigrid acceleration?

2. Develop a finite difference scheme of the second order for the equation

$$\frac{\partial}{\partial x}\left[a(x,y)\frac{\partial u}{\partial x}\right] + \frac{\partial}{\partial y}\left[b(x,y)\frac{\partial u}{\partial y}\right] = g(x,y).$$

Assume that a, b, and g are known functions of x and y, and use the five-point difference molecule on a uniform Cartesian grid (see Figure 8.1). *Hint*: Use the discretization approach applied to a one-dimensional heat equation with variable conductivity in Section 4.4.1. Use linear interpolation.

3. Rewrite the scheme developed in Problem 2 in matrix form. Develop the row equation and expressions for coefficients as in (8.9) and (8.10).

4. The two-dimensional heat conduction equation (8.15) is solved in the rectangular domain $0 \le x \le L_x$, $0 \le y \le L_y$ with the boundary conditions $u = 0$ at $x = 0$, $x = L_x$, $y = 0$, $y = L_y$. The solution uses the Crank–Nicolson central difference scheme on a structured, uniform, Cartesian grid (see (8.17) and (8.18)). Rewrite the problem in matrix form. Develop the row equation and expressions for the coefficients as in (8.9) and (8.10). Include the rows corresponding to the boundary conditions.

5. For the finite difference scheme in Problem 4:
   a) Write the formulas for the Gauss–Seidel algorithm in terms of the grid point values.
   b) Prove convergence of the Gauss–Seidel iteration procedure.

6. Derive the expressions for submatrices B_j, D_j, A_j in the block-tridiagonal form (8.38) of the matrix equation for the five-point discretization (8.7) of the two-dimensional Poisson equation. Consider only the blocks corresponding to internal points of the computational domain.

7. Consider the two-dimensional Poisson equation (8.6) in a rectangular domain $0 \le x \le L_x$, $0 \le y \le L_y$. The boundary conditions are

$$\frac{\partial u}{\partial x}(0,y) = g_1(y), \ u(L_x,y) = g_2(y), \ \frac{\partial u}{\partial y}(x,0) = g_3(x), \ u(x,L_y) = g_4(x).$$

The computational grid is uniform with steps $\Delta x = L_x/N_x$ and $\Delta y = L_y/N_y$. Develop the system of finite difference equations approximating the entire problem (PDE and boundary conditions) with the second order of accuracy. Write the scheme in the matrix form similar to (8.9) and (8.10). Take into account that the equations and expressions for coefficients are different for internal and boundary grid points.

8. Write the formulas for the Gauss–Seidel algorithm (similar to (8.57)) for the central difference scheme applied to the three-dimensional Poisson equation (8.14).

9. Develop a finite difference scheme for the PDE in problem 2. Use the same spatial discretization as in problem 2, as well the pseudo-transient approach with the first-order implicit discretization in pseudo-time.

10. Consider the steady-state version of the governing equations for an incompressible flow and convection heat transfer (2.22) and (2.29). Disregard the incompressibility condition (2.7) and pressure field p for the moment. Methods dealing with them will be discussed in Chapter 10. Can the sequential solution procedure be applied to the system? For each equation, determine the dominant variable, and write the symbolic linearized systems similar to (8.78).

Programming Exercises

1. Implement the Jacobi, Gauss–Seidel, and Gauss–Seidel with over- and underrelaxation iterative schemes for the five-point discretization scheme of the two-dimensional Poisson equation. Conduct computations to verify the data presented in Table 8.1. Perform additional numerical experiments to determine the value of relaxation parameter ω optimal for this particular problem.

2. Implement the generalized Thomas algorithm for a block-tridiagonal matrix. Use one of the freely downloadable matrix inversion subroutines (e.g. a subroutine from the *Lapack* linear algebra package at the www.netlib.org repository). Apply the algorithm to solve the two-dimensional Poisson problem illustrated in Table 8.1 and Figure 8.5.

9

UNSTEADY COMPRESSIBLE FLUID FLOWS AND CONDUCTION HEAT TRANSFER

9.1 INTRODUCTION

In the next two chapters, we extend the methods of finite difference and finite volume solution to the general case of unsteady multidimensional flows and heat transfer. It would be impossible to cover the entire variety of existing schemes, many of them quite complex, in a single book, especially in an introductory book like ours. Many methods are specialized and only used for certain types of problems. Furthermore, CFD is a rapidly evolving field. An algorithm may become obsolete soon after the book is published. For all these reasons, the following two chapters focus on the common principles of widely used methods and on introduction of several time-honored techniques that have formed the basis of modern CFD.

The main attention is given to calculation of incompressible flows. Chapter 10 is entirely devoted to this subject. This chapter contains a brief review of the issues arising in computations of unsteady multidimensional compressible flows and heat transfer. Description of several schemes is included for the sake of completeness and can be skipped with no loss of

Essential Computational Fluid Dynamics, Second Edition. Oleg Zikanov.
© 2019 John Wiley & Sons, Inc. Published 2019 by John Wiley & Sons, Inc.
Companion Website: www.wiley.com/go/zikanov/essential

continuity. A reader interested in these areas is encouraged to study the chapter and proceed to the books listed at the end for a deeper and more detailed discussion.

9.2 COMPRESSIBLE FLOWS

9.2.1 Equations, Mathematical Classification, and General Comments

We have finally arrived at the point where we can discuss the methods used to solve actual fluid flow equations. The discussion starts with compressible flows in this section and continues with incompressible flows in Chapter 10. The reasons for such a separation will be explained later.

The system of equations to be solved always includes the continuity (mass conservation) equation

$$\frac{D\rho}{Dt} + \rho \frac{\partial u_k}{\partial x_k} = 0 \tag{9.1}$$

and the Navier–Stokes (momentum conservation) equations

$$\rho \frac{Du_i}{Dt} = \rho f_i - \frac{\partial p}{\partial x_i} + \frac{\partial}{\partial x_j}\left[\mu\left(\frac{\partial u_i}{\partial x_j} + \frac{\partial u_j}{\partial x_i}\right) - \frac{2}{3}\mu\left(\frac{\partial u_k}{\partial x_k}\right)\delta_{ij}\right]. \tag{9.2}$$

Summation over repeating indices is assumed in (9.1) and (9.2).

If the flow is compressible, the system should also contain the thermodynamic equation of state connecting pressure and density (e.g. the ideal gas equation (2.32)). If significant heat transfer occurs, the energy conservation equation in some form (see Section 2.2.4) should be included.

We would like to stress that it is not always necessary to solve the full system of conservation equations. In many applications, the system can be replaced, with no significant loss of accuracy, by one of the asymptotic approximations (e.g. an incompressible flow, inviscid flow, or boundary layer). Moreover, the use of an approximation is often a requirement. Only in this way the specific properties of the solution can be utilized for developing more accurate and efficient numerical algorithms. An example illustrating this statement will be given in Chapter 10, when we discuss the implications of incompressibility.

Mathematical Classification of the Equations: We know that the partial differential equations of the second order can be classified into three basic types: parabolic, hyperbolic, and elliptic equations. As we have already demonstrated on the examples of model one-dimensional equations, each type implies certain underlying physical effect and certain type of solution behavior and requires certain methods of numerical solution. The situation is more complex for the three-dimensional nonlinear equations that describe fluid flows and heat transfer. The solutions of these equations may simultaneously demonstrate all three types of behavior. A rigorous mathematical treatment is far beyond the scope of this book. As an alternative, we present the simple analysis based on the analogies of various terms of the equations to the classifiable model equations. This approach was applied to the convection heat transfer equation in Section 3.2.5. Here it is extended to the Navier–Stokes equations.

First of all, the viscous terms with their spatial second derivatives of the velocity components render the Navier–Stokes equations parabolic. Leaving only these terms in the right-hand side of the momentum equations and keeping only the time derivative terms in the left-hand side, we obtain the truncated equations, which are purely parabolic. This can be illustrated by taking a one-dimensional version with u_i depending on only one coordinate and time

$$\rho \frac{\partial u_i}{\partial t} = \frac{\partial}{\partial x_j} \left[\mu \left(\frac{\partial u_i}{\partial x_j} \right) \right] \tag{9.3}$$

and applying the classification criterion introduced in Section 3.2.1. Identification of the equations as parabolic is also in agreement with the diffusion nature of the process of viscous dissipation.

Let us now keep only the material derivative terms of the continuity equation (9.1) and the momentum equation (9.2):

$$\frac{D\rho}{Dt} = \frac{\partial \rho}{\partial t} + u_k \frac{\partial \rho}{\partial x_k} = 0, \tag{9.4}$$

$$\frac{Du_i}{Dt} = \frac{\partial u_i}{\partial t} + u_k \frac{\partial u_i}{\partial x_k} = 0, \quad i = 1, 2, 3. \tag{9.5}$$

The system (9.4) and (9.5) is hyperbolic. This can be shown rigorously using the theory of classification of systems of PDE. Alternatively, we can use the same approach as in Section 3.2.5 and think of the truncated equations as three-dimensional variable-coefficient versions of the linear convection equation

$$\frac{\partial u}{\partial t} + c\frac{\partial u}{\partial x} = 0. \tag{9.6}$$

The system (9.4) and (9.5) has characteristic curves tangential to the local velocity $V(x, t)$. We should expect wavelike solutions. The analogy extends to the physical nature of the phenomenon described by the convection terms, which is the transport of mass and momentum by the flow velocity.

We see that the Navier–Stokes equations for compressible flows (9.1) and (9.2) can be characterized as *hyperbolic–parabolic*. The classification changes when we apply asymptotic approximations. Let us consider the most prominent cases. If the effect of viscosity is negligible, and the momentum equations are the Euler equations (2.23), the unsteady system becomes purely *hyperbolic*. In the steady-state case, the nature of the Euler system depends on the Mach number (the ratio $M = v/a$ between the typical magnitude of flow velocity and the typical speed of sound). The equations are *elliptic* for subsonic flow ($M < 1$) and *hyperbolic* for supersonic flows ($M > 1$). At last, if we retain the viscosity and assume that the fluid is incompressible, the unsteady Navier–Stokes system not only retains its hyperbolic–parabolic character but also acquires *elliptic* features because of the new role of pressure, which we will discuss in detail in Chapter 10.

Features of Compressible Flows: A flow is considered compressible if the Mach number is not small, approximately larger than 0.3. The effect of compressibility may need to be taken into account in many areas, for example, in high-speed aerodynamics, turbomachinery, and combustion. Led by industrial and military applications, the CFD analysis of compressible flows has long become an established discipline with its own specialized methods.

Several features of compressible flows have to be kept in mind, since they directly affect the choice of numerical approach. The first is that the Reynolds number is usually very high. This often means that the solution domain can be divided into thin boundary layers, where viscosity, heat transfer, and turbulence are important, and the outer domain, where the flow can be approximately considered ideal (nonviscous and nonconductive) or even potential (with zero vorticity $\boldsymbol{\omega} = \nabla \times \boldsymbol{v}$). The logical, historically established, and still used approach is to combine the ideal flow equations in the outer domain with physically justified models of the boundary layer behavior.

In the outer domain, the equations describing unsteady flows are hyperbolic. The numerical schemes should be stable and accurate (without excessive numerical dissipation and dispersion) for solutions dominated by waves. Our experience of finding such schemes for the linear convection equation (see Section 7.1) is evidently useful here.

In the flows or within the flow domains, where turbulence is present, we have to apply turbulence models, some of which are discussed in Chapter 11. The effects of heat transfer and temperature variability of fluid properties are, often, important in turbulent compressible flows. The energy conservation equation has to be solved.

Some compressible flows are supersonic ($M > 1$), which means existence of shock waves. This raises complex questions for CFD analysis. The shock waves are extremely thin (with the thickness of the same order of magnitude as the mean free path of the gas molecules) layers, across which the flow variables (pressure, density, temperature, velocity, internal energy, etc.) change strongly. Discretization of a shock wave in the standard CFD manner is impossible for practical reasons and also because the continuum model underlying the governing equations becomes invalid. The analysis usually relies on the idealized physical models, in which shock waves are treated as infinitely thin surfaces, at which the variables change abruptly, according to certain algebraic relations expressing the physical conservation laws. The presence of such discontinuities completely changes the landscape of available computational techniques. Many of them, in particular those based on the second-order central differences popular in other areas of CFD, cannot be used since they lead to spurious oscillations around the shock. To address this issue, specialized methods for flows with stationary and moving discontinuities have been developed.

All these and other issues are discussed in depth in the books dealing with the specific subject of CFD analysis of compressible flows. Several references are listed at the end of this chapter. Our discussion is limited to a few methods that illustrate common techniques and often encountered difficulties.

It is customary and convenient to present the compressible Navier–Stokes system in the standardized vector form (see Section 2.6)

$$\frac{\partial U}{\partial t} + \frac{\partial A}{\partial x} + \frac{\partial B}{\partial y} + \frac{\partial C}{\partial z} = 0. \tag{9.7}$$

For simplicity, we omit body forces and internal heat sources. The formal representation (9.7) is convenient in the sense that it is valid for all compressible flows. Details of a specific flow are hidden in the expressions (2.45) and (2.46) for the vector fields U, A, B, and C. A numerical scheme designed and analyzed for (9.7) can be applied to any flow, provided proper discretizations of (2.45) and (2.46) have been used. The form (9.7) explicitly shows the hyperbolic character of the equations (compare with the model hyperbolic equation $u_t + cu_x = 0$, which can be rewritten as $u_t + v_x = 0$ with $v = cu$). We would like to stress that the seeming simplicity of (9.7) does not change the complex three-dimensional nonlinear character of the equations.

9.2.2 MacCormack Scheme

The finite difference methods designed for the linear convection equation (see Section 7.1) can be generalized to the case of compressible flow equations (9.7). Some of them are, in fact, quite efficient. An example of such a method is the explicit MacCormack scheme introduced in Section 7.1.5. Every time step is split into two substeps:

$$\text{Predictor:} \quad U^*_{i,\,j,k} = U^n_{i,\,j,k} - \frac{\Delta t}{\Delta x}(A^n_{i+1,\,j,k} - A^n_{i,\,j,k})$$

$$-\frac{\Delta t}{\Delta y}(B^n_{i,\,j+1,k} - B^n_{i,\,j,k})$$

$$-\frac{\Delta t}{\Delta z}(C^n_{i,\,j,k+1} - C^n_{i,\,j,k}), \tag{9.8}$$

$$\text{Corrector:} \quad U^{n+1}_{i,\,j,k} = \frac{1}{2}\left[U^n_{i,\,j,k} + U^*_{i,\,j,k} - \frac{\Delta t}{\Delta x}\left(A^*_{i,\,j,k} - A^*_{i-1,\,j,k}\right)\right.$$

$$-\frac{\Delta t}{\Delta y}\left(B^*_{i,\,j,k} - B^*_{i,\,j-1,k}\right)$$

$$\left. -\frac{\Delta t}{\Delta z}\left(C^*_{i,\,j,k} - C^*_{i,\,j,k-1}\right)\right], \tag{9.9}$$

where A^*, B^*, and C^* are computed using the predicted velocity U^*. We assume that the computational grid is rectangular and uniform with steps Δx, Δy, and Δz.

The scheme has the second order of approximation in space and time if the following rules are followed. For the x-derivatives in A, we use one-sided differences in the direction opposite to the direction of the difference used at this particular substep for $\partial A/\partial x$. Central differences are applied for the y- and z-derivatives. Similarly, one-sided differences are used for the y-derivatives in B and z-derivatives in C. The directions of these differences are opposite to the directions used for $\partial B/\partial y$ and $\partial C/\partial z$, respectively. For the other derivatives in B and C, central differences are applied. As an illustration, we discretize the third component of C (see (2.46)):

$$C_3 = \rho v w - \sigma_{yz} = \rho v w - \mu \left(\frac{\partial v}{\partial z} + \frac{\partial w}{\partial y} \right).$$

For the predictor step, this term is approximated as

$$C_{i,j,k}^n = (\rho v w)_{i,j,k}^n - \mu \left(\frac{v_{i,j,k}^n - v_{i,j,k-1}^n}{\Delta z} + \frac{w_{i,j+1,k}^n - w_{i,j-1,k}^n}{2\Delta y} \right),$$

while for the corrector step, the approximation is

$$C_{i,j,k}^* = (\rho v w)_{i,j,k}^* - \mu \left(\frac{v_{i,j,k+1}^* - v_{i,j,k}^*}{\Delta z} + \frac{w_{i,j+1,k}^* - w_{i,j-1,k}^*}{2\Delta y} \right).$$

As it is typical for schemes applied to full Navier–Stokes equations, a rigorous stability analysis of the MacCormack method is impossible. One can, however, use the empirical formula (see Tannehill et al. 1997)

$$\Delta t \leq \frac{\sigma(\Delta t)_{CFL}}{1 + 2/Re_\Delta}. \tag{9.10}$$

The CFL limit for the MacCormack scheme is

$$(\Delta t)_{CFL} = \left[\frac{|u|}{\Delta x} + \frac{|v|}{\Delta y} + \frac{|w|}{\Delta z} + a \left(\frac{1}{(\Delta x)^2} + \frac{1}{(\Delta y)^2} + \frac{1}{(\Delta z)^2} \right)^{1/2} \right]^{-1}, \tag{9.11}$$

where a is the local speed of sound. The effect of viscosity on stability is accounted for by the minimum mesh Reynolds number

$$Re_\Delta = \min(Re_{\Delta x}, Re_{\Delta y}, Re_{\Delta z}) \tag{9.12}$$

with

$$Re_{\Delta x} = \frac{\rho|u|\Delta x}{\mu}, \quad Re_{\Delta y} = \frac{\rho|v|\Delta y}{\mu}, \quad Re_{\Delta z} = \frac{\rho|w|\Delta z}{\mu}. \tag{9.13}$$

Several important and interesting observations can be made in regard to (9.10). We will use them to illustrate the common problems arising in the time integration of the Navier–Stokes equations. First, the stability condition is usually inexact. As a result, the scheme can sometimes become unstable at Δt close to the stability limit even if the condition is formally satisfied. To deal with this problem and avoid the instability, the *safety factor* σ is introduced. Although (9.10) is supposed to guarantee a stable time step at $\sigma = 1$, it is common to use $\sigma \approx 0.9$ or smaller.

Second, the stability condition is a combination of the inviscid (Courant–Friedrichs–Lewy or CFL) limit and the viscous limit. This reflects the fact that the equations combine hyperbolic (convective) and parabolic (viscous) features.

Third, as we see in (9.11) and (9.13), the stability limits are derived on the basis of the "freezing" assumption. The variable velocity components u, v, and w are used as if they were constants. This means that (9.10) provides a *local* stability criterion valid at a given space point and a given moment of time. In calculations, the time step that provides stability for the entire solution is evaluated using some estimates of u, v, and w. Alternatively, if the variable time step approach is followed, the lower bound of (9.10) is calculated after each time step on the basis of the current velocity field and used to determine the stable $\Delta t^n = t^{n+1} - t^n$.

The last comment concerns the problem faced by the MacCormack scheme in flows at high Reynolds numbers. Such flows are characterized by thin boundary layers at solid walls, within which the flow has sharp gradients and, thus, the computational grid must be refined (see Chapter 12 for a discussion of refinement). The small grid steps would require small time step Δt. This can be easily seen. Let, for example, the requirements of the resolution of boundary layers be such that $\Delta x \ll \Delta y, \Delta z$. From (9.11) and (9.13) we have $(\Delta t)_{CFL} \sim \Delta x$ and $Re_\Delta \sim \Delta x$. This means (see (9.10)) that for stability we have to use the time step $\Delta t \sim (\Delta x_{min})^2$. This may require an unacceptably large number of time steps. The problem is not unique and concerns other explicit methods. Having said that we should also mention that in CFD applications the effect of turbulent boundary layers on the rest of the flow is sometimes determined using an approximate physical model, in which case excessively fine grids are not required.

9.2.3   Beam–Warming Scheme

The stability limit on the time step can be avoided if we apply an implicit method. Such methods are usually unconditionally stable, so larger time steps can be used. The downside is, of course, the larger amount of computations needed to complete each step. This is significant since the governing equations (9.7) are multidimensional and nonlinear (remember that the vector fields A, B, and C shown in (2.45)–(2.46) are quadratic in terms of the flow variables). The high computational cost of a time step can make an implicit solution completely unfeasible, unless we find a way to linearize the equations and solve the resulting discretized system efficiently.

The historically first efficient implicit method was designed by Beam and Warming (1976). A family of schemes following the same principles was later developed. The common features of these schemes are that they are *noniterative* (do not require iterations to complete one time step), use a truncated Taylor expansion for *linearization*, and avoid dealing with nontridiagonal matrices that appear in multidimensional problems using *approximate factorization*.

We will give a brief description of the classical Beam–Warming scheme for the case of two-dimensional compressible ideal (nonviscous) flow. The equations are written in the vector form

$$\frac{\partial U}{\partial t} + \frac{\partial A}{\partial x} + \frac{\partial B}{\partial y} = 0. \tag{9.14}$$

An extension to the three-dimensional case is straightforward. A more detailed discussion of the method can be found, for example, in Tannehill et al. (1997).

To derive the scheme, we start with the implicit Crank–Nicolson time integration of (9.14):

$$U^{n+1} = U^n - \frac{\Delta t}{2} \left[\left(\frac{\partial A}{\partial x} + \frac{\partial B}{\partial y} \right)^n + \left(\frac{\partial A}{\partial x} + \frac{\partial B}{\partial y} \right)^{n+1} \right], \tag{9.15}$$

which generates the truncation error $O\left((\Delta t)^2\right)$. Here, U^n, U^{n+1}, A^n, B^n, A^{n+1}, and B^{n+1} are the vector functions of x and y, which are obtained after the time discretization but before we conduct the spatial one.

The terms in (9.15) that include A^{n+1} and B^{n+1} are nonlinear with respect to the unknown U^{n+1}. We approximate them by local linearizations in the form of Taylor approximations around the time level t^n:

$$A^{n+1} \approx A^n + \mathbf{F}^n \cdot (U^{n+1} - U^n), \tag{9.16}$$

$$B^{n+1} \approx B^n + \mathbf{G}^n \cdot (U^{n+1} - U^n), \tag{9.17}$$

where $\mathbf{F}^n$ and $\mathbf{G}^n$ are the Jacobian matrices

$$\mathbf{F}^n = \left(\frac{\partial A}{\partial U}\right)^n, \quad \mathbf{G}^n = \left(\frac{\partial B}{\partial U}\right)^n. \tag{9.18}$$

The error introduced by the linearization is $O\left((\Delta t)^2\right)$. To see this, we consider that the lowest order term of the reminders of the Taylor series omitted in (9.16) and (9.17) are

$$\sim (U^{n+1} - U^n)^2 \sim \left(\frac{\partial U}{\partial t}\Delta t\right)^2 \sim (\Delta t)^2.$$

Substituting the expansions (9.16)–(9.17) into (9.15) and rearranging the equation so that the terms with the unknown U^{n+1} are in the left-hand side and all the other terms are in the right-hand side, we obtain

$$\left[\mathbf{I} + \frac{\Delta t}{2}\left(\frac{\partial}{\partial x}\mathbf{F}^n + \frac{\partial}{\partial y}\mathbf{G}^n\right)\right] \cdot U^{n+1}$$
$$= \left[\mathbf{I} + \frac{\Delta t}{2}\left(\frac{\partial}{\partial x}\mathbf{F}^n + \frac{\partial}{\partial y}\mathbf{G}^n\right)\right] \cdot U^n - \Delta t\left(\frac{\partial A}{\partial x} + \frac{\partial B}{\partial y}\right)^n, \tag{9.19}$$

where $\mathbf{I}$ is the identity matrix. The equation is written in the operator form. The operator

$$\mathbf{I} + \frac{\Delta t}{2}\left(\frac{\partial}{\partial x}\mathbf{F}^n + \frac{\partial}{\partial y}\mathbf{G}^n\right)$$

stands for the combination of matrix multiplications and additions, and differentiations in x and y, such that the result of its action on a vector field, for example, on U^{n+1}, is

$$\left[\mathbf{I} + \frac{\Delta t}{2}\left(\frac{\partial}{\partial x}\mathbf{F}^n + \frac{\partial}{\partial y}\mathbf{G}^n\right)\right] \cdot U^{n+1}$$
$$= U^{n+1} + \frac{\Delta t}{2}\frac{\partial}{\partial x}(\mathbf{F}^n \cdot U^{n+1}) + \frac{\Delta t}{2}\frac{\partial}{\partial y}(\mathbf{G}^n \cdot U^{n+1}). \tag{9.20}$$

Equation (9.19) is linear for the unknown U^{n+1}. The spatial discretization applied at this stage would result in a system of linear algebraic equations (a matrix equation). Because of multidimensionality of the problem, the coefficient matrix of the system would be nontridiagonal. This can be easily verified by applying the second-order central difference discretization to

(9.20), which results in the matrix of the structure similar to the structure in Figure 8.2.

To solve the linearized system, direct or iterative methods discussed in Chapter 8 can be applied. Another efficient approach is to apply the approximation of the multidimensional operator by a sequence of one-dimensional operators. There exist several versions, including the *approximate factorization* method explained here.

The operator in the left-hand side of (9.20) is approximated as

$$\left[\mathbf{I} + \frac{\Delta t}{2}\left(\frac{\partial}{\partial x}\mathbf{F}^n + \frac{\partial}{\partial y}\mathbf{G}^n\right)\right] \cdot \mathbf{U}^{n+1} \approx \left(\mathbf{I} + \frac{\Delta t}{2}\frac{\partial}{\partial x}\mathbf{F}^n\right)$$

$$\cdot \left(\mathbf{I} + \frac{\Delta t}{2}\frac{\partial}{\partial y}\mathbf{G}^n\right)\cdot \mathbf{U}^{n+1}. \quad (9.21)$$

Direct matrix multiplication shows that the right-hand side and left-hand side differ by the term

$$\left(\frac{\Delta t}{2}\right)^2 \left(\frac{\partial}{\partial x}\mathbf{F}^n \cdot \frac{\partial}{\partial y}\mathbf{G}^n\right)\cdot \mathbf{U}^{n+1}, \quad (9.22)$$

which constitutes the error $O\left((\Delta t)^2\right)$. After a similar factorization of the operator in the right-hand side, (9.19) becomes

$$\left(\mathbf{I} + \frac{\Delta t}{2}\frac{\partial}{\partial x}\mathbf{F}^n\right)\cdot \left(\mathbf{I} + \frac{\Delta t}{2}\frac{\partial}{\partial y}\mathbf{G}^n\right)\cdot \mathbf{U}^{n+1}$$

$$= \left(\mathbf{I} + \frac{\Delta t}{2}\frac{\partial}{\partial x}\mathbf{F}^n\right)\cdot \left(\mathbf{I} + \frac{\Delta t}{2}\frac{\partial}{\partial y}\mathbf{G}^n\right)\cdot \mathbf{U}^n - \Delta t\left(\frac{\partial \mathbf{A}}{\partial x} + \frac{\partial \mathbf{B}}{\partial y}\right)^n. \quad (9.23)$$

A simpler and computationally more efficient version is obtained if we introduce the vector field $\Delta \mathbf{U}^n = \mathbf{U}^{n+1} - \mathbf{U}^n$ and rewrite the equation as

$$\left(\mathbf{I} + \frac{\Delta t}{2}\frac{\partial}{\partial x}\mathbf{F}^n\right)\cdot \left(\mathbf{I} + \frac{\Delta t}{2}\frac{\partial}{\partial y}\mathbf{G}^n\right)\cdot \Delta \mathbf{U} = -\Delta t\left(\frac{\partial \mathbf{A}}{\partial x} + \frac{\partial \mathbf{B}}{\partial y}\right)^n. \quad (9.24)$$

The advantage of the factorized forms (9.23) and (9.24) in comparison to the original form (9.19) is that now the left-hand side is a product of two operators, which are one-dimensional in the sense that each involves a derivative with respect to only one coordinate. The time step can be arranged as a sequence of substeps, which only require solving matrix equations with tridiagonal matrices if central differences of second order are applied for discretization. For example, the sequence for (9.24) is:

Step 1: Solve

$$\left(\mathbf{I} + \frac{\Delta t}{2} \frac{\partial}{\partial x} \mathbf{F}^n \right) \cdot \widetilde{\Delta U} = -\Delta t \left(\frac{\partial \mathbf{A}}{\partial x} + \frac{\partial \mathbf{B}}{\partial y} \right)^n .$$

Step 2: Solve

$$\left(\mathbf{I} + \frac{\Delta t}{2} \frac{\partial}{\partial y} \mathbf{G}^n \right) \cdot \Delta U = \widetilde{\Delta U} .$$

Step 3: Update the solution $U^{n+1} = U^n + \Delta U^n$.

The Beam–Warming algorithm presented here is unconditionally stable. Since the additional discretization errors introduced by the linearization (9.16) and (9.17) and the approximate factorization (9.23) and (9.24) are $O\left((\Delta t)^2\right)$, the scheme retains the second order in time introduced by the Crank–Nicolson method. The order of approximation in space is determined by the scheme used for spatial discretization. Other versions of the method, some with higher order of time integration, have been developed. They are discussed, for example, in Tannehill et al. (1997).

9.2.4 Upwinding

We already saw an example of upwinding in Section 7.1.1 when we solved the linear convection equation

$$\frac{\partial u}{\partial t} + c \frac{\partial u}{\partial x} = 0, \quad c > 0. \tag{9.25}$$

It was found, perhaps surprisingly, that a simple explicit scheme is only useful if it is based on the backward difference for the space derivative:

$$\frac{u_i^{n+1} - u_i^n}{\Delta t} + c \frac{u_i^n - u_{i-1}^n}{\Delta x} = 0. \tag{9.26}$$

The other two schemes we analyzed, namely, those based on the forward and central differences, were found unconditionally unstable.

The results can be generalized to the rule valid for an arbitrary sign of c. Considering the amplification factors (7.12)–(7.14), it is easy to see that the only meaningful simple explicit scheme is the *upwind scheme*, in which the space derivative is approximated as

$$\left. \frac{\partial u}{\partial x} \right|_i^n \approx \begin{cases} (u_i^n - u_{i-1}^n)/\Delta x & \text{if} \quad c > 0, \\ (u_{i+1}^n - u_i^n)/\Delta x & \text{if} \quad c < 0. \end{cases} \tag{9.27}$$

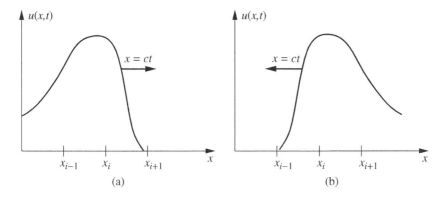

Figure 9.1 Wavelike solutions of the linear convection equation (9.25) at $c > 0$ (a) and $c < 0$ (b).

The same scheme is obtained if we follow the finite volume approach and apply the upwind interpolation of the first order (5.16) (see Section 5.3.1).

The clear advantage of the upwind approximation of spatial derivatives (or, simply, *upwinding*) is associated with the physical properties of the solutions of (9.25). The equation is the simplest representative of hyperbolic equations, solutions of which are dominated by waves propagating along characteristics. In the case of (9.25), there is only one family of characteristics. The solution propagates along the x-axis to the right if $c > 0$ and to the left if $c < 0$ (see Figure 9.1). The behavior can also be described in terms of the domain of dependence of the solution at the grid point x_i: the interval $x < x_i$ at $c > 0$ and the interval $x > x_i$ at $c < 0$.

The upwind scheme is consistent with the physical nature of the solution, since it draws the information exclusively from the domain of dependence. For example, the scheme (9.26) at $c > 0$ leads us to compute u_i^{n+1} as a function of u_i^n and u_{i-1}^n. On the contrary, the forward and central difference schemes (7.6) and (7.8) use u_{i+1}^n, which is outside the domain of dependence (at the location, to which the wave has not arrived yet). As a result, the upwind scheme produces an acceptable approximation, while central and downwind schemes do not.

The idea of upwinding extends to many areas of CFD as a broad concept: If a solution has strong wavelike features propagating in certain directions, better stability and accuracy can often be achieved through the use of upwind differentiation or, in the case of finite volume methods, upwind interpolation. The upwinding is understood here not as a particular approximation, such as (9.27), but as the general approach, according to which we design a difference or interpolation formula so that it either uses only the grid points on the upwind side or gives to these points larger weights than to the downwind points (the second strategy is also called "upwind bias"). It is important to mention that the first-order schemes, such as (9.27), have very strong numerical dissipation and low accuracy (see Section 7.1.1). For this reason, the modern upwind or upwind-bias methods mostly use multipoint approximations of the second order or higher.

One important situation, in which the upwinding can be helpful, is the approximation of convection terms of the Navier–Stokes equations. We have already discussed in Section 9.2.1 that these terms are analogous to the linear convection equation (9.6) and responsible for hyperbolic features of the solution. Of course, the Navier–Stokes equations are not purely hyperbolic. The waves are affected by viscosity and pressure and do not have clearly defined domains of influence and dependence. In many cases, schemes with central difference approximation of convection terms produce good results. Still, there are cases in which the hyperbolic effects are quite strong. This happens, in particular, in the *convection-dominated* flows, in which the amplitude of the convection terms is much larger than the amplitude of viscous terms. The central difference formulas are known not to work in an optimal way in such flows. The situation becomes even worse when central discretization formulas of high order (fourth of higher) are applied. The main troubles are the low stability limit and spurious small-scale oscillations in the numerical solution. Adding a certain amount of upwinding usually eliminates or reduces these effects and makes the computational schemes more robust. There are various techniques of doing that. Typically, an upwind-bias discretization or interpolation formula can be considered as a combination of weighted symmetric and purely upwind formulas.

9.2.5   Methods for Purely Hyperbolic Systems: TVD Schemes

The need for special discretization approach is critical in the case of purely hyperbolic systems. In CFD, such systems are often associated with inviscid gas dynamics and with supersonic flows, although hyperbolic equations may describe processes in many other areas: acoustics, optics, electrodynamics, population balance dynamics, and so on. The difficulty of numerical analysis of purely hyperbolic systems is often increased by the presence of discontinuities in the solution. The shock waves in supersonic flows provide the best known, but not unique, example. A numerical scheme should be able to reproduce the shape and the motion of a shock correctly and to maintain its structure as a discontinuity.

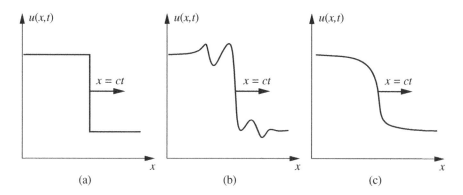

Figure 9.2 Discontinuous solution of the linear convection equation (9.6) at $c > 0$. (a) Exact solution. (b) Non-TVD numerical solution with spurious oscillations. (c) Numerical solution with smearing due to the first order of approximation.

The common problem is the development of spurious oscillations near the areas of strong gradients and discontinuities of the solution. As an example, consider the step wave solution of the linear convection equation (9.6) shown in Figure 9.2a. The exact solution has the form of a wave moving to the right (we assume that $c > 0$) with constant speed and without changing its shape. The spurious oscillations leading to the grossly incorrect shape of the wave illustrated in Figure 9.2b appear when we use schemes, which are stable but not purely upwind, i.e. draw information from outside the domain of dependence of the solution. An example of such a scheme is the Lax–Wendroff method (7.24).

Interestingly, the upwind character of the scheme by itself does not guarantee the absence of spurious oscillations. For example, if we attempt to reproduce the shock wave solution of the linear convection equation using the second-order upwind discretization

$$\frac{\partial u}{\partial x}\bigg|_i^n \approx \begin{cases} (3u_i^n - 4u_{i-1}^n + u_{i-2}^n)/2\Delta x & \text{if} \quad c > 0, \\ (-3u_i^n + 4u_{i+1}^n - u_{i+2}^n)/2\Delta x & \text{if} \quad c \leq 0, \end{cases} \tag{9.28}$$

the solution will still develop the spurious oscillations as in Figure 9.2b.

The condition that guarantees that the spurious oscillations do not appear is called *total variation diminishing* or TVD condition. We will introduce it for the one-dimensional solutions, for which the total variation is defined as

$$TV(u) \equiv \int_\Omega \left|\frac{\partial u}{\partial x}\right| \Delta x \tag{9.29}$$

or, in terms of numerical solution,

$$TV(u) \equiv \sum_{i=1}^{N} |u_{i+1} - u_i|. \qquad (9.30)$$

The integration and summation are over the entire solution domain and all grid points, respectively. In a solution consisting of propagating waves, the shape of u does not change, so the total variation $TV(u)$ remains constant. A good numerical scheme should, ideally, reproduce this property. In practice, we require that the scheme does not allow the total variation to grow. The scheme is TVD if it satisfies

$$TV\left(u^{n+1}\right) \leq TV\left(u^n\right). \qquad (9.31)$$

The importance of the TVD requirement becomes clear if we consider the illustration in Figure 9.2b. The schemes that generate spurious oscillations around a shock increase the total variation. They are not TVD and would be banned if the TVD requirement were imposed.

The first-order scheme (9.27) for the linear convection equation is TVD. Other TVD first-order methods have been developed over the years to calculate solutions with discontinuities in simple and in multidimensional complex systems. They all, however, have a serious flaw. The first-order approximation means very strong numerical dissipation. The sharp gradient features of the solution, including the shocks, are smeared out. This is illustrated in Figure 9.2c.

One popular way to improve the situation is to modify the higher-order schemes so that they become TVD, thus preventing the oscillations. The modifiers (also called limiters) are specifically designed so as to limit the possible variations of the solution. The second order of approximation can be lost near a discontinuity but maintained in the areas where the solution is smooth. The approach is very effective. It is, in fact, so effective that it allows us to avoid oscillations in nonupwind schemes – such as, the Beam–Warming or the Lax–Wendroff method. A detailed discussion of these high-resolution schemes for hyperbolic systems is beyond the scope of this book but can be found elsewhere – for example, in Leveque (2002).

9.3   UNSTEADY CONDUCTION HEAT TRANSFER

9.3.1   Overview

We solve the equation

$$\frac{\partial u}{\partial t} = a^2 \nabla^2 u, \tag{9.32}$$

which describes the process of diffusion of field $u(x, t)$ with constant diffusivity coefficient a^2. The physical nature of the diffusion can vary, although the most common application is the conduction heat transfer in a motionless medium.

The discussion is focused on the finite difference methods of solution. There are other techniques, numerical and analytical, to solve the equation. In particular, many widely used engineering finite element tools have the heat transfer capability, which provides a relatively simple and straightforward solution. This method is especially effective in the case of complex geometries. The more academic Green's function approach should also be mentioned. A reader interested in these and other methods is urged to consult more comprehensive or more specialized books. Some references are provided at the end of the chapter.

We have already considered finite difference schemes for (9.32), but only in the special cases of a steady-state problem (in Chapter 8) or a one-dimensional unsteady problem (in Chapter 7). Here, we address the general unsteady multidimensional form of the equation. For simplicity, the schemes are presented on the example of the two-dimensional equation discretized on a uniform Cartesian grid with steps Δx and Δy.

9.3.2   Simple Methods for Multidimensional Heat Conduction

The schemes derived earlier for the one-dimensional heat equation can be easily generalized to the two-dimensional case. The five-point central difference approximation is used for spatial derivatives. The simple explicit, simple implicit, and Crank–Nicolson methods become

$$\frac{u_{i,j}^{n+1} - u_{i,j}^{n}}{\Delta t} = a^2 \left[\frac{u_{i+1,j}^{n} - 2u_{i,j}^{n} + u_{i-1,j}^{n}}{(\Delta x)^2} + \frac{u_{i,j+1}^{n} - 2u_{i,j}^{n} + u_{i,j-1}^{n}}{(\Delta y)^2} \right], \tag{9.33}$$

$$\frac{u_{i,j}^{n+1} - u_{i,j}^n}{\Delta t} = a^2 \left[\frac{u_{i+1,j}^{n+1} - 2u_{i,j}^{n+1} + u_{i-1,j}^{n+1}}{(\Delta x)^2} + \frac{u_{i,j+1}^{n+1} - 2u_{i,j}^{n+1} + u_{i,j-1}^{n+1}}{(\Delta y)^2} \right],$$

(9.34)

and

$$\frac{u_{i,j}^{n+1} - u_{i,j}^n}{\Delta t} = \frac{a^2}{2} \left[\frac{u_{i+1,j}^{n+1} - 2u_{i,j}^{n+1} + u_{i-1,j}^{n+1}}{(\Delta x)^2} + \frac{u_{i+1,j}^n - 2u_{i,j}^n + u_{i-1,j}^n}{(\Delta x)^2} \right.$$

$$\left. + \frac{u_{i,j+1}^{n+1} - 2u_{i,j}^{n+1} + u_{i,j-1}^{n+1}}{(\Delta y)^2} + \frac{u_{i,j+1}^n - 2u_{i,j}^n + u_{i,j-1}^n}{(\Delta y)^2} \right]. \quad (9.35)$$

The truncation error of these schemes is of the same order as the error of the one-dimensional version: $O\left(\Delta t, (\Delta x)^2, (\Delta y)^2\right)$ for the simple explicit and simple implicit methods and $O\left((\Delta t)^2, (\Delta x)^2, (\Delta y)^2\right)$ for the Crank–Nicolson method.

Generalization of the stability results obtained for the one-dimensional equation to the multidimensional case must be done with caution. Implicit schemes (9.34) and (9.35) remain unconditionally stable. For the explicit scheme (9.33), a naive generalization of the criterion $r \leq 1/2$ derived in Section 7.1.1 would lead to

$$\frac{a^2 \Delta t}{(\Delta x)^2} \leq \frac{1}{2}, \quad \frac{a^2 \Delta t}{(\Delta y)^2} \leq \frac{1}{2}.$$

In fact, the Fourier–Neumann analysis of the numerical stability of (9.33) gives the more restrictive criterion

$$a^2 \Delta t \left(\frac{1}{(\Delta x)^2} + \frac{1}{(\Delta y)^2} \right) \leq \frac{1}{2}. \quad (9.36)$$

Similarly, if the simple explicit method is applied to the three-dimensional heat equation, the stability condition changes to

$$a^2 \Delta t \left(\frac{1}{(\Delta x)^2} + \frac{1}{(\Delta y)^2} + \frac{1}{(\Delta z)^2} \right) \leq \frac{1}{2}. \quad (9.37)$$

9.3.3 Approximate Factorization

The stability requirements of explicit methods, such as (9.36) and (9.37), often impose severe limitations on the size of the time step Δt. This typically happens because at least one of the grid steps Δx, Δy, Δz must be small, at least in some areas, to resolve strong variations of u. The implicit schemes, such as (9.34) and (9.35), do not have this disadvantage, since they are unconditionally stable and allow us to take arbitrarily large time steps. In the multidimensional case, however, the implicit schemes present the difficulty of having to solve matrix equations with large nontridiagonal matrices.

Direct and iterative methods of solving such equations were discussed in Chapter 8. Here we present an alternative approach based on approximate factorization. The implicit step is split into two or more substeps, each requiring solution of a tridiagonal system. The basic technique is the same as in the Beam–Warming method applied to compressible flows in Section 9.2.3, but the realization is simpler and can be presented directly in terms of discretization equations.

We will work with the increments of the solution variable at the grid points

$$\Delta u_{i,j} = u_{i,j}^{n+1} - u_{i,j}^n \qquad (9.38)$$

and use the shorthand notation for the operators of central difference discretization on the five-point molecule:

$$L_x u_{i,j}^n \equiv \frac{u_{i+1,j}^n - 2u_{i,j}^n + u_{i-1,j}^n}{(\Delta x)^2}, \quad L_y u_{i,j}^n \equiv \frac{u_{i,j+1}^n - 2u_{i,j}^n + u_{i,j-1}^n}{(\Delta y)^2}. \qquad (9.39)$$

As in our earlier discussion of the Beam–Warming method, we will illustrate the method on the example of the Crank–Nicolson time discretization. The finite difference formula (9.35) can be rewritten as

$$\frac{1}{\Delta t}\Delta u_{i,j} = \frac{a^2}{2}(L_x u_{i,j}^{n+1} + L_x u_{i,j}^n + L_y u_{i,j}^{n+1} + L_y u_{i,j}^n). \qquad (9.40)$$

Replacing $u_{i,j}^{n+1}$ by $u_{i,j}^n + \Delta u_{i,j}$, we obtain

$$\frac{1}{\Delta t}\Delta u_{i,j} = \frac{a^2}{2}(L_x \Delta u_{i,j} + L_y \Delta u_{i,j} + 2L_x u_{i,j}^n + 2L_y u_{i,j}^n). \qquad (9.41)$$

Moving the terms with the unknowns $\Delta u_{i,j}$ into the left-hand side and all the other terms into the right-hand side produces

$$\left(1 - \frac{a^2 \Delta t}{2} L_x - \frac{a^2 \Delta t}{2} L_y\right) \Delta u_{i,j} = a^2 \Delta t (L_x + L_y) u_{i,j}^n. \tag{9.42}$$

The system of such equations can be written as a matrix equation with non-tridiagonal matrix and is as costly to solve as the system associated with the original Crank–Nicolson scheme. The procedure of approximate factorization is applied to reduce the solution to a sequence of applications of the Thomas algorithm. Similarly to (9.21), we approximate the left-hand side as

$$\left(1 - \frac{a^2 \Delta t}{2} L_x - \frac{a^2 \Delta t}{2} L_y\right) \Delta u_{i,j} \approx \left(1 - \frac{a^2 \Delta t}{2} L_x\right)$$
$$\times \left(1 - \frac{a^2 \Delta t}{2} L_y\right) \Delta u_{i,j}. \tag{9.43}$$

The approximation introduces the error $(1/4)a^4(\Delta t)^2 L_x L_y \Delta u_{i,j}$, which is of the second order in Δt and can, therefore, be tolerated.

Algorithmically, the factorized operator is a consecutive application of the central difference operators (9.39) in the x- and y-directions. We introduce the intermediate increment $\widetilde{\Delta u}_{i,j}$ such that

$$\left(1 - \frac{a^2}{2} \Delta t L_y\right) \Delta u_{i,j} = \widetilde{\Delta u}_{i,j}. \tag{9.44}$$

The intermediate increment is a solution of

$$\left(1 - \frac{a^2}{2} \Delta t L_x\right) \widetilde{\Delta u}_{i,j} = a^2 \Delta t (L_x + L_y) u_{i,j}^n. \tag{9.45}$$

Equation (9.44) is a system of N_x decoupled matrix equations, one for each $i = 1, \ldots, N_x$, with tridiagonal matrices of size $N_y \times N_y$. Similarly, (9.45) is a system of N_y decoupled matrix equations with tridiagonal matrices of size $N_x \times N_x$.

The procedure can be summarized as follows: solve, using the Thomas algorithm, all the equations (9.45) for $\widetilde{\Delta u}_{i,j}$ and then all the equations (9.44) for $\Delta u_{i,j}$. After that, use the calculated $\Delta u_{i,j}$ to find $u_{i,j}^{n+1}$ according to (9.38). The scheme is of the second order in time and space and unconditionally stable. These properties are retained by the three-dimensional version, which requires factorization into three substeps.

9.3.4 ADI Method

The approximate factorization concept was, in fact, originally developed for the transient multidimensional heat equation. The first scheme was published more than 50 years ago by Peaceman and Rachford (1955) under the name of alternating direction implicit (ADI) scheme. We will briefly describe its simplest version.

In the ADI method, the solution for the implicit values $u_{i,j}^{n+1}$ is split into two successive substeps. On the first, an intermediate solution is found using the scheme, which is implicit for the x-derivative term and explicit for the y-derivative term:

$$\frac{\tilde{u}_{i,j} - u_{i,j}^n}{\Delta t/2} = a^2 \left[\frac{\tilde{u}_{i+1,j} - \tilde{u}_{i,j} + \tilde{u}_{i-1,j}}{(\Delta x)^2} + \frac{u_{i,j+1}^n - u_{i,j}^n + u_{i,j-1}^n}{(\Delta y)^2} \right]. \qquad (9.46)$$

On the second substep, the solution is updated using the scheme, which is implicit for the y-term and explicit for the x-term:

$$\frac{u_{i,j}^{n+1} - \tilde{u}_{i,j}}{\Delta t/2} = a^2 \left[\frac{\tilde{u}_{i+1,j} - \tilde{u}_{i,j} + \tilde{u}_{i-1,j}}{(\Delta x)^2} + \frac{u_{i,j+1}^{n+1} - u_{i,j}^{n+1} + u_{i,j-1}^{n+1}}{(\Delta y)^2} \right]. \qquad (9.47)$$

Rearranging the equations so that the quantities unknown at each particular substep are in the left-hand side and the known quantities are in the right-hand side, we obtain two systems, which are expressed in terms of one-dimensional differential operators as

$$\left(1 - \frac{a^2 \Delta t}{2} L_x\right) \tilde{u}_{i,j} = \left(1 + \frac{a^2 \Delta t}{2} L_y\right) u_{i,j}^n, \qquad (9.48)$$

$$\left(1 - \frac{a^2 \Delta t}{2} L_y\right) u_{i,j}^{n+1} = \left(1 + \frac{a^2 \Delta t}{2} L_x\right) \tilde{u}_{i,j}. \qquad (9.49)$$

Similarly to (9.44) and (9.45), the systems (9.48) and (9.49) consist of decoupled small systems with tridiagonal matrices, which can be solved using the Thomas algorithm.

The ADI method has the truncation error $O\left((\Delta t)^2, (\Delta x)^2, (\Delta y)^2\right)$. The scheme is unconditionally stable when applied to a two-dimensional problem. Its extension to three dimensions, which would include three substeps, each implicit in one direction, retains the second-order accuracy and efficiency but becomes only conditionally stable. It requires that $r_x = a^2 \Delta t / (\Delta x)^2$, $r_y = a^2 \Delta t / (\Delta y)^2$, and $r_z = a^2 \Delta t / (\Delta z)^2$ are all smaller than 3/2.

An important aspect of the approximate factorization methods is that the intermediate solution, such as $\tilde{u}_{i,j}$, should be viewed as a preliminary approximation of $u_{i,j}^{n+1}$. Generally, it does not represent the solution at some intermediate time level, say, at $t^{n+1/2} = t_n + \Delta t / 2$. This has a significant implication that the intermediate solution does not have to satisfy the physical boundary conditions imposed on the solution $u(x, y, t)$ of the PDE problem. Moreover, imposing such conditions may introduce additional truncation error and compromise the accuracy of the entire scheme. For example, imposing physical boundary conditions on $\tilde{u}_{i,j}$ in the ADI method may result in a truncation error $\sim O(\Delta t)$.

The loss of accuracy can be avoided if special numerical boundary conditions are designed from the factorized equations themselves. As an example, to obtain the necessary conditions on $\tilde{u}$ in the ADI method, we should subtract (9.47) from (9.46) and evaluate the resulting equation at the boundary points while substituting the physical boundary conditions for u^n and u^{n+1}.

As a last comment, the approximate factorization methods can also be used to solve the steady-state elliptic problems. The application is based on the pseudo-transient approach. The elliptic problem is converted into a parabolic problem through a fictitious time derivative term (see Section 8.3.7).

BIBLIOGRAPHY

Beam, R.M. and Warming, R.F. (1976). An implicit finite-difference algorithm for hyperbolic systems in conservation law form. *J. Comput. Phys.* **22**: 87–110.

Cole, K.D., Beck, J.V., Haji-Sheikh, A., and Litkouhi, B. (2010). *Heat Conduction Using Green's Functions*. Boca Raton, FL: CRC Press.

Fletcher, C.A.J. (1996). *Computational Techniques for Fluid Dynamics: Specific Techniques for Different Flow Categories*, vol. **2**. Berlin: Springer-Verlag.

Jaluria, Y. and Torrance, K.E. (2003). *Computational Heat Transfer*. London: Taylor & Francis.

Leveque, R.J. (2002). *Finite Volume Methods for Hyperbolic Problems*. Cambridge: Cambridge University Press.

Peaceman, D. and Rachford, M. (1955). The numerical solution of parabolic and elliptic differential equations. *J. SIAM* **3**: 28–41.

Samarskii, A.A. and Vabischhevich, P.N. (1995). *Computational Heat Transfer*, vols. I and II. New York: Wiley.

Tannehill, J.C., Anderson, D.A., and Pletcher, R.H. (1997). *Computational Fluid Mechanics and Heat Transfer*. Philadelphia, PA: Taylor & Francis.

PROBLEMS

1. For a two-dimensional compressible flow with all variables depending on (x, y, t), the equations are written in vector form as

$$\frac{\partial U}{\partial t} + \frac{\partial A}{\partial x} + \frac{\partial B}{\partial y} = 0.$$

   Rewrite the expressions (2.45) and (2.46) of the vector fields U, A, and B for this case.

2. Can the incompressible flow equations be written in the vector form (9.7)? What are the expressions for the vector fields U, A, B, and C in this case?

3. For the two-dimensional system of compressible flow equations in Problem 1, write the complete MacCormack scheme of the second order in space and time. Follow the rules described in Section 9.2.2 to derive the discretization of the internal derivatives in A and B.

4. A three-dimensional flow of air is modeled using the MacCormack scheme. The flow velocity is estimated to be between 1 and 200 m/s. The computational grid has $\Delta x = \Delta y = 10^{-2}$ m and variable step in the z-direction $10^{-3} \leq \Delta z \leq 10^{-2}$ m. Taking the air properties as $\rho = 1.2$ kg/m^3, $\mu = 1.81 \times 10^{-5}$ kg/m s, and $a = 340$ m/s (approximately the properties at 20 °C), find the maximum time step that guarantees numerical stability of solution.

5. Considering that the Beam–Warming scheme is unconditionally stable, would it be justified to use very large time steps?

6. For the Beam–Warming scheme, what is the order of error introduced by linearization and by approximate factorization? Does the accuracy of the scheme deteriorate when we use these approximations?

7. Use matrix algebra to prove that the error introduced by the approximate factorization in the two-dimensional Beam–Warming method is as given by (9.22).

8. Is the leapfrog scheme (7.21) a good choice for calculation of solutions of the linear convection equation, in which the initial state has a discontinuity, such as in Figure 9.2?

9. A rectangular bar of dimensions $L_x \times L_y \times L_z$, constant material properties κ, ρ, C, and initial temperature T_0 is immersed in cold water maintained at constant temperature $T_w < T_0$.

   a) Write the complete PDE problem (heat equation and initial and boundary conditions) for conduction heat transfer within the bar. Assume that the boundaries of the bar have the same temperature as water at all times.

   b) Develop the simple explicit and Crank–Nicolson schemes. Include the finite difference approximations of the boundary conditions.

   c) What are the truncation error and stability conditions for each scheme?

10. Rewrite the finite difference approximations developed in Problem 9.9 for the case when the material properties κ, ρ, C are functions of temperature.

11. Rewrite the finite difference approximations developed in Problem 9 for the situation when the body has the shape of a cylindrical ring of inner and outer radii R_i and R_o and height H.

12. How would you proceed with analyzing the heat transfer in Problems 9–11 if your interest were only to find the asymptotic equilibrium temperature at $t \to \infty$.

13. Rewrite the approximate factorization method of Section 9.3.3 for the case of three-dimensional heat conduction equation.

14. Compare the properties of the approximate factorization method, Crank–Nicolson method (9.35), and simple explicit method (9.33), all applied to solution of two-dimensional heat conduction problems. Discuss relative advantages and disadvantages of each method.

15. Show that the ADI scheme for two-dimensional heat equation can be obtained from the Crank–Nicolson scheme (9.35) by approximate factorization and that the factorization error is $O\left((\Delta t)^3\right)$.

16. The two-dimensional heat equation is solved in a rectangular domain $0 < x < A$ and $0 < y < B$ with Dirichlet boundary conditions

$u(x = 0, y, t) = f(y, t)$, $u(x = A, y, t) = g(y, t)$, $u(x, y = 0, t) = h(x, t)$, $u(x, y = B, t) = p(y, t)$. The ADI scheme is applied. Derive the numerical boundary conditions for the intermediate solution $\tilde{u}$. See the text for a hint how this can be done.

Programming Exercises

1. Implement the simple explicit algorithm for two-dimensional heat equation, and apply it to solve unsteady heat conduction in a rectangular plate of dimensions $L_x \times L_y = 0.1 \times 0.1$ m. The initial temperature is uniform and equal to 293 K. The boundary conditions are constant temperature of 293 K at $x = L_x$ and $y = 0$, perfectly insulated boundary at $x = 0$, and constant heat flux 1000 W/m^2 into the plate at $y = L_y$. Use the material properties of pure aluminum ($\rho \approx 2.7$ kg/m^3, $C \approx 900$ J/kg K, $\kappa \approx 204$ W/m K). Conduct the solution until the asymptotic steady state is found. Use uniform grids of 10×10, 50×50, and 100×100 points. Compare the results.

10

INCOMPRESSIBLE FLOWS

10.1 GENERAL CONSIDERATIONS

10.1.1 Introduction

No fluid is perfectly incompressible. Even in the most carefully controlled situations, minute variations of density are present because of inevitable variations of temperature and for other reasons. The incompressible fluid model is only an approximation in which we assume that the variations of density are negligibly weak. It is shown in texts on fluid mechanics that the typical relative magnitude of density variation $\Delta\rho/\rho$ is estimated by the square of the Mach number $M = U/a$, where U is the typical flow velocity and a is the typical local speed of sound. The assumption of incompressibility is equivalent to the asymptotic limit $M \equiv U/a \to 0$. The approximation is, thus, accurate for flows, in which M is substantially smaller than one. The condition $M < 0.3$ is often used for practical purposes.

In this section, we consider methods developed specifically for the equations that describe flows in the approximation of incompressible fluid.

Essential Computational Fluid Dynamics, Second Edition. Oleg Zikanov.
© 2019 John Wiley & Sons, Inc. Published 2019 by John Wiley & Sons, Inc.
Companion Website: www.wiley.com/go/zikanov/essential

In addition to constant density, we assume that the fluid properties, such as viscosity or heat conductivity, are constant.

The governing Navier–Stokes and continuity equations simplify to

$$\rho\frac{DV}{Dt} = -\nabla p + \mu\nabla^2 V + \rho f \tag{10.1}$$

$$\nabla \cdot V = 0, \tag{10.2}$$

where f is an external body force and

$$\frac{DV}{Dt} \equiv \frac{\partial V}{\partial t} + \frac{\partial(uV)}{\partial x} + \frac{\partial(vV)}{\partial y} + \frac{\partial(wV)}{\partial z} \equiv \frac{\partial V}{\partial t} + N(V, V) \tag{10.3}$$

is the material derivative in conservation form, with N being the shorthand notation for the nonlinear term. The energy equation can be reduced to the equation of convection heat transfer with, possibly, a heat source due to viscous dissipation and other effects:

$$\rho C\frac{DT}{Dt} = \kappa\nabla^2 T + Q. \tag{10.4}$$

In the incompressible fluid model, the heat equation is decoupled from the momentum and incompressibility equations and can be solved separately.[1] We will focus on schemes designed to compute fluid flows, that is, to solve (10.1) and (10.2).

10.1.2   Role of Pressure

The incompressible flow model has certain distinctive mathematical and physical properties that require special computational techniques. The most important is the new role of the pressure field. In compressible flows, pressure is defined by the equation of state as a function of other variables (e.g. density and temperature), which, in turn, are found as solutions of evolutionary equations. When a flow is incompressible, the situation is entirely

[1]An exception is the Boussinesq model of natural and mixed (combined natural and forced) thermal convection, in which the incompressibility condition is relaxed to allow density variations caused by temperature variations, and the resulting Archimedes forces. Evidently, the heat and momentum equations are coupled in that case. We mention the Boussinesq model here because apart from the Archimedes force, the momentum and incompressibility equations are the same as in the purely incompressible flow. They can be solved by the numerical methods described in this chapter.

different. There is no equation of state, except for $\rho = $ const. The pressure, which still remains a function of space and time and has to be computed at every time step, is now determined as a part of the general solution of the momentum and incompressibility equations (10.1) and (10.2).

The main difficulty arises here. We cannot find the pressure field by simple advancement of a marching scheme or by evaluating a function of other thermodynamic variables. Furthermore, the mass conservation condition (10.2) is not an independent evolutionary equation for density, as in the case of compressible flows. In fact, it is not an evolutionary equation at all, but a constraint on velocity that has to be satisfied by any state of the flow. How can we find a velocity field, which follows the evolution prescribed by the momentum equation (10.1) and satisfies this constraint at the same time?

The answer to these questions is that the pressure field adjusts itself to the instantaneous state of the flow so that the gradient ∇p in (10.1) enforces the incompressibility. Let us see how this pressure field can be found. We apply the divergence operator $(\nabla \cdot)$ to the momentum equation (10.1). Requiring that $\nabla \cdot V = 0$ and taking into account that the operator $(\nabla \cdot)$ commutes with $\partial/\partial t$ in the time derivative term and with ∇^2 in the viscous term, we find that the pressure field satisfies the Poisson equation

$$\nabla \cdot (\nabla p) \equiv \nabla^2 p = \rho \nabla \cdot [f - N(V, V)] = \rho \nabla \cdot F, \qquad (10.5)$$

where we use the notation $F = f - N(V, V)$.

From the mathematical viewpoint, this means that the incompressible flow equations have some features of an elliptic system. We can say that the equations are of the mixed hyperbolic (convective terms), parabolic (viscous terms), and elliptic (pressure and incompressibility) type. The elliptic properties of the solution have physical meaning. They show that the pressure field in the entire flow domain adjusts instantaneously to any, however localized, perturbation. This is in perfect agreement with the fact that the speed of propagation of weak perturbations (the speed of sound) tends to infinity in the asymptotic limit of the incompressible fluid approximation.

The special features of incompressible flows require special computational tools. An attempt to dismiss the peculiar role of pressure as a mere technicality and apply the methods developed for compressible flows is likely to lead to disappointing results. For example, let us assume that we could adapt the MacCormack method (9.8) and (9.9) to the case of an incompressible flow. The CFL stability criterion (9.11) would then require zero time step Δt since the speed of sound a is infinite. In fact, many established explicit schemes for compressible flows fail when compressibility is weak (the Mach number M is much smaller than 1). The required time step becomes too small when $M \to 0$.

The pressure equation (10.5) poses new questions that have to be answered by any scheme designed for incompressible flows. Some of them are of familiar kind and refer to discretization of (10.5). The others are quite unique. They concern the organization of steps of a time-marching or iteration procedure. Every such step should combine solutions of two equations: the momentum equation (10.1) and the pressure equation (10.5). As we will see in this chapter, the task is nontrivial.

10.2 DISCRETIZATION APPROACH

We start with a general discussion of the discretization approach. Finite difference approximations of the momentum and pressure equations can be obtained using the formulas derived earlier for the model equations. For example, in the momentum equations, we can use central or upwind-biased schemes for convective terms and central differences for viscous terms. The pressure gradient term in (10.1) can be approximated by various schemes, not necessarily coinciding with the schemes used for the other terms and not necessarily on the same set of grid points (we will discuss this later).

In finite volume methods, the surface integrals corresponding to convective and viscous terms of the momentum equation are interpolated as convective and diffusive flux integrals, respectively (see Section 5.2.2). The pressure gradient in the momentum equation is usually treated as a surface force. For example, for the x-momentum equation, we use

$$\int_\Omega \frac{\partial p}{\partial x} d\Omega = \int_{\partial\Omega} p e_x \cdot n \, dS = \int_{\partial\Omega} p n_x \, dS \qquad (10.6)$$

and apply an interpolation scheme to approximate the surface integral.

The pressure equation in the integral form is

$$\int_\Omega \nabla \cdot (\nabla p) \, d\Omega = \int_\Omega \nabla \cdot F \, d\Omega. \qquad (10.7)$$

The volume integrals are transformed into surface integrals using the divergence theorem,

$$\int_{\partial\Omega} \nabla p \cdot n \, dS = \int_{\partial\Omega} F \cdot n \, dS, \qquad (10.8)$$

and then approximated using interpolation formulas.

10.2.1 Conditions for Conservation of Mass by Numerical Solution

One important aspect of the system should be taken into account if a numerical scheme with exact conservation of mass is desired. The discretization schemes for the Laplace operator in (10.5), the pressure gradient term in the momentum equation (10.1), and the divergence operator in the right-hand side of (10.5) cannot be chosen independently of each other.

Let us use the symbolic notation for the discretized operators: $\delta p / \delta x_i$ for the components of ∇p, $\delta V_i / \delta x_i$ for $\nabla \cdot V$, and $\delta^2 p / \delta x_i^2$ for $\nabla^2 p$. The discretized pressure equation (10.5) is written as

$$\frac{\delta^2 p}{\delta x_i^2} = \rho \frac{\delta F_i}{\delta x_i}. \tag{10.9}$$

Mathematically, the Laplace operator and the right-hand side of (10.5) or (10.9) are equivalent to the results of application of the divergence operator to the momentum equation (10.1) and imposing the incompressibility condition (10.2). In order to assure that the velocity solution of the discretized equations satisfy the discretized incompressibility condition

$$\frac{\delta V_i}{\delta x_i} = 0, \tag{10.10}$$

we need to maintain the same equivalency for the discretized operators, i.e. that

$$\frac{\delta^2 p}{\delta x_i^2} = \frac{\delta}{\delta x_i} \left(\frac{\delta p}{\delta x_i} \right) \tag{10.11}$$

and that identical discretization schemes are applied to the divergence operators in (10.10) and the right-hand sides of (10.9) and (10.11), and to the gradients of pressure in (10.11) and the momentum equation (10.1).

If there is no such equivalency, the incompressibility condition (10.10) is satisfied by the computed velocity field with an error. The error has the order of magnitude of the truncation error of the scheme. The effect is as if a source (positive or negative) of mass were created by the numerical scheme. However small, this artificial mass source is, in many cases, undesirable, since its effect may accumulate with time and lead to significant mass imbalance and to poor conservation of kinetic energy by the solution. There is an additional rather unpleasant effect. In many cases, the errors of the energy balance are not strictly dissipative, which results in numerical instability.

10.2.2 Colocated and Staggered Grids

When designing a finite difference or finite volume scheme, we have to choose whether to use the same or different sets of grid points for velocity and pressure. The obvious choice seems to have a single set of points, at which all the variables and all the equations are discretized. Such a grid has the name of a *colocated grid*. Albeit simple and easy in operation, the colocated grids were out of favor for long time because of their tendency to generate spurious oscillations in the solution.

Let us understand the origins of the oscillations. The phenomenon itself is quite general. It appears for colocated variable arrangements on two- and three-dimensional, structured and unstructured, or uniform and nonuniform grids. For simplicity, we illustrate it on the example of a finite volume grid with uniform rectangular cells shown in Figure 10.1. The grid points are the cell midpoints marked by capital letters: P, E, N, etc. The same points can be considered as forming a rectangular finite difference grid. It will be clear from the following discussion that the discretized equations resulting from the two approaches are equivalent.

We start by approximating the divergence operator. The finite volume approach gives, by the divergence theorem,

$$\int_\Omega \nabla \cdot \boldsymbol{V} \, d\Omega = \int_{\partial\Omega} \boldsymbol{V} \cdot \boldsymbol{n} \, dS. \tag{10.12}$$

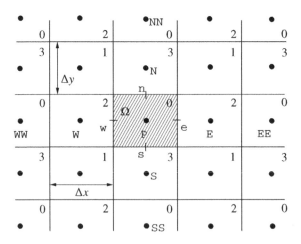

Figure 10.1 Finite volume colocated grid.

On each of the faces marked by e, w, s, and n, the outward-facing normal $\boldsymbol{n}$ is given by a positive or negative unit vector in the x- or y-direction. Using the mean value theorem, we represent the surface integral in (10.12) as

$$\int_{\partial\Omega} \boldsymbol{V} \cdot \boldsymbol{n} \, dS \approx u_e \Delta y - u_w \Delta y + v_n \Delta x - v_s \Delta x. \tag{10.13}$$

Dividing by the cell volume $\Delta x \Delta y$, we obtain the approximation of the second order:

$$\frac{1}{\Delta x \Delta y} \int_{\Omega} \nabla \cdot \boldsymbol{V} \, d\Omega \approx \frac{u_e - u_w}{\Delta x} + \frac{v_n - v_s}{\Delta y} = \left. \frac{\delta u}{\delta x} \right|_P + \left. \frac{\delta v}{\delta y} \right|_P, \tag{10.14}$$

where the symbolic expressions $\delta u/\delta x$ and $\delta v/\delta y$ now denote the specific discretizations of the partial derivatives $\partial u/\partial x$ and $\partial v/\partial y$. The velocities u and v are only defined at the grid points. Their values at the face points have to be obtained by interpolation. Willing to retain the second order of accuracy, we use the linear interpolation:

$$u_e \approx \frac{u_E + u_P}{2}, \quad u_w \approx \frac{u_W + u_P}{2}, \quad v_n \approx \frac{v_N + v_P}{2}, \quad v_s \approx \frac{v_S + v_P}{2}. \tag{10.15}$$

Substitution into (10.14) gives the familiar formula

$$\left. \frac{\delta u}{\delta x} \right|_P + \left. \frac{\delta v}{\delta y} \right|_P = \frac{u_E - u_W}{2\Delta x} + \frac{v_N - v_S}{2\Delta y}, \tag{10.16}$$

which could be obtained as a finite difference approximation by direct application of the central difference formulas to the derivatives.

The next step is to approximate the pressure gradient ∇p. The second-order approximation of integrals of pressure derivatives such as (10.6) can be derived similarly to (10.13) and (10.14):

$$\frac{1}{\Delta x \Delta y} \int_{\partial\Omega} p n_x \, dS \approx \frac{p_e - p_w}{\Delta x} = \left. \frac{\delta p}{\delta x} \right|_P, \tag{10.17}$$

$$\frac{1}{\Delta x \Delta y} \int_{\partial\Omega} p n_y \, dS \approx \frac{p_n - p_s}{\Delta y} = \left. \frac{\delta p}{\delta y} \right|_P, \tag{10.18}$$

where we use the symbolic notation as in (10.14).

The values at the face points are obtained by the linear interpolation as in (10.15), which results in

$$\left. \frac{\delta p}{\delta x} \right|_P = \frac{p_E - p_W}{2\Delta x}, \quad \left. \frac{\delta p}{\delta y} \right|_P = \frac{p_N - p_S}{2\Delta y}. \tag{10.19}$$

This could also be obtained in the result of the central difference approximation of the components of $\nabla p|_P$ on the finite difference grid.

We now consider the pressure equation (10.5) or (10.8) and recall the condition for the exact mass conservation by a numerical solution derived in Section 10.2.1. They require that the pressure equation, which is (10.8) in the finite volume method, is discretized using a combination of the same discretizations as in (10.16) and (10.19) of the divergence and gradient operators. Applying the analog of (10.16) to (10.8), we obtain

$$\frac{(\delta p/\delta x)_E - (\delta p/\delta x)_W}{2\Delta x} + \frac{(\delta p/\delta y)_N - (\delta p/\delta y)_S}{2\Delta y} = \frac{F_{xE} - F_{xW}}{2\Delta x} + \frac{F_{yN} - F_{yS}}{2\Delta y}.$$
$$(10.20)$$

The pressure derivatives are approximated as in (10.19), except that the formulas should be taken for the cells with the grid points E, W, S, and N instead of P. For example, we should use

$$\left.\frac{\delta p}{\delta x}\right|_E = \frac{p_{EE} - p_P}{2\Delta x}.$$

The final approximation of the pressure equation is

$$\frac{p_{EE} - 2p_P + p_{WW}}{(2\Delta x)^2} + \frac{p_{NN} - 2p_P + p_{SS}}{(2\Delta y)^2} = \frac{F_{xE} - F_{xW}}{2\Delta x} + \frac{F_{yN} - F_{yS}}{2\Delta y}. \quad (10.21)$$

The expression on the left-hand side is the familiar five-point operator, which has the second order of accuracy in x and y. It has a strange feature, however, that the grid steps of double size $2\Delta x$ and $2\Delta y$ are used. Albeit not especially troubling at first glance, this may lead to dangerous behavior in the form of spurious oscillations of the pressure field. The reason is the splitting of the system of discretization equations into four subsystems, which are only coupled with each other in weak sense via the right-hand sides. Each such subsystem contains the equations that connect the pressure values at cells marked by only one of the numbers 0, 1, 2, or 3 in Figure 10.1. For example, it is easy to see that the left-hand side of (10.21) connects the values of p at the 0-cells. Any equation written for any other 0-cell connects only 0-cells. Similarly, equations for a 1-cell connect 1-cells but not the others.

As a particularly striking example of the splitting effect, let us assume that the numbers 0, 1, 2, and 3 in Figure 10.1 represent actual values of pressure in the corresponding cells. Such a bizarre distribution would be a solution of the pressure equation with zero right-hand side. The approximations (10.19) would ignore the pressure variations and show the gradient ∇p as identically zero.

The source of trouble is, of course, the use of the double grid steps $2\Delta x$ and $2\Delta y$ in the five-point formula in (10.21) that stems from our use of such steps in the approximations (10.16) and (10.19). Can this be avoided? The positive answer was first given by Harlow and Welch (1965) in the form of a *staggered grid*.

The idea is simple and based on understanding that we do not have to use the same set of grid points for all variables. In particular, we avoid double-step differences by evaluating velocity components directly at the corresponding face points instead of the grid points located at the centers of the cells. The u-component is calculated at the midpoints e and w of the eastern and western faces of each cell, while the midpoints n and s of the northern and southern faces are used for the v-component.

The same arrangement is used for the components of vector F. The grid points at the cell centers are still used for the pressure. The formulas (10.14) can now be applied directly without need for an interpolation. The main benefit, however, is removing the splitting problem. If we use (10.14) instead of (10.16) as an approximation of the divergence operator, the consistent approximation of the pressure equation becomes, instead of (10.20),

$$\frac{(\delta p/\delta x)_e - (\delta p/\delta x)_w}{\Delta x} + \frac{(\delta p/\delta y)_n - (\delta p/\delta y)_s}{\Delta y} = \frac{F_{xe} - F_{xw}}{\Delta x} + \frac{F_{yn} - F_{ys}}{\Delta y}.$$
(10.22)

The partial derivatives of pressure at the face points can be approximated by single-step differences:

$$\left.\frac{\delta p}{\delta x}\right|_e = \frac{p_E - p_P}{\Delta x}, \quad \left.\frac{\delta p}{\delta x}\right|_w = \frac{p_P - p_W}{\Delta x},$$

$$\left.\frac{\delta p}{\delta y}\right|_n = \frac{p_N - p_P}{\Delta y}, \quad \left.\frac{\delta p}{\delta y}\right|_s = \frac{p_P - p_S}{\Delta y}.$$

Substitution into (10.22) results in the standard five-point scheme:

$$\frac{p_E - 2p_P + p_W}{(\Delta x)^2} + \frac{p_N - 2p_P + p_S}{(\Delta y)^2} = \frac{F_{xe} - F_{xw}}{\Delta x} + \frac{F_{yn} - F_{ys}}{\Delta y}.$$
(10.23)

Equations for every grid point are now strongly coupled by their left-hand sides. Unphysical oscillations like those shown in Figure 10.1 are registered by the pressure gradient and corrected by the solution.

To be able to evaluate the components of V and F directly at the face points, we have to introduce additional sets of grid points or, in finite volume methods, cells. This leads to the staggered grid arrangements illustrated by two-dimensional examples in Figure 10.2. Generalization to the three-dimensional case is trivial.

In finite difference methods (see Figure 10.2a), we evaluate pressure and approximate the pressure equation (satisfy the incompressibility condition) at the integer grid points (x_i, y_j) shown by solid circles. These points are also used for other scalar fields (e.g. temperature). The velocity components are evaluated and momentum equations are solved at points of the half-integer grids obtained from the integer grid by shifting in the x- or y-direction by half a grid step. The points $(x_{i+1/2}, y_j)$ shown by hollow circles are used to evaluate u, while v is computed at points $(x_i, y_{j+1/2})$ shown by squares.

In the finite volume method, additional sets of cells are created. Examples of the cells used to solve the pressure or other scalar equation, u-momentum

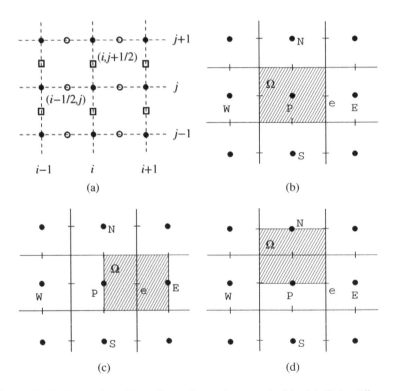

Figure 10.2 Examples of two-dimensional staggered grids. (a) Finite difference grid. (b)–(d) Finite volume grid. Separate figures present cells used to satisfy mass balance (b), x-momentum balance (c), and y-momentum balance (d).

equation, and v-momentum equation are shown in Figure 10.2b, c, and d, respectively.

The staggered arrangement increases complexity of a scheme. Programming becomes somewhat more difficult, since it requires accounting for three (or four in the three-dimensional case) indexing systems. Interpolations must be used to compute the nonlinear terms of the momentum equations. Further complications arise when the grid is nonuniform. All these difficulties, however, can be relatively easily handled in computations with structured grids such as those shown in Figure 10.2. For this reason and because of the benefit of removing the splitting problem, the staggered arrangement was a popular choice during the early years of CFD.

The difficulties of handling a staggered grid arrangement increase significantly when unstructured grids are used. When such grids started to be broadly applied in the general-purpose CFD codes in recent years, colocated arrangements returned to favor. This area of CFD is still evolving and, in general, requires discussion on a more advanced level than appropriate for this book. We only mention that methods have been developed to cure the splitting problem. Some of them are based on filtering out the oscillating component of pressure field or periodically averaging the pressure values at neighboring points. Others avoid the problem by fully or partially neglecting the requirement that $\nabla^2 p$ in the pressure equation, $\nabla \cdot V$ in the incompressibility condition, and ∇p in the momentum equation are discretized consistently. The property of exact mass and kinetic energy conservation is lost if this approach is taken. It was, however, shown, for example, by Morinishi et al. (1998), that the schemes can be arranged so that the error is relatively small and does not induce numerical instability.

10.3 PROJECTION METHOD FOR UNSTEADY FLOWS

In this section, we consider the procedure commonly used to compute time-dependent flows of incompressible flows. The procedures for steady-state equations, which we review in the next section, follow the same basic principle, although they are implemented differently. The details of spatial discretization are of little importance at this moment and, therefore, omitted. We assume in the following discussion that the spatial discretization is completed using a finite difference, finite volume, or, perhaps, spectral scheme.

The problem of pressure calculation in unsteady solutions is formulated as follows: *Given the solution p^n, V^n at the previous time layer t^n, find the next time-layer pressure p^{n+1} and velocity V^{n+1} such that they together satisfy the momentum equation and the velocity is divergence-free $\nabla \cdot V^{n+1} = 0$ and satisfies the boundary conditions.*

A widely used and well-established approach is based on the *pressure correction* or *projection* method. The general strategy is to decompose each time step into two substeps. On the first substep, the momentum equation is solved for the velocity components. The pressure gradient is either removed from the equation or approximated by an available estimate. The obtained velocity field cannot be considered a solution at the new time level since it does not satisfy the incompressibility condition. The second substep is, therefore, needed, at which the correct pressure distribution is found and the correction of velocity is made. The term *projection* reflects the fact that we find a preliminary solution, which is not divergence-free, and then project it onto the space of divergence-free vector functions. Versions of the projection method were proposed by Harlow and Welch (1965) (the marker-and-cell method), Chorin (1968), Temam (1969), and Yanenko (1971) (the fractional-step method) and further developed in the 1980s and 1990s.

10.3.1   Explicit Schemes

As the first illustration, we solve a marching problem using the simple explicit scheme

$$\frac{V^{n+1} - V^n}{\Delta t} = -\frac{1}{\rho}\nabla p^{n+1} + [-N(V, V) + \nu\nabla^2 V + f]^n$$

$$= -\frac{1}{\rho}\nabla p^{n+1} + F^n, \tag{10.24}$$

$$\nabla \cdot V^{n+1} = 0, \tag{10.25}$$

where F^n is the shorthand notation for the combination of all terms of the right-hand side except the pressure gradient. The pressure term is marked as belonging to the time layer t^{n+1} to stress its role in ensuring incompressibility at that time layer.

As we have seen, the incompressibility condition is fulfilled if the pressure field satisfies the Poisson equation (10.5). In the case of the simple explicit scheme (10.24)–(10.25), the equation takes the form

$$\nabla^2 p^{n+1} = \rho\nabla \cdot F^n, \tag{10.26}$$

which is derived by applying the divergence operator directly to the discretized momentum equation (10.24) and requiring that $\nabla \cdot V^n = \nabla \cdot V^{n+1} = 0$.

The solution procedure is as follows. Instead of the original system (10.24) and (10.25), we solve the mathematically equivalent system (10.24) and (10.26). To advance from the time layer t^n to the layer t^{n+1}, we first calculate the right-hand side of the momentum equation F^n. It is based on the known velocity V^n and does not include the pressure term. We then solve the pressure equation (10.26) and use ∇p^{n+1} and F^n to update the velocity according to (10.24).

The procedure can be presented as a two-step cycle of the form common for all projection methods. To do so, we split the velocity update into two substeps. On the first, F^n is taken into account to produce the intermediate velocity field V^*:

$$\text{Predictor:} \quad \frac{V^* - V^n}{\Delta t} = F^n \text{ or } V^* = V^n + \Delta t F^n. \tag{10.27}$$

The pressure gradient is added at the second substep to satisfy the incompressibility constraint:

$$\text{Corrector:} \quad \frac{V^{n+1} - V^*}{\Delta t} = -\frac{1}{\rho}\nabla p^{n+1} \text{ or } V^{n+1} = V^* - \frac{\Delta t}{\rho}\nabla p^{n+1}. \tag{10.28}$$

It is easy to verify that (10.27) and (10.28) added together give the original scheme (10.24). The pressure equation is solved either in the beginning of the cycle or between the predictor and corrector steps. In the latter case, it can be rewritten as

$$\nabla^2 p^{n+1} = \frac{\rho}{\Delta t}\nabla \cdot V^*, \tag{10.29}$$

which follows from (10.26) and (10.27) or from direct application of the divergence operator to (10.28).

An interesting and important question arises as to what boundary conditions should be used for the pressure field. Such conditions are required at every point of the boundary for the Poisson problem (10.26) or (10.29) to be well posed. The conditions, however, do not naturally follow from the flow physics for the boundaries between a fluid and solid walls, unless a full fluid–structure interaction problem with stresses in the wall as variables is solved. Since the latter option is, in most cases, an unnecessary complication, we have to find a way to derive the pressure boundary conditions from the flow equations themselves.

To do so, we observe that the velocity field should satisfy the impermeability condition $V \cdot n|_{\partial\Omega} = 0$, where n is the normal to the solid wall boundary $\partial\Omega$. In the case of an explicit scheme, this results in a simple boundary condition for pressure. Taking the projection of the momentum equation (10.24) on $\boldsymbol{n}$, we find

$$\left(-\frac{1}{\rho}\nabla p^{n+1} \cdot \boldsymbol{n} + \boldsymbol{F}^n \cdot \boldsymbol{n} \right)_{\partial\Omega} = 0 \text{ or}$$

$$\left.\frac{\partial p^{n+1}}{\partial n}\right|_{\partial\Omega} = \rho \boldsymbol{F}^n \cdot \boldsymbol{n}|_{\partial\Omega} = \frac{\rho}{\Delta t} \boldsymbol{V}^* \cdot \boldsymbol{n}\Big|_{\partial\Omega}. \tag{10.30}$$

The wall-normal component of $\boldsymbol{F}^n$ is, generally, nonzero. For example, let us consider a solid wall located at $x = 0$. Using the no-slip condition $u^n = v^n = w^n = 0$ at $\partial\Omega$, we find the wall-normal component

$$F_x^n|_{x=0} = \left(v\frac{\partial^2 u^n}{\partial x^2} + f_x^n \right)_{x=0} \neq 0.$$

We note that the no-slip conditions for the wall-parallel components of velocity $\boldsymbol{V}^{n+1}$ are not enforced in the course of the explicit scheme procedure (10.27)–(10.30). The conditions have to be imposed after every time step.

The scheme outlined here remains valid, with technical modifications, for other fully explicit methods based, for example, on the Adams–Bashforth or Runge–Kutta time-integration algorithms.

Solving the Poisson equation for pressure is a computationally expensive part of the solution, since it requires solution of a matrix equation (see Chapter 8). This is particularly true in the case of explicit methods, where the solution of the Poisson equation is responsible for up to 90% of the total number of computer operations. Realizing that and taking into account that the Poisson equation is often solved by one of the iterative methods (see Section 8.3), we immediately see a way to improve the computational efficiency of the projection algorithm. The number of iterations needed to achieve the desired accuracy of the solution of the matrix equation is reduced if a good initial guess of the solution is employed as a starting point. In our case, such a guess is readily available in the form of the pressure field from the previous time step $p_0 = p^n$ or an interpolation from several previous steps, such as $p_0 = 2p^n - p^{n-1}$. It can be used in two different ways. First, we can directly apply p_0 to start the first iteration. Alternatively, we can include the effect of p_0 into $\boldsymbol{F}^n$. The two-step version of the algorithm (10.27) and (10.28) is modified as

$$\text{Predictor:} \quad \boldsymbol{V}^* = \boldsymbol{V}^n + \Delta t \boldsymbol{F}^n - \frac{\Delta t}{\rho}\nabla p_0, \tag{10.31}$$

$$\text{Corrector:} \quad \boldsymbol{V}^{n+1} = \boldsymbol{V}^* - \frac{\Delta t}{\rho}\nabla(\delta p), \tag{10.32}$$

where $\delta p = p^{n+1} - p_0$ is the pressure correction. The Poisson equation and boundary conditions for δp are expressed in terms of V^* in the same way as in (10.29) and (10.30), but the right-hand sides are closer to 0 in terms of an appropriate norm. One can expect (and in many cases the expectations are fulfilled) that the number of iterations needed to solve the Poisson equation decreases.

10.3.2 Implicit Schemes

Talking about implicit methods, we have to distinguish between two approaches: the fully implicit approach, according to which all terms including the nonlinear term are approximated at the new time layer t^{n+1}, and the semi-implicit approach, in which the implicit treatment is limited to the viscous term and pressure.

The fully implicit approach leads to a system of nonlinear discretization equations, which has to be solved at every time step. Such systems have already been discussed in Section 8.4. We have found that their direct solution, for example, by Newton's method, is inefficient. Efficiency can be achieved using an iteration procedure and linearization. The same general approach can be applied, after some modification, to fully implicit schemes for incompressible flows. Because of importance and certain peculiar features (primarily related to the role of pressure), it is worthwhile to present a detailed description of the method.

As an illustration, we consider the solution by the simple implicit method:

$$\frac{V^{n+1} - V^n}{\Delta t} = -\frac{1}{\rho}\nabla p^{n+1} - N(V^{n+1}, V^{n+1}) + \nu\nabla^2 V^{n+1} + f^{n+1} \qquad (10.33)$$

$$\nabla \cdot V^{n+1} = 0. \qquad (10.34)$$

As before, some kind of spatial discretization is assumed but not shown.

The general Poisson equation for pressure (10.5) can be easily specified for the implicit scheme. Applying the divergence operator to (10.33), we obtain, assuming that $\nabla \cdot V^n = 0$ and requiring that $\nabla \cdot V^{n+1} = 0$,

$$\nabla^2 p^{n+1} = -\rho\nabla \cdot [N(V^{n+1}, V^{n+1}) + f^{n+1}] = -\rho\nabla \cdot F^{n+1}. \qquad (10.35)$$

We see the problem now. Unlike the explicit version (10.26), the right-hand side of the pressure equation now uses the velocity V^{n+1} that must already include the pressure correction ∇p^{n+1}. The momentum and pressure equations (10.33), (10.35) are fully coupled and have to be solved simultaneously. The matter is further complicated by the nonlinearity of the discretized momentum equation. We have to apply iterations and linearization to solve the system.

The iterative procedures are discussed in the next section. Here, we consider the version of the method valid in the case of sufficiently small Δt. Convergence of iterations is not required in this method. In fact, we only perform one iteration at a time step. This introduces additional error, which we will estimate later. The method uses linearization around the solution obtained at the previous time layer and is arranged as a predictor–corrector procedure similar to the procedure introduced earlier for the explicit method.

We describe the linearization first. The unknown velocity and pressure are represented as sums of the known values found at t^n and perturbations:

$$V^{n+1} = V^n + \delta V, \quad p^{n+1} = p^n + \delta p. \tag{10.36}$$

The nonlinear term is quadratic in velocity components. It can be rewritten as

$$N(V^{n+1}, V^{n+1}) = N(V^n, V^n) + N(V^n, \delta V) + N(\delta V, V^n) + N(\delta V, \delta V). \tag{10.37}$$

If the time step Δt is small, the typical amplitude of the velocity perturbations can be estimated according to

$$\delta V \sim \frac{\partial V}{\partial t} \Delta t.$$

The last term in the right-hand side of (10.37) is quadratic in δV and, thus, is of the order of $O\left((\Delta t)^2\right)$. It can be neglected in comparison with other terms, which are either $O(1)$ or $O(\Delta t)$. Dropping the quadratic term resolves the issue of nonlinearity at the price of introducing the error $\sim O\left((\Delta t)^2\right)$ into the solution. The price is fair, since this error is of the higher order than the truncation error of the simple implicit formula (10.33). The linearization would still be justified if we used a second-order time discretization scheme.

The linearized version of (10.33) and (10.34) is

$$\frac{V^{n+1} - V^n}{\Delta t} = -\frac{1}{\rho}\nabla p^n - \frac{1}{\rho}\nabla \delta p - \tilde{N}(V^n, V^{n+1})$$
$$+ \nu\nabla^2 V^{n+1} + f^{n+1}, \tag{10.38}$$

$$\nabla \cdot V^{n+1} = 0, \tag{10.39}$$

where $\tilde{N}(V^n, V^{n+1}) = N(V^n, V^n) + N(V^n, \delta V) + N(\delta V, V^n)$ is the linearized convection term. We assume that the body force is a constant, a function of space and time, or a linear function of V.

Now we will deal with the problem of coupling between the momentum and pressure equations. The solution follows the predictor–corrector scheme. On the predictor substep, the linear system (10.38) is solved with ∇p^n used as an estimate of the pressure gradient. The perturbation δp is omitted. This produces the intermediate velocity field V^*.

On the corrector substep, the velocity field is updated as $V^{n+1} = V^* - \Delta t \nabla \delta p$. The requirement of incompressibility of V^{n+1} gives the compact form of the pressure equation:

$$\nabla^2 \delta p = \frac{\rho}{\Delta t} \nabla \cdot V^*. \tag{10.40}$$

This equation has to be solved between the predictor and corrector substeps.

The predictor part of the procedure is computationally costly, significantly more so than in the case of explicit methods, since it requires solution of a large linear system (the matrix size is three times the total number of grid points). Methods have been developed to achieve computational efficiency, for example, the methods based on the approximate factorization approach (see Chapter 9) or the efficient direct or iterative techniques for solution of matrix equations (see Chapter 8).

An alternative to the fully implicit schemes is the semi-implicit approach, in which only the viscous, pressure, and body force terms are treated implicitly. The nonlinear term is treated explicitly, so no linearization is needed. The simple semi-implicit scheme is

$$\frac{V^{n+1} - V^n}{\Delta t} = -\frac{1}{\rho} \nabla p^{n+1} - N(V^n, V^n) + \nu \nabla^2 V^{n+1} + f^{n+1} \tag{10.41}$$

$$\nabla \cdot V^{n+1} = 0. \tag{10.42}$$

The projection procedure is organized as a sequence of the predictor substep, on which we solve the momentum equation with ∇p^n or without pressure at all, solution of the pressure equation (10.40) or (10.29), and the correction substep. Similarly to the fully implicit linearized solution, the predictor substep requires solving a large linear system for the components of the intermediate velocity V^*.

One important difference between the fully implicit and semi-implicit methods is in their stability characteristics. The fully implicit schemes are typically unconditionally stable, while semi-implicit schemes have stability limits on the time step. This does not necessarily mean that the fully implicit schemes are more efficient. The time step should be reasonably small anyway in order to keep the linearization and truncation errors under control.

The linear elliptic problems for V^* encountered in both versions of the implicit method need boundary conditions on the boundaries $\partial\Omega$ of the computational domain. Correct formulation of these conditions and of the boundary conditions for pressure is a nontrivial matter, which is discussed in the research literature and cannot be adequately addressed here.

10.4 PROJECTION METHODS FOR STEADY-STATE FLOWS

There is a commonality between the solutions of steady-state problems and transient problems solved by fully implicit methods. Almost identical systems of nonlinear discretization equations have to be solved, once for a steady-state problem and at every time step for a transient problem.[2] The same numerical procedure can be applied.

We will give an overview of several popular methods. For the sake of consistency with other textbooks, for example, by Patankar (1980) and Ferziger and Perić (2001) and with documentation of many CFD codes, the discussion will use the equations written in spatially discretized form. This should not obscure the fact that the methods are based on the same basic principles as the schemes described in Section 10.3: projection and linearization of convection term.

The discretized momentum equation can be written in the general form as

$$a_{\mathrm{P}}(\boldsymbol{u})u_{i,\mathrm{P}} + \sum_{\ell} a_{\ell,\mathrm{P}}(\boldsymbol{u})u_{i,\ell} = Q_{\mathrm{P}}(\boldsymbol{u}) - \left(\frac{\delta p}{\delta x_i}\right)_{\mathrm{P}}. \qquad (10.43)$$

We have already considered a similar general form of a partial differential equation (PDE) in Section 4.4.3. In (10.43), u_i is the velocity component, conservation of which is described. The equation is a result of spatial discretization of the fully implicit formula (10.33) or of a steady-state momentum equation. The discretization is conducted for the grid point P (in finite difference methods) or cell P (in finite volume methods). The right-hand

---

[2]We assume here that the simplified solution algorithm developed in Section 10.3 for transient problems cannot be utilized because we cannot afford small time steps Δt.

side includes the source term $Q_P(\boldsymbol{u})$ combining the body forces, which may linearly depend on velocity, and the terms with variables other than velocity (e.g. temperature). It also contains the discretized pressure gradient $\delta p / \delta x_i$. The left-hand side represents the discretized nonlinear and viscous terms. The summation index ℓ runs over all neighboring nodes used by the discretization formulas. Some of the coefficients $a_P(\boldsymbol{u})$ and $a_{\ell,P}(\boldsymbol{u})$ depend on the velocity components (all three, not just u_i), which reflects the quadratic nonlinearity of the momentum equation.

Equation (10.43) is written in the general way that does not imply any particular grid or any particular method of discretization. Since the methods discussed here are commonly applied by the general-purpose CFD codes, however, it is useful to view (10.43) as a result of discretization on an unstructured finite volume grid.

The solution $\boldsymbol{u}$ has to satisfy the discretized incompressibility condition

$$\left.\frac{\delta u_i}{\delta x_i}\right|_P = 0, \tag{10.44}$$

where summation over repeating indices is assumed.

The system of coupled equations (10.43) and (10.44) is solved in a sequence of converging approximations, the so-called outer iterations. An iteration is based on the projection method and linearization and consists of the following principal substeps:

Step 1: Use the best currently available approximations of velocity $u_i^{(m)}$ and pressure $p^{(m)}$ to evaluate the coefficients a_P and $a_{\ell,P}$ and the terms Q_P and $(\delta p / \delta x_i)_P$ in (10.43).

Step 2: Solve the linearized momentum equations

$$a_P(\boldsymbol{u}^{(m)})u_{i,P}^* + \sum_\ell a_{\ell,P}\left(\boldsymbol{u}^{(m)}\right) u_{i,\ell}^* = Q_P\left(\boldsymbol{u}^{(m)}\right) - \left(\frac{\delta p^{(m)}}{\delta x_i}\right)_P. \tag{10.45}$$

This is a linear system usually characterized by a sparse coefficient matrix. It can be efficiently solved by an iterative method. The steps of this procedure are called *inner iterations*. At the end, we obtain the intermediate velocity field u_i^*, which does not satisfy the incompressibility condition.

Step 3: Solve the pressure equation to find the new pressure field $p^{(m+1)}$. The exact form of the pressure equation varies with the implementation of the method. Several forms are presented in Sections 10.4.1–10.4.3.

Step 4: Use the new pressure field to update velocity. The result is the new approximation $u_i^{(m+1)}$.

Step 5: Test convergence. If the desired accuracy is not achieved, repeat the procedure starting at step 1.

Several variations of this general procedure have been developed. They are widely used today. We will discuss the most popular among them.

10.4.1 SIMPLE

One of the first versions of the method is SIMPLE (Semi-Implicit Method for Pressure-Linked Equations) (see Caretto et al. 1973; Patankar and Spalding 1972). The method was originally designed for finite volume approximation with staggered grid arrangement but can be straightforwardly extended to other discretization techniques. Our description is not tied to any particular discretization.

In the method, the next iteration values of velocity and pressure found in Steps 3 and 4 given above are represented as

$$u_i^{(m+1)} = u_i^* + u_i', \quad p^{(m+1)} = p^{(m)} + p', \tag{10.46}$$

where u_i^* is solution of the linearized momentum equation (10.45) and $p^{(m)}$ is used as an estimate of pressure.

To find the relation between the unknown corrections u_i' and p', we require that $u_i^{(m+1)}$ and $p^{(m+1)}$ satisfy the linearized momentum equation

$$a_P(\boldsymbol{u}^{(m)})u_{i,P}^{(m+1)} + \sum_\ell a_{\ell,P}(\boldsymbol{u}^{(m)})u_{i,\ell}^{(m+1)} = Q_P(\boldsymbol{u}^{(m)}) - \left(\frac{\delta p^{(m+1)}}{\delta x_i}\right)_P. \tag{10.47}$$

Subtracting (10.45) from (10.47), we find

$$a_P(\boldsymbol{u}^{(m)})u_{i,P}' + \sum_\ell a_{\ell,P}(\boldsymbol{u}^{(m)})u_{i,\ell}' = -\left(\frac{\delta p'}{\delta x_i}\right)_P. \tag{10.48}$$

This is, again, a linearized system, which, at the moment, cannot be solved for either $u_{i,P}'$ or p_P', since both are yet unknown. We formally write the solution as

$$u_{i,P}' = -\frac{\sum_\ell a_{\ell,P}(\boldsymbol{u}^{(m)})u_{i,\ell}'}{a_P(\boldsymbol{u}^{(m)})} - \frac{1}{a_P(\boldsymbol{u}^{(m)})}\left(\frac{\delta p'}{\delta x_i}\right)_P = \tilde{u}_{i,P}' - \frac{1}{a_P(\boldsymbol{u}^{(m)})}\left(\frac{\delta p'}{\delta x_i}\right)_P,$$

$$\tag{10.49}$$

where the first term in the right-hand side is denoted as $\tilde{u}'_{i,\mathrm{P}}$. It is a function of u'_i and, thus, unknown at this moment.

We are now in a position to formally derive the pressure equation. Substituting the first expansion of (10.46) into the incompressibility condition

$$\left.\frac{\delta u_i^{(m+1)}}{\delta x_i}\right|_{\mathrm{P}} = 0 \tag{10.50}$$

and using (10.49), we obtain

$$\left(\frac{\delta u_i^*}{\delta x_i}\right)_{\mathrm{P}} + \left(\frac{\delta \tilde{u}'_i}{\delta x_i}\right)_{\mathrm{P}} - \frac{\delta}{\delta x_i}\left(\frac{1}{a_{\mathrm{P}}(\boldsymbol{u}^{(m)})}\frac{\delta p'}{\delta x_i}\right)_{\mathrm{P}} = 0$$

or

$$\frac{\delta}{\delta x_i}\left(\frac{1}{a_{\mathrm{P}}(\boldsymbol{u}^{(m)})}\frac{\delta p'}{\delta x_i}\right)_{\mathrm{P}} = \left(\frac{\delta u_i^*}{\delta x_i}\right)_{\mathrm{P}} + \left(\frac{\delta \tilde{u}'_i}{\delta x_i}\right)_{\mathrm{P}}. \tag{10.51}$$

We see a familiar problem. The equation contains $\tilde{u}'_i$ and p', both of which are unknown. In the SIMPLE method, the problem is resolved in a bold move. The unknown terms with $\tilde{u}'_i$ are simply removed from (10.49) and (10.51). There is no fully satisfactory justification of such a drastic simplification except that the SIMPLE schemes have been found to converge in many cases. It has to be noted, however, that the convergence is slower than in the case of more sophisticated methods discussed next.

To summarize, an outer iteration of the SIMPLE method consists of the following substeps:

Step 1: Evaluate the coefficients a_{P} and $a_{\ell,\mathrm{P}}$ and the terms of the right-hand side of the linearized momentum equation (10.45) using the velocity and pressure fields from the previous iteration.

Step 2: Solve (10.45) to find the intermediate velocity u_i^*.

Step 3: Solve the approximate pressure equation (note the difference with the full equation (10.51)):

$$\frac{\delta}{\delta x_i}\left[\frac{1}{a_{\mathrm{P}}(\boldsymbol{u}^{(m)})}\frac{\delta p'}{\delta x_i}\right]_{\mathrm{P}} = \frac{\delta}{\delta x_i}[u_i^*]_{\mathrm{P}}. \tag{10.52}$$

Step 4: Update velocity and pressure fields as

$$u'_{i,\mathrm{P}} = -\frac{1}{a_{\mathrm{P}}(\boldsymbol{u}^{(m)})}\left(\frac{\delta p'}{\delta x_i}\right)_{\mathrm{P}},$$

$$u_{i,\mathrm{P}}^{(m+1)} = u_{i,\mathrm{P}}^* + u'_{i,\mathrm{P}}, \quad p_{\mathrm{P}}^{(m+1)} = p_{\mathrm{P}}^{(m)} + p'_{\mathrm{P}}. \tag{10.53}$$

Step 5: Check convergence and start the next iteration if needed.

The convergence of SIMPLE algorithm can be accelerated by using successive underrelaxation, that is, by updating pressure and velocity as

$$u_{i,\mathrm{P}}^{(m+1)} = u_{i,\mathrm{P}}^{*} + \alpha_u u_{i,\mathrm{P}}', \quad p_{\mathrm{P}}^{(m+1)} = p_{\mathrm{P}}^{(m)} + \alpha_p p_{\mathrm{P}}', \tag{10.54}$$

where $0 < \alpha_u \leq 1$ and $0 < \alpha_p \leq 1$ are the underrelaxation coefficients. Rather small value of α_p is recommended on the basis of empirical studies. It can be shown that the fastest convergence is obtained when $\alpha_u = 1 - \alpha_p$.

10.4.2 SIMPLEC and SIMPLER

The drastic simplification of completely neglecting the terms with $\tilde{u}_i'$ results in slow convergence of SIMPLE. Improved versions of the algorithm have been developed and are widely applied in modern CFD codes.

One such version is SIMPLEC (SIMPLE Consistent) proposed by van Doormaal and Raithby (1984). The unknown terms with $\tilde{u}_i'$ are approximated rather than neglected. To derive the approximation, we start with the definition (see (10.49))

$$\tilde{u}_{i,\mathrm{P}}' = -\frac{\sum_{\ell} a_{\ell,\mathrm{P}}(\boldsymbol{u}^{(m)}) u_{i,\ell}'}{a_{\mathrm{P}}(\boldsymbol{u}^{(m)})}. \tag{10.55}$$

As a next step, we approximate u_i' at every grid point by a weighted average of the values at neighboring points as

$$u_{i,\mathrm{P}}' \approx \frac{\sum_{\ell} a_{\ell,\mathrm{P}}(\boldsymbol{u}^{(m)}) u_{i,\ell}'}{\sum_{\ell} a_{\ell,\mathrm{P}}(\boldsymbol{u}^{(m)})} \tag{10.56}$$

or

$$\sum_{\ell} a_{\ell,\mathrm{P}}(\boldsymbol{u}^{(m)}) u_{i,\ell}' \approx u_{i,\mathrm{P}}' \sum_{\ell} a_{\ell,\mathrm{P}}(\boldsymbol{u}^{(m)}).$$

Substitution of the last expression into the right-hand side of (10.55) gives the desired approximation of $\tilde{u}_i'$:

$$\tilde{u}_{i,\mathrm{P}}' \approx -u_{i,\mathrm{P}}' \frac{\sum_{\ell} a_{\ell,\mathrm{P}}(\boldsymbol{u}^{(m)})}{a_{\mathrm{P}}(\boldsymbol{u}^{(m)})}. \tag{10.57}$$

It can be used in the velocity correction formula (10.49) and in the pressure equation (10.51), which become, respectively,

$$u'_{i,\mathrm{P}} = -\frac{1}{a_{\mathrm{P}}(\boldsymbol{u}^{(m)}) + \sum_{\ell} a_{\ell,\mathrm{P}}(\boldsymbol{u}^{(m)})} \left(\frac{\delta p'}{\delta x_i}\right)_{\mathrm{P}} \qquad (10.58)$$

and

$$\frac{\delta}{\delta x_i}\left[\frac{1}{a_{\mathrm{P}}(\boldsymbol{u}^{(m)}) + \sum_{\ell} a_{\ell,\mathrm{P}}(\boldsymbol{u}^{(m)})}\left(\frac{\delta p'}{\delta x_i}\right)\right]_{\mathrm{P}} = \frac{\delta}{\delta x_i}[u_i^*]_{\mathrm{P}}. \qquad (10.59)$$

The iteration strategy remains the same as in SIMPLE.

Another approach is to add extra corrector substeps to the SIMPLE routine. One such method called SIMPLER (SIMPLE Revised) was proposed by Patankar (1980). Every outer iteration is implemented as follows:

Step 1 and 2: Evaluate the coefficients and solve the approximate momentum equation, which is obtained by deleting the pressure gradient term $(\delta p^{(m)}/\delta x_i)_{\mathrm{P}}$ from (10.45). The result is the intermediate velocity $\hat{u}_i^{(m+1)}$ different from u_i^*.

Step 3: Solve the pressure equation

$$\frac{\delta}{\delta x_i}\left[\frac{1}{a_{\mathrm{P}}(\boldsymbol{u}^{(m)})}\frac{\delta p^{(m+1)}}{\delta x_i}\right]_{\mathrm{P}} = \frac{\delta}{\delta x_i}[\hat{u}_i^{(m+1)}]_{\mathrm{P}}. \qquad (10.60)$$

The solution is the pressure field at the next iteration level, so no further pressure correction is necessary.

Step 4: Substitute $p^{(m+1)}$ into the momentum equation (10.45) and solve them to obtain the new intermediate velocity u_i^*.

Step 5: Solve yet another pressure equation for p' identical to equation (10.52) of SIMPLE.

Step 6: Use p' to update velocity u_i^* to $u_i^{(m+1)}$ as in (10.53). Pressure is not updated.

Every iteration of SIMPLER requires roughly twice the number of operations of the original SIMPLE algorithm, since two elliptic equations for pressure and two sets of momentum equations have to be solved. This disadvantage is often outweighed by significantly faster convergence, so SIMPLER is, in general, more efficient.

Unlike the original SIMPLE algorithm, SIMPLEC and SIMPLER do not need underrelaxation.

10.4.3 PISO

The last version to be reviewed is the popular PISO (Pressure Implicit with Splitting Operators) algorithm. It was originally proposed by Issa (1986) for noniterative (with small time steps) solution of unsteady compressible flow problems. The version described here is the successful adaptation to the iterative solution of steady-state or fully implicit unsteady problems for incompressible flows. The algorithm can be viewed as a version of SIMPLE, in which we use two instead of one pressure corrections.

The first three steps of the outer iteration are the same as in SIMPLE:

Step 1: Evaluate a_P and $a_{\ell,P}$ and the terms of the right-hand side of (10.45) using the best available approximation of velocity and pressure fields, typically the values $u^{(m)}$ and $p^{(m)}$ from the previous iteration.

Step 2: Solve (10.45) to find the intermediate velocity u_i^*.

Step 3: Solve the pressure equation (10.52) to find the pressure correction p'.

The next step is different from that in the SIMPLE procedure. We assume that the corrected pressure and velocity are not the final outcomes of the iteration, but preliminary estimates to be corrected further:

Step 4: Compute the first corrections of velocity and pressure as

$$u_{i,P}' = -\frac{1}{a_P(u^{(m)})}\left(\frac{\delta p'}{\delta x_i}\right)_P,$$

$$u_{i,P}^{**} = u_{i,P}^* + u_{i,P}', \quad p_P^{**} = p_P^{(m)} + p_P'. \tag{10.61}$$

The rest of the procedure is entirely new. It consists of the second correction and the final update of velocity and pressure. We start with evaluating the term $\tilde{u}_i'$ omitted in the SIMPLE algorithm:

Step 5: Compute

$$\tilde{u}_{i,P}' = -\frac{\sum_\ell a_{\ell,P}(u^{(m)})u_{i,\ell}'}{a_P(u^{(m)})}. \tag{10.62}$$

The finally updated fields are

$$u_{i,P}^{(m+1)} = u_{i,P}^{**} + u_{i,P}'', \quad p_P^{(m+1)} = p_P^{**} + p_P'', \tag{10.63}$$

where $u''_{i,\mathrm{P}}$ is the second velocity correction given by the full formula (10.49), i.e.

$$u''_{i,\mathrm{P}} = \tilde{u}'_{i,\mathrm{P}} - \frac{1}{a_\mathrm{P}} \left(\frac{\delta p''}{\delta x_i} \right)_\mathrm{P}. \qquad (10.64)$$

Applying the incompressibility condition to (10.63), we obtain the second pressure equation

$$\frac{\delta}{\delta x_i} \left[\frac{1}{a_\mathrm{P}(\boldsymbol{u}^{(m)})} \frac{\delta p''}{\delta x_i} \right]_\mathrm{P} = \frac{\delta}{\delta x_i} [\tilde{u}'_i]_\mathrm{P}. \qquad (10.65)$$

The remaining steps of the iteration are, thus, the following:

Step 6: Solve the second pressure equation (10.65).

Step 7: Update velocity and pressure according to (10.63), (10.64).

Step 8: Check convergence and start the next iteration if needed.

The PISO algorithm does not require underrelaxation and allows numerically stable solutions of unsteady problems with larger time steps than SIMPLE and SIMPLEC. At the same time, similarly to SIMPLER, it requires approximately double amount of computations per iteration.

10.5 OTHER METHODS

10.5.1 Vorticity–Streamfunction Formulation for Two-Dimensional Flows

The methods discussed so far can be applied to arbitrary incompressible flows. If the flow is two-dimensional, with velocity and pressure fields being functions of only two space coordinates and time, $V = u(x, y, t)\boldsymbol{i} + v(x, y, t)\boldsymbol{j}$, $p = p(x, y, t)$, the governing equations become

$$\frac{\partial u}{\partial x} + \frac{\partial v}{\partial y} = 0 \qquad (10.66)$$

$$\frac{\partial u}{\partial t} + u\frac{\partial u}{\partial x} + v\frac{\partial u}{\partial y} = -\frac{1}{\rho}\frac{\partial p}{\partial x} + v\nabla^2 u \qquad (10.67)$$

$$\frac{\partial v}{\partial t} + u\frac{\partial v}{\partial x} + v\frac{\partial v}{\partial y} = -\frac{1}{\rho}\frac{\partial p}{\partial y} + v\nabla^2 v. \qquad (10.68)$$

Formulation: The problem can be solved in a new simpler way if instead of the *primitive variables* V and p, we use *streamfunction* and *vorticity*. The streamfunction $\psi(x, y, t)$ can always be introduced for an incompressible two-dimensional flow so that

$$u = \frac{\partial \psi}{\partial y}, \quad v = -\frac{\partial \psi}{\partial x}. \tag{10.69}$$

The name of ψ refers to the fact that the velocity vector at every point of space and every moment of time is tangential to the line ψ =const. passing through this point. The lines ψ =const. thus represent the *streamlines* of the flow.

Vorticity is a vector field (in general, three-dimensional) $\boldsymbol{\omega} = \nabla \times V$. For a two-dimensional flow in the x–y-plane, only the z-component of vorticity is nonzero:

$$\omega = \frac{\partial v}{\partial x} - \frac{\partial u}{\partial y}. \tag{10.70}$$

An important result of using the streamfunction instead of the velocity components is that the incompressibility condition (10.66) is satisfied automatically. This is easy to verify substituting (10.69) into (10.66).

We can now transform the system of governing equations (10.66)–(10.68). Applying the operator $(\nabla \times)$ to the momentum equation, which in the two-dimensional case means calculating $(\partial / \partial x)(10.68)–(\partial / \partial y)\,(10.67)$, and taking into account the incompressibility (10.66) and the mathematical identity

$$\frac{\partial}{\partial y}\left(\frac{\partial p}{\partial x}\right) - \frac{\partial}{\partial x}\left(\frac{\partial p}{\partial y}\right) = 0,$$

we obtain the *transport equation for vorticity*:

$$\frac{\partial \omega}{\partial t} + u\frac{\partial \omega}{\partial x} + v\frac{\partial \omega}{\partial y} = \nu\left(\frac{\partial^2 \omega}{\partial x^2} + \frac{\partial^2 \omega}{\partial y^2}\right) \tag{10.71}$$

or, in a short form,

$$\frac{D\omega}{Dt} = \nu\nabla^2\omega. \tag{10.72}$$

The second equation is obtained by substituting (10.69) into (10.70). This results in the connection between the vorticity and streamfunction:

$$\nabla^2\psi = -\omega. \tag{10.73}$$

Note that (10.72) and (10.73) form a coupled system. The coupling occurs via appearance of velocity components in the material derivative in the left-hand side of (10.72).

We have replaced the original system of three partial differential equations (10.66) and (10.68) by just two equations (10.72) and (10.73). Furthermore, the incompressibility condition is satisfied automatically by the velocity field (10.69). The pressure field does not explicitly appear in (10.72) and (10.73) and, in principle, is not needed for the solution. If, for some reason, knowledge of pressure field is required, p can be evaluated after the velocity field is found by solving the pressure equation (10.5).

The system (10.72) and (10.73) requires boundary conditions on ψ and ω. For the streamfunction, imposing physically plausible conditions is not difficult. One has to write the proper boundary conditions for the velocity components and use (10.69) to represent them as conditions for ψ and its first derivatives. The situation is more difficult in the case of vorticity. There are no natural conditions on ω. They can, however, be derived from the conditions on ψ applying Eq. (10.73) at the boundary points. Since this typically results in expressions containing second derivatives, special numerical treatment is required.

Methods of Solution: The vorticity transport equation (10.72) and Poisson equation (10.73) can be discretized using the schemes developed for parabolic and elliptic equations, respectively. For example, we can apply finite difference discretization and use the simple explicit scheme for (10.72):

$$\frac{\omega_{i,j}^{n+1} - \omega_{i,j}^n}{\Delta t} = -u_{i,j}^n \frac{\omega_{i+1,j}^n - \omega_{i-1,j}^n}{2\Delta x} - v_{i,j}^n \frac{\omega_{i,j+1}^n - \omega_{i,j-1}^n}{2\Delta y}$$
$$+ v \left[\frac{\omega_{i+1,j}^n - 2\omega_{i,j}^n + \omega_{i-1,j}^n}{(\Delta x)^2} + \frac{\omega_{i,j+1}^n - 2\omega_{i,j}^n + \omega_{i,j-1}^n}{(\Delta y)^2} \right].$$

$$(10.74)$$

For (10.73), the standard five-point scheme of the second order gives

$$\frac{\psi_{i+1,j}^{n+1} - 2\psi_{i,j}^{n+1} + \psi_{i-1,j}^{n+1}}{(\Delta x)^2} + \frac{\psi_{i,j+1}^{n+1} - 2\psi_{i,j}^{n+1} + \psi_{i,j-1}^{n+1}}{(\Delta y)^2} = -\omega_{i,j}^{n+1}. \quad (10.75)$$

We also need discretized expressions for u and v. Keeping the second order of approximation, we use

$$u_{i,j}^{n+1} = \frac{\psi_{i,j+1}^{n+1} - \psi_{i,j-1}^{n+1}}{2\Delta y}, \quad v_{i,j}^{n+1} = -\frac{\psi_{i+1,j}^{n+1} - \psi_{i-1,j}^{n+1}}{2\Delta x}. \quad (10.76)$$

The system of coupled equations (10.74)–(10.76) can be solved by a multistep procedure conceptually similar to the projection method developed in Section 10.3 for explicit methods. Each time step consists of the following substeps:

Step 1: Use the known values of velocity, vorticity, and streamfunction at $t = t^n$ to solve (10.74) and find $\omega_{i,j}^{n+1}$ at the interior points of the computational domain.

Step 2: Solve the Poisson equation (10.75) together with the boundary conditions on ψ to find $\psi_{i,j}^{n+1}$. This can be done by a direct or iterative method.

Step 3: Find values $u_{i,j}^{n+1}$ and $v_{i,j}^{n+1}$ of the velocity components at the new time layer using (10.76) and velocity boundary conditions.

Step 4: Update $\omega_{i,j}^{n+1}$ at the boundary points using $\psi_{i,j}^{n+1}$ and the discretization of the vorticity boundary conditions derived from (10.73).

The scheme (10.74)–(10.76) has the truncation error $\text{TE} = O\left((\Delta x)^2, (\Delta y)^2, \Delta t\right)$ and is stable if

$$\Delta t \le \frac{1}{2v}\left[\frac{1}{(\Delta x)^2} + \frac{1}{(\Delta y)^2}\right]^{-1} \quad \text{and } \Delta t \le \frac{2v}{(u_{i,j}^2 + v_{i,j}^2)}. \tag{10.77}$$

The second of these conditions must be satisfied at all grid points (x_i, y_j). This means that the maximum of $u^2 + v^2$ must be either estimated a priori or evaluated at every time layer with subsequent corresponding adjustment of Δt.

The algorithm described above is only one of many possible solutions. Other methods may use implicit schemes for the vorticity transport equation or a pseudo-transient representation of the Poisson equation (10.73).

If the flow is steady state, the vorticity equation becomes

$$\mathbf{V} \cdot \nabla\omega - v\nabla^2\omega = 0. \tag{10.78}$$

The resulting system (10.73) and (10.78) has to be solved as an equilibrium problem. Iterative methods for systems of coupled nonlinear equations, such as the method of sequential iterations discussed in Section 8.4.3, can be applied.

10.5.2 Artificial Compressibility

Yet another approach to solution of incompressible flow equations is to modify them so that the methods developed for compressible flows can be applied. The approach can be used to compute steady-state flows. Introducing fictitious time τ and adding the term $\partial V / \partial \tau$ to the left-hand side of the momentum equation, we also replace the incompressibility condition

$$\nabla \cdot V = 0 \tag{10.79}$$

by

$$\frac{\partial p}{\partial \tau} + a^2 \nabla \cdot V = 0. \tag{10.80}$$

Because in the limit $\tau \to \infty$ p converges to a steady field independent of τ, (10.80) converges to (10.79). The incompressibility is, thus, satisfied by the final steady state. Equation (10.80) reminds the continuity equation in the case of isentropic compressible flows, with a playing the role of the speed of sound. For this reason, this method introduced by Chorin (1967) bears the name of the *artificial compressibility* approach.

BIBLIOGRAPHY

Caretto, L.S., Gosman, A.D., Patankar, S.V., and Spalding, D.B. (1973). Two calculation procedures for steady, three-dimensional flows with recirculation. In: *Proceedings of the 3rd International Conference on Numerical Methods in Fluid Mechanics*, Lecture Notes in Physics, vol. **19** (ed. H. Cabannes and R. Temam), 60–68. Berlin, Heidelberg: Springer-Verlag.

Chorin, A.J. (1967). A numerical method for solving incompressible viscous flow problems. *J. Comput. Phys.* **2**: 12–26.

Chorin, A.J. (1968). Numerical solution of the Navier–Stokes equations. *Math. Comput.* **22**: 745–762.

Ferziger, J.H. and Perić, M. (2001). *Computational Methods for Fluid Dynamics*, 3e. New York: Springer.

Fletcher, C.A.J. (1996). *Computational Techniques for Fluid Dynamics: Specific Techniques for Different Flow Categories*, vol. **2**. Springer.

Ghia, U., Ghia, K.N., and Shin, C.T. (1982). High-Re solutions for incompressible flow using the Navier–Stokes equations and a multigrid method. *J. Comput. Phys.* **48**: 387–411.

Harlow, F.H. and Welch, J.E. (1965). Numerical calculation of time dependent viscous incompressible flow with free surface. *Phys. Fluids* **8**: 2182–2189.

Issa, R.I. (1986). Solution of the implicitly discretised fluid flow equations by operator-splitting. *J. Comput. Phys.* **62**: 40–65.

Morinishi, Y., Lund, T.S., Vasilyev, O.V., and Moin, P. (1998). Fully conservative higher order finite difference schemes for incompressible flows. *J. Comput. Phys.* **143**: 90–124.

Patankar, S.V. (1980). *Numerical Heat Transfer And Fluid Flow*, 1e. London: Taylor & Francis.

Patankar, S.V. and Spalding, D. (1972). A calculation procedure for heat, mass and momentum transfer in three-dimensional parabolic flows. *Int. J. Heat Mass Transf.* **15**: 1787–1807.

Temam, R. (1969). Sur l'approximation de la solution des equations de Navier–Stokes par la méthode des pas fractionnaires. *Arch. Ration. Mech. Anal.* **32**: 135–153 and **33**: 377–385.

Van Doormaal, J.P. and Raithby, G.D. (1984). Enhancement of the SIMPLE method for predicting incompressible fluid flows. *Numer. Heat Transfer, Part B* **7**: 147–163.

Yanenko, N.N. (1971). *The Method of Fractional Steps*. Berlin: Springer-Verlag.

PROBLEMS

1. What is the difference between colocated and staggered grid arrangements? Discuss the comparative advantages of each approach.

2. If your course involves exercises with a CFD code, study the manual to determine whether the discretization uses staggered or colocated grids. Does the manual say anything about the exact mass conservation when the schemes are applied to incompressible flows?

3. Derive the approximations (10.17) and (10.18) of the pressure gradient terms.

4. Describe the staggered grid arrangement of finite difference and finite volume structured grids in the three-dimensional case.

5. Write the finite difference formulas for the nonlinear term of the x-momentum equation for a two-dimensional flow discretized on a staggered Cartesian uniform grid. The discretization must be of the second order and based on central differences and linear interpolations. Solve the problem for the equation written in:

a) Conservation form

b) Nonconservation form.

6. Write the complete set of formulas for the finite difference discretization of two-dimensional Navier–Stokes equation system for incompressible viscous unsteady flow with zero body forces. Use the projection method, simple explicit scheme of the first order for time discretization, and approximation of the second order based on the central differences for spatial discretization. Use a colocated Cartesian uniform grid. Write the discretization that conserves mass exactly.

7. The simple explicit scheme and the projection method (see Section 10.3.1) are applied to compute the flow of an incompressible viscous fluid in a rectangular box $0 < x < A$, $0 < y < B$, $0 < z < C$. Write the boundary conditions for pressure. Why do we need them? Would we still need them if the flow were compressible?

8. For the flow in Problem 7, write the boundary conditions for pressure when the flow is incompressible and inviscid and there is no body force.

9. Modify the predictor–corrector formulas (10.27)–(10.30) for the method that uses the second-order Adams–Bashforth time-integration scheme (see Section 7.4.1) instead of the simple explicit scheme.

10. Show that the decomposition (10.37) of the nonlinear term is correct. Use direct substitution of (10.36) into the expression for one component of vector N.

11. Consider the explicit, fully implicit, and semi-implicit methods discussed in Section 10.3. Order them by the amount of computations required at every time step. Explain your answer.

12. Develop the formula (10.43) for the case of the x-momentum equation of a two-dimensional steady-state incompressible flow discretized on a structured uniform finite difference grid. Use central differences of the second-order and staggered grid arrangement. Derive expressions for all the coefficients $a_P(u)$, $a_{\ell,P}(u)$ $Q_P(u)$.

13. Repeat the solution of Problem 12 for the case when the flow is unsteady and the problem is solved using the fully implicit time discretization (the simple implicit scheme).

14. Compare the four algorithms described in Section 10.4 (SIMPLE, SIMPLER, SIMPLEC, PISO) by the amount of computations required for one iteration. Does the smaller amount necessarily mean smaller time needed to solve the CFD problem?

15. If your course involves exercises with a CFD code, study the manual to determine which of the projection schemes discussed in Section 10.4 (SIMPLE, SIMPLEC, SIMPLER, PISO) are implemented. Are there other schemes available for incompressible flows? Does the manual provide any recommendations concerning the choice of the scheme?

Programming exercise The lid-driven cavity flow has long been used as a benchmark for numerical methods. The simplest version is the two-dimensional flow of an incompressible fluid in a square cavity $x < 0 < L$, $0 < y < L$. The walls at $x = 0$, $x = L$, and $y = 0$ are stationary, while the wall at $y = L$ (the lid) is moving with constant velocity U in the tangential direction. If your course involves exercises with a CFD code, calculate the flow at several values of the Reynolds number $Re = UL/\nu$. For example, try $Re = 10, 100$, and 1000. Experiment with different projection schemes and different grid sizes. Compare your results with the results available in literature, for example, in the paper of Ghia et al. (1982).

Part III

ART OF CFD

11

TURBULENCE

11.1 INTRODUCTION

Most flows that we observe in nature, technology, and everyday life are turbulent. As examples, we name, quite arbitrarily, flows in the Earth's atmosphere (the weather) and liquid core (this flow causes the terrestrial magnetic field by the dynamo effect), wakes behind moving bodies (such as airplanes and automobiles), a flow within a cylinder of an internal combustion engine, and a flow in a cup of coffee, to which stirring by a spoon is applied to facilitate mixing and dissolution of sugar. These flows are generated by different mechanisms and occur in domains of different sizes ranging from the size of a coffee cup to the size of the Earth. Still, they have one feature in common. The Reynolds number

$$Re \equiv UL/\nu \tag{11.1}$$

is sufficiently large, so that the flow has turbulent form. In (11.1) U and L are the typical scales of velocity and length in the flow, and $\nu = \mu/\rho$ is the kinematic viscosity of the fluid.

Essential Computational Fluid Dynamics, Second Edition. Oleg Zikanov.
© 2019 John Wiley & Sons, Inc. Published 2019 by John Wiley & Sons, Inc.
Companion Website: www.wiley.com/go/zikanov/essential

11.1.1 A Few Words About Turbulence

A complete, rigorous, and brief definition of turbulence seems impossible. The usual approach is to define turbulence through its following main traits.

Irregularity, Time Dependence, and Three Dimensionality: A turbulent flow field may have a component with smooth large-scale regular structure. It is usually identified as the mean flow with velocity $\langle u \rangle$, where brackets stand for statistical averaging or some form of time or space averaging (a more precise definition of the mean flow is given in Section 11.3). In addition to the mean component, however, the flow field necessarily contains fluctuations:

$$u' = u - \langle u \rangle, \tag{11.2}$$

which are irregular, time dependent, and three-dimensional. Figure 11.1 shows an example of a turbulent flow field and the corresponding mean flow.

(Pseudo-)chaoticity: A seemingly obvious conclusion can be obtained from a plain visual observation of a turbulent flow. The flow behaves chaotically.[1] The change of flow pattern in space and time seems random. However, a turbulent flow is still a solution of the Navier–Stokes equations, which do not contain any stochastic terms. The flow should follow a fully predictable evolution determined by the equations and the boundary and initial conditions. The reason for the observed pseudorandomness is that small perturbations that are constantly added to velocity and pressure fields are enhanced exponentially in time in a turbulent solution. Practical CFD analysis usually disregards the subtleties and treats the turbulent fluctuations as truly chaotic.

Broad Range of Length and Time Scales: A turbulent flow consists of motions with typical length and time scales that continuously fill a very broad range. This phenomenon can be explained by the nonlinearity of the Navier–Stokes equations or, from another viewpoint, by hydrodynamic instabilities and interaction between flow structures. It is formalized using the concept of energy cascade introduced by L.F.

[1]Drawings indicating such observations were made by Leonardo da Vinci. These drawings have been repeatedly used by the turbulence research community in a bid to add age and respectability to the discipline.

Richardson (1922). According to the concept, large flow structures (also called *eddies* or *vortices* to stress the role of vorticity in their dynamics) constantly generated by the hydrodynamic instability of the flow are unstable themselves and generate smaller eddies. The smaller eddies are also unstable and break into even smaller ones. The kinetic energy is thus constantly transferred from large-scale to small-scale motions. The cascade stops at the level where the structures are so small that strong velocity gradients associated with them lead to complete dissipation of transferred kinetic energy into heat. The pattern is schematically represented in Figure 11.2.

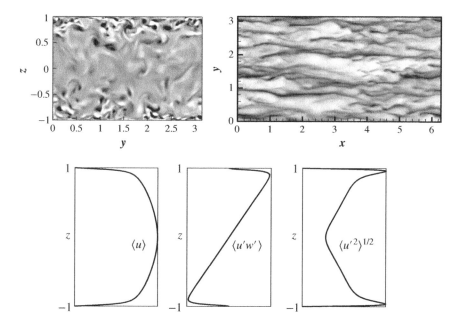

Figure 11.1 Example of a computed turbulent flow in a channel. The Reynolds number based on the channel width and the channel-averaged velocity is 13,333. x, y, and z are the nondimensional coordinates obtained from the physical coordinates by scaling with the half of the channel width. The top row shows instantaneous snapshots of the x-component of vorticity $\omega = \nabla \times u$ in the channel cross section (on the left-hand side) and of the x-component of velocity in the wall-parallel plane at $z = 0.95$ (on the right-hand side). The bottom row from left to right shows profiles of horizontally and time-averaged characteristics: mean flow velocity $\langle u \rangle$, turbulent Reynolds stress $\langle u'w' \rangle$, and root-mean-square amplitude of velocity fluctuation component $\langle u'^2 \rangle^{1/2}$. Source: Courtesy of D. Krasnov, Ilmenau University of Technology.

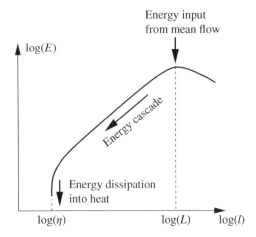

Figure 11.2 Schematic representation of the distribution of energy of velocity fluctuations over the length scales in a turbulent flow. L is the typical size of the largest and most energetic fluctuations. η is the Kolmogorov scale, the typical size of the smallest fluctuations, in which the viscous dissipation into heat primarily occurs.

The cascade concept and certain physical assumptions led A.N. Kolmogorov (1941) to develop the phenomenological picture of turbulence, which remains one of the most profound results of the turbulence theory. Essentially for us, the Kolmogorov phenomenology also provides the basis for understanding the numerical methods used to compute turbulent flows. In particular, it allows us to estimate the ranges of active scales of the flow. The estimates are strictly valid for isotropic homogeneous turbulence but are qualitatively correct in general case. The typical length and time scales of the smallest eddies η and τ are related to the typical length and time scales of the largest eddies as

$$\eta/L \sim Re^{-3/4}, \quad \tau/T \sim Re^{-1/2}. \tag{11.3}$$

The kinetic energy is distributed very unevenly over the active length scales. As illustrated in Figure 11.2, the small-scale fluctuations have much lower (orders of magnitude lower) energy than the large-scale ones.

Strong Mixing: The turbulent fluctuations provide a much more efficient mechanism of mixing than molecular diffusion. This concerns all kinds of mixing: of momentum, internal energy (in flows with heat transfer), or dissolved admixture. The actual mixing – that is, the transport of any of these physical fields between two neighboring fluid particles – is still accomplished by molecular diffusion. The role of turbulent fluctuations is to add an intensive stirring action that brings particles with different concentrations of the mixed field into contact with each other.

Coherent Structures: One of the significant advances of the turbulence research is the understanding of the role played by localized *coherent structures*, which have identifiable shapes, are repeatedly generated by the flow, and persist for relatively long times. The structures have various typical sizes and take various forms, such as regions of strong vorticity or high- or low-velocity streaks (see the top two plots in Figure 11.1 for an illustration). They play an important, sometimes dominant, role in turbulent mixing. We should note that the existence of coherent structures does not annul the pseudostochastic nature of turbulence. Size, time of occurrence, location, and specific features of the shape of the structures follow the pseudorandom pattern.

This description of the traits of turbulence is deliberately brief. Our intention is merely to introduce the concepts needed for the following discussion of computational methods and to draw the reader's attention to the fascinating and challenging subject of turbulence. Several of the many good books available for a thorough study of the subject are listed at the end of this chapter.

11.1.2 Why Is the Computation of Turbulent Flows Difficult?

Accurate numerical solution describing a turbulent flow is either difficult or impossible. Before we start explaining why this is so, it is necessary to say that, so far, nobody has found a way to describe and predict turbulent flows mathematically in the form of explicitly written solutions of Navier–Stokes or some other equations. A complete characterization of a turbulent flow by experimental techniques is also impossible. The situation with numerical methods seems, at first glance, more promising. Here, at least, we can rely on constantly growing computational power. The optimism is, unfortunately, not fully justified, as shown by the following simple analysis.

Let us estimate the size of the computational grid needed to accurately calculate a turbulent flow. It is obvious that the grid step should not be larger than the size η of the smallest turbulent eddy existing in the flow. If this condition is not satisfied, the velocity fluctuations corresponding to this eddy "fall through the gaps" between the grid points. They are not detected and their effect on the flow is ignored. We assume, for simplicity, that the grid step is about the same in every direction and require that $\Delta x \sim \Delta y \sim \Delta z \sim \eta$.

The computational domain does not have to cover the entire flow domain. Using artificial boundary conditions, we can limit computations to a fraction of the flow (see the discussion in Section 2.7). However, to reproduce the flow dynamics, the computational domain has to be at least several times larger than the typical size of the largest turbulent eddies. For simplicity, we assume that the dimensions of the computational domain are $L_x \sim L_y \sim L_z \sim L$. The number of the grid points in every direction is estimated using the Kolmogorov scaling (11.3) as

$$N_x \sim N_y \sim N_z \sim N = L/\eta \sim Re^{3/4}. \tag{11.4}$$

In three dimensions, the total size of the computational grid is

$$N^3 \sim Re^{9/4}. \tag{11.5}$$

This appears a stunningly large number if we consider that the Reynolds number is usually large, anywhere between 10^4 and 10^{12} or even higher. Even at the lower end, the required grid consists of 10^9 points or cells.

An estimate of the required computational effort should also take into account that the time step has to be not larger than the typical time scale of the smallest turbulent eddies τ. The result, discussed, for example, in Pope (2000), is that even in the simplest case of an incompressible flow without boundary layers, the product of the number of grid nodes and the number of time steps needed for a simulation representing a typical evolution of a turbulent flow is approximately $160Re^3$.

The number of floating point operations per node per time step varies, depending on the effectiveness of the method, but is unlikely to be less than 100. Assuming this lower bound, we obtain the final estimate that the total number of floating point operations needed for a meaningful simulation of a turbulent flow is, in any case, not smaller than $\sim 10^4 Re^3$. To complete the discussion, we consider an example of a moderately high Reynolds number $Re = 10^6$ (exceeded in many common flows). The total number of operations is, at least, $\sim 10^{22}$. Assuming that we work on a hypothetical multiprocessor workstation with peak performance 1 Tflops (10^{12} floating point operations per second), the computations would take at least 10^{10} s, i.e. more than 300 years.

The example illustrates the main difficulty of numerical simulations of turbulent flows. The requirement of computational accuracy leads to unrealistically large computational grids. Even if we adopt the most optimistic predictions of the future growth of computational power, direct computations of realistic turbulent flows will remain unfeasible for a long time.

11.1.3 Overview of Numerical Approaches

There are different ways to overcome the computational challenge and acquire useful information on turbulent flows from numerical solutions. Let us overview the landscape before moving to descriptions of particular methods.

The approaches to numerical analysis of turbulence can be divided into two groups: *simulations* and *modeling*. In simulations, we calculate an actual realization of the flow (simulate it). The methods of this group are the direct numerical simulation (DNS) and large eddy simulation (LES) methods. The DNS, which we discuss in Section 11.2, is the most honest approach. We solve the Navier–Stokes equations without any modifications or modeling assumptions. The result is a complete picture of the evolution of a time-dependent flow field $u(x, t)$, $p(x, t)$. This is the approach we have assumed to follow so far in this book. The disadvantage of this approach in the case of turbulent flows is that, as we have just discovered, the requirement of accurate approximation of the flow features at small length scales leads to unrealistically large grids.

In the LES approach introduced in Section 11.4, equations for spatially filtered variables $\overline{u}(x, t)$, $\overline{p}(x, t)$ are solved. The variables represent the time-dependent behavior of the flow features with large and moderate length scales. The effect of small-scale fluctuations appears in the equations in the form of additional terms, which cannot be calculated directly and have to be substituted by model approximations.

In modeling, we do not try to compute an actual realization of the flow. Instead, the problem is recast as a system of equations for mean flow quantities, such as the mean velocity and pressure $\langle u \rangle$, $\langle p \rangle$, Reynolds stresses $\langle u_i u_j \rangle$, and so on. The results correspond to our expectations of the flow characteristics that would be obtained after averaging over many realizations. This approach is called the Reynolds-averaged Navier-Stokes (RANS) method. As discussed in Section 11.3, the method is computationally very efficient. However, the results are often inaccurate because of the large error introduced by the approximations of the additional terms that appear in the RANS equations.

Which of the three methods – DNS, LES, or RANS – should we use? The answer depends on the purpose of the analysis and the kind of the flow. The factors to be considered are:

Accuracy: Extra terms that appear in the LES and RANS equations have to be approximated. As we will discuss in Sections 11.3 and 11.4, the models used for the approximation are inherently imprecise and based on rather weak physical assumptions. They introduce the model error, whose magnitude varies depending on the type of the model, grid step, and flow characteristics. In general, the amplitudes of the error are related as $0 = \epsilon_{model,DNS} < \epsilon_{model,LES} < \epsilon_{model,RANS}$.

Level of description: The three approaches describe the flow on different levels. In DNS, we find a complete flow field $u(x, t), p(x, t)$, which can be used to find all the desired characteristics related to the behaviors at all length scales. LES also produces a flow field, but the computed variables $\bar{u}(x, t), \bar{p}(x, t)$ do not provide information on motions at small scales. At last, RANS methods only produce the mean flow $\langle u \rangle, \langle p \rangle$. Characteristics of turbulent fluctuations remain unknown, except for the basic properties, such as their kinetic energy and dissipation rate, which are estimated (with an error) by the turbulence models.

Computational cost: The computational cost is highest for DNS and lowest for RANS. This is in accordance with the size of the structures that have to be resolved by a computational grid: the smallest turbulent eddies in DNS and large-scale features of mean flow in RANS. The computational cost of LES varies with the model and the desired accuracy. In general, it lies between the costs of DNS and RANS.

It is easy to see that each of the three methods is optimal for applications in certain areas. The ability of DNS to accurately simulate the complete flow behavior makes it a powerful and irreplaceable tool for fundamental turbulence research. Of course, high computational cost means that DNS analysis is limited to flows with small-to-moderate Reynolds numbers. Also, as we explain in Section 11.2, the method is much better suited and usually applied for flow in simple domains. The majority of actual flows in technology and environment are beyond reach for DNS.

The practical engineering analysis is traditionally conducted using RANS models. The advantages are obvious. Mean flow characteristics are often sufficient for engineering problems. The computations can be conducted in a few hours or even minutes. Unfortunately, RANS provides no data on turbulent fluctuations and predicts the mean flow with significant error.

The LES approach occupies an intermediate position between DNS and RANS. It can be used for fundamental science, albeit with due awareness of the model error it introduces. The trend of modern CFD development is the growing role of LES in practical engineering computations, often in combination with RANS. Although more expensive to conduct than RANS (this problem is being gradually taken care of by increasing computer power), LES analysis adds important, sometimes crucial, information on large- and moderate-scale fluctuations and is more accurate than RANS in predicting mean flow properties.

11.2   DIRECT NUMERICAL SIMULATION (DNS)

In principle, any numerical method – for example, finite difference or finite volume – can be used in DNS. Another popular choice is the spectral method (see Section 3.3.1). While limited to domains of simple geometric shapes, such as a rectangular box, a cylinder, or a sphere, the spectral methods are computationally very efficient. This makes them attractive for analysis of fundamental properties of turbulent flows, where the use of a simple flow domain is allowed and even recommended. For this reason and also because the spectral methods have not received much attention in this book so far, the discussion in this section is focused on them.

11.2.1   Homogeneous Turbulence

The first DNS of a flow with realistic turbulent features appeared in the paper by Orszag and Patterson (1972). The numerical approach pioneered in that paper still serves, with modifications, as an important tool of fundamental turbulence research. The method is based on the assumption that turbulent fluctuations are *spatially homogeneous*, which means that their statistically averaged properties are the same at every point of the flow domain. The assumption excludes flows with solid walls and other realistic boundaries but still leaves plenty of interesting behavior to look at. The situation nearest to homogeneous is achieved in wind tunnel experiments, where turbulence is generated by a uniform honeycomb or grid positioned at the entrance of the test section. As an approximation, the state of turbulence in many other flows can be considered homogeneous, if we consider a small zone far from the walls.

In DNS, a homogeneous turbulent flow is simulated by a flow in a rectangular box of dimensions $L_x \times L_y \times L_z$ with periodic (cyclic) boundary conditions in all three directions:

$$u(x,t) = u(x + L_x e_x, t) = u(x + L_y e_y, t) = u(x + L_z e_z, t). \qquad (11.6)$$

The periodicity allows us to apply the efficient Fourier spectral method introduced by Orszag and Patterson and later developed by others, most notably by Rogallo (1981). The velocity field is approximated by a three-dimensional Fourier series:

$$u(x,t) = \sum_k \hat{u}(k,t) e^{ik \cdot x}, \qquad (11.7)$$

where $k = k_x e_x + k_y e_y + k_z e_z$ is the wavenumber vector and $\hat{u}(k,t)$ is the complex-valued expansion coefficient. The exponential term is the shorthand notation $e^{ik \cdot x} \equiv e^{ik_x x} e^{ik_y y} e^{ik_z z}$, where, for example, $e^{ik_x x} = \cos(k_x x) + i\sin(k_x x)$.

The wavenumbers are related to the dimensions of the box, which are also the periodicity lengths, as

$$k_x = \frac{2\pi m_x}{L_x}, \quad k_y = \frac{2\pi m_y}{L_y}, \quad k_z = \frac{2\pi m_z}{L_z}. \qquad (11.8)$$

The sum in (11.7) is over the values of k_x, k_y, and k_z corresponding to the integer indices $m_x = -N_x/2, \ldots, N_x/2$, $m_y = -N_y/2, \ldots, N_y/2$, and $m_z = -N_z/2, \ldots, N_z/2$. Since $u(x,t)$ is a real vector function, the complex Fourier expansion coefficients must satisfy the conjugate symmetry $\hat{u}(k,t) = \hat{u}^*(-k,t)$.

An important and nontrivial question concerns the values of N_x, N_y, and N_z. How many terms of the expansion (11.7) should we use in a simulation? One necessary condition is that (11.7) accurately reproduces the turbulent fluctuations of the smallest length scale, which is the Kolmogorov scale η. The numerical resolution of a spectral method based on (11.7) is determined by the smallest wavelength used in the expansion. Specifically, the resolution approximately corresponds to the half of such a wavelength in each direction, i.e. to

$$\Delta x \sim \frac{L_x}{N_x} = \frac{\pi}{k_{x,max}}, \quad \Delta y \sim \frac{L_y}{N_y} = \frac{\pi}{k_{y,max}}, \quad \Delta z \sim \frac{L_z}{N_z} = \frac{\pi}{k_{z,max}}, \qquad (11.9)$$

where

$$k_{x,max} = 2\pi \frac{N_x}{2} \frac{1}{L_x} = \frac{\pi N_x}{L_x},$$

etc. are the maximum wavenumbers in the expansion. The small-scale resolution requirement is $\Delta x \sim \Delta y \sim \Delta z \sim \eta$, which implies $k_{x,max} \sim k_{y,max} \sim k_{z,max} \sim \pi/\eta$. Numerical studies have shown that this formula is a slight overestimation. Unless the dynamics of turbulent fluctuations with the length scales approaching η is of special interest, the resolution requirement can be relaxed to

$$k_{x,max} \sim k_{y,max} \sim k_{z,max} \sim \frac{1.5}{\eta}. \tag{11.10}$$

There is another numerical resolution requirement related to the fact that the periodic box is an artificial construction. The periodicity introduces unphysical perfect correlations over the distances L_x, L_y, and L_z. Meaningful flow dynamics can, therefore, only be observed if the box dimensions are significantly (at least seven to eight times) larger than the typical size of the largest turbulent flow structures.

Combining the two requirements and using the Kolmogorov scaling $L/\eta \sim Re^{3/4}$, we see that the numerical resolution is fully determined by the Reynolds number. The first periodic box DNS were conducted by Orszag and Patterson (1972) using $N_x = N_y = N_z = 32$ for a flow with a very low Reynolds number. At the time of writing this book, ambitious simulations use values of N_x, N_y, and N_z between 4,096 and 16,348 and achieve the Reynolds numbers comparable to those in high-Re laboratory experiments.

A peculiar aspect of the periodic box turbulence is that the flow does not have a mechanism to sustain itself. The energy is constantly transferred into small-scale structures and dissipated into heat, so the flow decays. The decay is a legitimate and interesting situation, which is similar to the situation in wind tunnel experiments, except that the role of the distance from the turbulence-generating grid in a wind tunnel is played by time in DNS. Alternatively, we can maintain the flow in a statistically steady state by adding energy to large-scale motions. This can be done in a purely artificial way (the artificial forcing) or by imitating a natural mechanism, for example, by imposing a mean flow with constant shear.

The implementation of the Fourier spectral method in the case of a periodic box flow is not difficult and, in its main features, is similar to the procedure outlined in Section 3.3.1 for a one-dimensional model equation. We substitute the expansion (11.7) and the similar expansion for pressure into the Navier–Stokes equations and derive a system of ordinary differential equations for the Fourier coefficients $\hat{u}(k, t)$. The equations are solved using one of the established time-integration schemes, for example, the Runge–Kutta or Adams–Bashforth method.

The main difficulty and computational challenge of the procedure are due to the nonlinearity of the momentum equation. To see the problem, let us consider one product in the nonlinear term, for example, uv. After substituting the Fourier expansions for u and v, we obtain

$$\sum_{k_1} \sum_{k_2} \hat{u}(k_1, t)\hat{v}(k_2, t)e^{\iota(k_1 \cdot x + k_2 \cdot x)}, \tag{11.11}$$

which should be expressed as the Fourier expansion over the same basis as the linear terms of the equation, i.e. as

$$\sum_k \hat{f}(k, t)e^{\iota k \cdot x},$$

with $\hat{f}$ being some expression in terms of $\hat{u}$ and $\hat{v}$. This can be done by direct calculation of the convolution sums in (11.11), but at the cost of approximately N^6 operations, where N is an estimate of N_x, N_y, and N_z. Much more efficient way is to compute the sums (11.7) for u and v at certain grid points, calculate the product uv at these points, and use the resulting field to compute its Fourier coefficients $\hat{f}(k, t)$. With appropriate choice of grid points, the evaluation of grid point values and Fourier coefficients can be performed using direct and inverse fast Fourier transforms. Each transform requires approximately $N^3 \log N$ operations. The linear part of the momentum equations requires $\sim N^3$ operations per time step. If an explicit time discretization is used, the total number of operations per time step is, thus, estimated as $\sim N^3 \log N$.

There are other nontrivial aspects of the procedure, which we do not consider here. An interested reader can find a detailed description of the method in research literature, for example, in the papers listed at the end of this chapter.

11.2.2 Inhomogeneous Turbulence

If turbulence is inhomogeneous in one or several directions – for example, if the flow domain has realistic boundaries, such as solid walls – the spectral method based on the three-dimensional Fourier expansion cannot be used. We can apply finite difference or finite volume methods. Schemes of the second or higher order are desirable in this case, since strong numerical dissipation of the first-order schemes leads to unphysical suppression of small-scale turbulent fluctuations.

Another approach is possible if the flow is inhomogeneous in only one or two directions and homogeneous in the others. A classical example is the fully developed turbulent flow in a channel (see Figure 11.1). We can assume homogeneity and apply periodic boundary conditions in the streamwise (x) and spanwise (y) directions. This means that the Fourier spectral scheme can be used for discretization in x and y. The Fourier expansion becomes

$$u(x, y, z, t) = \sum_{k_x, k_y} \hat{u}(k_x, k_y, z, t) e^{i(k_x x + k_y y)}. \tag{11.12}$$

Substituting into the Navier–Stokes equations and performing the standard spectral method transformations in the x- and y-coordinates, we obtain a system of partial differential equations for coefficients $\hat{u}\left(k_x, k_y, z, t\right)$ considered as functions of z and t. The discretization in the inhomogeneous (z) direction can be achieved using a finite difference scheme with proper grid clustering near the walls (we discuss clustering in Chapter 12) or by a different version of spectral method. The expansion over Chebyshev polynomials $T_{m_z}(z) = \cos(m_z \arccos z)$ is particularly convenient for the latter, since the polynomials satisfy the no-slip boundary conditions at the walls and provide good resolution of boundary layers and their series can be computed and inverted by a fast Fourier transform.

The first high-resolution DNS of the channel flow published in 1987 by Kim, Moin, and Moser used the Fourier–Chebyshev spectral method. The results have had strong and lasting impact on the turbulence research and CFD in general by creating a complete and reliable set of flow characteristics. The flow presented in Figure 11.1 was also computed using a version of this method. The number of the expansion terms in the series was $N_x = N_y = 512$ and $N_z = 256$.

11.3 REYNOLDS-AVERAGED NAVIER–STOKES (RANS) MODELS

RANS is a traditional method of turbulence modeling, which still remains a primary tool of practical CFD analysis. Its advantages are the simplicity, low computational cost (relative to DNS and LES), broad selection of models readily available in the general-purpose CFD codes, and significant experience accumulated in its use for different types of turbulent flows. Its disadvantages are the low level of description (no information is available beyond the mean flow characteristics), necessity to fine-tune the models to specific features of the flow, and relatively large modeling error. The first disadvantage is often acceptable in engineering analysis. The second and third, however, must be taken very seriously. Two questions should always be asked when applying an RANS model: *How well does the model capture the flow physics, and how accurate are the quantitative predictions obtained in the analysis?*

11.3.1 Mean Flow and Fluctuations

We begin with the precise definition of the mean flow fields. The universally applicable definition is based on the *ensemble averaging*:

$$\langle u \rangle(x, t) = \lim_{M \to \infty} \frac{1}{M} \sum_{m=1}^{M} u^{(m)}(x, t), \qquad (11.13)$$

where $u^{(m)}$ are the realizations of the flow in M identical experiments.

If the flow conditions are time independent, the mean flow can be defined as a result of time averaging:

$$\langle u \rangle(x) = \lim_{T \to \infty} \frac{1}{T} \int_{0}^{T} u^{(m)}(x, t) dt. \qquad (11.14)$$

The definition remains valid, with a small modification, for flows that experience slow nonturbulent variations due to time-dependent boundary conditions, a time-dependent body force, or some internal hydrodynamic mechanisms (for example, separation of boundary layer). It is necessary that the typical time T_V of the variation is much larger than the typical time scale T_t of the slowest (corresponding to the largest eddies) turbulent fluctuations. In this case, we can apply

$$\langle u \rangle(x, t) = \frac{1}{T} \int_{t}^{t+T} u^{(m)}(x, \tau) d\tau, \qquad (11.15)$$

where $T_t \ll T \ll T_V$.

The averaging operations (11.13)–(11.15) are linear and commute with space derivatives. They also commute with time derivative, which is obvious for (11.13)–(11.14) and can be derived for (11.15) as an approximation assuming that T is the new infinitesimal time step. Other relevant properties include

$$\langle\langle f\rangle\rangle = \langle f\rangle, \quad \langle f\langle g\rangle\rangle = \langle f\rangle\langle g\rangle, \tag{11.16}$$

where f and g are arbitrary functions.

The flow fields can be seen as sums of mean and fluctuating parts:

$$\boldsymbol{u}(\boldsymbol{x}, t) = \langle\boldsymbol{u}\rangle(\boldsymbol{x}, t) + \boldsymbol{u}'(\boldsymbol{x}, t), \quad p(\boldsymbol{x}, t) = \langle p\rangle(\boldsymbol{x}, t) + p'(\boldsymbol{x}, t). \tag{11.17}$$

Taking the mean of the left-hand and right-hand sides of this equation and using the first formula of (11.16), we immediately find that the fluctuations always have zero means:

$$\langle\boldsymbol{u}'\rangle = 0, \quad \langle p'\rangle = 0. \tag{11.18}$$

11.3.2   Reynolds-Averaged Equations

Applying the averaging operation to the Navier–Stokes system results in the *Reynolds-averaged* equations. We will show the derivation for the incompressible and Newtonian flow equations in conservation form:

$$\rho\frac{\partial u_i}{\partial t} + \rho\frac{\partial}{\partial x_j}(u_i u_j) = -\frac{\partial p}{\partial x_i} + \mu\nabla^2 u_i, \quad \frac{\partial u_i}{\partial x_i} = 0, \tag{11.19}$$

where, as usual, we assume summation over repeating indices. Our ultimate goal is to write the system of equations for the mean flow fields $\langle\boldsymbol{u}\rangle$, $\langle p\rangle$. Taking the average and using the properties of this operation described in Section 11.3.1, we obtain

$$\rho\frac{\partial\langle u_i\rangle}{\partial t} + \rho\frac{\partial}{\partial x_j}\langle u_i u_j\rangle = -\frac{\partial\langle p\rangle}{\partial x_i} + \mu\nabla^2\langle u_i\rangle, \quad \frac{\partial\langle u_i\rangle}{\partial x_i} = 0. \tag{11.20}$$

The goal is not achieved, since $\langle u_i u_j\rangle$ is not expressed in terms of the mean flow quantities. It can be transformed as

$$\rho\langle u_i u_j\rangle = \rho\langle(\langle u_i\rangle + u_i')(\langle u_j\rangle + u_j')\rangle = \rho\langle u_i\rangle\langle u_j\rangle + \rho\langle u_i' u_j'\rangle, \tag{11.21}$$

i.e. as a combination of the product of mean velocities and the term expressing the effect of turbulent fluctuations, namely, the symmetric *Reynolds stress tensor*

$$\tau_{ij} \equiv -\rho\langle u_i' u_j'\rangle = -\rho\langle u_i u_j\rangle + \rho\langle u_i\rangle\langle u_j\rangle. \tag{11.22}$$

Using the new notation, we rewrite the RANS equations (11.20) as

$$\rho\frac{\partial\langle u_i\rangle}{\partial t} + \rho\frac{\partial}{\partial x_j}(\langle u_i\rangle\langle u_j\rangle) = -\frac{\partial\langle p\rangle}{\partial x_i} + \mu\nabla^2\langle u_i\rangle + \frac{\partial\tau_{ij}}{\partial x_j}, \quad \frac{\partial\langle u_i\rangle}{\partial x_i} = 0.$$

(11.23)

The goal is still not achieved because the components of the Reynolds stress tensor are unknown and cannot be expressed as functions of $\langle u\rangle$ and $\langle p\rangle$. The system (11.23) is not closed since there are ten variables (three components of mean velocity, mean pressure, and six independent components of τ_{ij}) and just four equations.

11.3.3   Reynolds Stresses and Turbulent Kinetic Energy

We can derive differential equations for the components of the Reynolds stress tensor (11.22). The procedure is discussed in detail in books on turbulence modeling including those listed at the end of this chapter. We multiply the momentum equations by velocity fluctuations u_i' and take the average of the result. After a straightforward albeit tedious transformation based on the properties of the averaging operation, we obtain

$$\frac{\partial\tau_{ij}}{\partial t} + \langle u_k\rangle\frac{\partial\tau_{ij}}{\partial x_k} = -\tau_{ik}\frac{\partial\langle u_j\rangle}{\partial x_k} - \tau_{jk}\frac{\partial\langle u_i\rangle}{\partial x_k} + E_{ij} - \Pi_{ij} + \frac{\partial}{\partial x_k}\left[\mu\frac{\partial\tau_{ij}}{\partial x_k} + C_{ijk}\right],$$

(11.24)

where we use summation from 1 to 3 over repeated indices and the notation

$$E_{ij} = 2\mu\left\langle\frac{\partial u_i'}{\partial x_k}\frac{\partial u_j'}{\partial x_k}\right\rangle$$

(11.25)

$$\Pi_{ij} = \left\langle p'\left(\frac{\partial u_i'}{\partial x_j} + \frac{\partial u_j'}{\partial x_i}\right)\right\rangle$$

(11.26)

$$C_{ijk} = \langle\rho u_i'u_j'u_k'\rangle + \langle p'u_i'\rangle\delta_{jk} + \langle p'u_j'\rangle\delta_{ik}.$$

(11.27)

Equation (11.24) can be used to derive the equation for the turbulent kinetic energy

$$k \equiv \frac{1}{2}\langle u_i'u_i'\rangle = \frac{1}{2}\langle(u_x')^2 + (u_y')^2 + (u_z')^2\rangle.$$

(11.28)

The trace of the residual tensor (11.22) is

$$\tau_{ii} = -\rho\langle u_i'u_i'\rangle = -2\rho k.$$

(11.29)

Adding together Eqs. (11.24) for τ_{11}, τ_{22}, and τ_{33}, we obtain, after simple transformations, the transport equation for k:

$$\frac{\partial(\rho k)}{\partial t} + \langle u_j \rangle \frac{\partial(\rho k)}{\partial x_j} = \tau_{ij} \frac{\partial \langle u_i \rangle}{\partial x_j} - \mu \left\langle \frac{\partial u_i'}{\partial x_k} \frac{\partial u_i'}{\partial x_k} \right\rangle$$
$$+ \frac{\partial}{\partial x_j} \left(\mu \frac{\partial k}{\partial x_j} - \frac{1}{2} \langle \rho u_i' u_i' u_j' \rangle - \langle p' u_j' \rangle \right). \quad (11.30)$$

The terms in the left-hand side form the material derivative of k, which is the rate of change of k in a fluid particle transported by the mean flow. The terms in the right-hand side represent the mechanisms, by which the energy within a particle can be changed. The first term is the rate of energy production (the rate, at which energy is transferred to fluctuations from the mean flow). The second term

$$\rho \epsilon = \mu \left\langle \frac{\partial u_i'}{\partial x_k} \frac{\partial u_i'}{\partial x_k} \right\rangle \quad (11.31)$$

is the rate of viscous dissipation of turbulent velocity taken per unit volume. Finally, the third term in the right-hand side is the rate of diffusion of the turbulent kinetic energy by molecular viscosity and turbulent fluctuations of velocity and pressure.

Equations (11.24) and (11.30) illustrate the very important property of turbulence, namely, the generally non-local relation between the mean flow and turbulent fluctuations. At any given moment, the local properties of the mean flow determine the instantaneous local generation of turbulent fluctuations. Similarly, the local instantaneous characteristics of the fluctuations determine the components of the local Reynolds stress tensor and, thus, the effect of the fluctuations on the mean flow. The non-local character of the relation appears because the fields of fluctuation properties are transported by the mean flow and diffused by molecular viscosity and turbulence. This can only be described by transport equations, such as (11.24) or (11.30).

We will illustrate the property by a simple example. Let particularly strong turbulence be generated by some mechanism, e.g. strong mean shear, at a certain location. In all but the simplest parallel flows, the fluid particles carrying strong turbulence are transported to other areas of the flow domain and diffused. As a result, we have a strong effect of fluctuations on the mean flow in the areas, where this could not be predicted by a model ignoring the transport and diffusion, and based on the assumption of local equilibrium between the mean flow and the fluctuations.

11.3.4 Eddy Viscosity Hypothesis

The often used first step toward closure models is based on an apparent albeit superficial similarity between the molecular viscosity and the effect of turbulent fluctuations on the mean flow. In both cases, quasi-random motion of small objects (molecules in one and turbulent eddies in the other) results in redistribution of the average momentum of the fluid. One can hypothesize that the similarity also exists on the level of functional relations and assume that the turbulent transport depends on the mean velocity gradients in the same way as molecular transport depends on the gradients of the full velocity field.[2] This is formalized as the *eddy viscosity hypothesis*. The Reynolds stress tensor is assumed to satisfy

$$\tau_{ij} \equiv -\rho\langle u_i' u_j' \rangle = 2\mu_t \langle S_{ij} \rangle - \frac{2}{3}\rho\delta_{ij}k, \tag{11.32}$$

where

$$\langle S_{ij} \rangle = \frac{1}{2}\left(\frac{\partial\langle u_i \rangle}{\partial x_j} + \frac{\partial\langle u_j \rangle}{\partial x_i} \right) \tag{11.33}$$

is the rate of strain tensor of the mean flow, $\mu_t(\mathbf{x}, t)$ is the scalar function called *eddy viscosity*, and $k(\mathbf{x}, t)$ is the turbulent kinetic energy per unit mass defined by (11.28).

The use of (11.32) reduces the number of unknowns associated with turbulent fluctuations to just two: μ and k. To convert the Reynolds-averaged momentum equations (11.23) into a closed system of equations for the mean flow variables, μ and k have to be modeled as functions of these variables. We note that full such modeling is only needed for compressible flows. When the fluid is assumed incompressible, substitution of (11.32) into (11.23) results in

$$\rho\frac{\partial\langle u_i \rangle}{\partial t} + \rho\frac{\partial}{\partial x_j}(\langle u_i \rangle\langle u_j \rangle) = -\frac{\partial\langle p \rangle}{\partial x_i} + \mu\nabla^2\langle u_i \rangle + \frac{\partial}{\partial x_j}\left[\mu_t\left(\frac{\partial\langle u_i \rangle}{\partial x_j} + \frac{\partial\langle u_j \rangle}{\partial x_i} \right) \right]$$
$$- \frac{2}{3}\rho\frac{\partial k}{\partial x_i}, \tag{11.34}$$

$$\frac{\partial\langle u_i \rangle}{\partial x_i} = 0. \tag{11.35}$$

---

[2]The hypothesis has two components. One is the assumption that the stress tensor depends solely on the local turbulence conditions. Another component is the actual functional form (11.32) of the dependency.

We can include k into the modified mean pressure

$$\bar{p} = \langle p \rangle + \frac{2}{3}\rho k,$$

which is found using the projection method described in Chapter 10.

The justification of the eddy viscosity hypothesis is questionable at best. There is no theoretical explanation beyond the generally unfounded analogy between the motion of molecules and motion of turbulent eddies. The formula (11.32) and the assumption that μ_t is a scalar coefficient imply that the tensors in the left-hand and right-hand sides of (11.32) are aligned (proportional). Experiments and DNS studies show that the alignment is good in simple parallel shear flows but disappears in flows of more complex structure. Despite these serious problems, the eddy viscosity model is widely applied in CFD analysis. The main reasons seem to be tradition, model's simplicity, and the fact that reasonably accurate results can be obtained with an appropriately chosen expression for $\mu_t(\boldsymbol{x}, t)$.

On the simplest and often accepted level of description, local turbulence is assumed to be fully characterized by the kinetic energy of fluctuations k or their root-mean-square velocity $q \equiv k^{1/2}$, and by a certain typical length scale ℓ. Dimensionality π-theorem analysis shows that in this case

$$\mu_t = C_\mu \rho q \ell, \tag{11.36}$$

where C_μ is a dimensionless proportionality constant.

11.3.5   Closure Models

Equations (11.23), (11.24), and (11.30) illustrate the fundamental problem of RANS modeling. Mathematical manipulations alone cannot produce a closed system of equations for the averaged flow variables. The number of unknowns is always larger than the number of equations. For example, the system (11.23) has four equations for ten unknowns (three velocity components $\langle u_i \rangle$, $\langle p \rangle$, and six independent components of the symmetric tensor τ_{ij}). If we decide to add the six equations (11.24) for the Reynolds stress tensor, new unknowns $E_{ij}(\boldsymbol{x}, t)$, $\Pi_{ij}(\boldsymbol{x}, t)$, $C_{ijk}(\boldsymbol{x}, t)$ appear. Differential equations can be derived for these fields, but these equations contain new unknowns – the higher-order statistical moments of u'_i, p', etc.

The model equations can only be made solvable if we find a way to express the higher-order unknown terms via the variables for which the system is to be solved. For example, the system (11.23) becomes solvable when the components of τ_{ij} are expressed through $\langle u_i \rangle$ and $\langle p \rangle$. Other solutions are possible, for example, expressing τ_{ij} through a smaller number of flow variables and adding a required number of *closed* equations for these variables.

Such solutions are called *closure models*. They are not mathematically rigorous. Rather, they are approximations based on physical reasoning, accumulated experimental and numerical (DNS) data, and certain, often crude and poorly justified assumptions about the behavior of turbulence.

A brief discussion of RANS closure models is provided in Sections 11.3.6–11.3.8. Since we cannot afford to go deeply into the physics of turbulence, the discussion is inevitably superficial. The focus is on the models for τ_{ij} in (11.23). A reader interested in the models for the unknown terms in the Reynolds stress equation (11.24) or, in general, a more thorough and detailed discussion is referred to specialized books, including the books listed at the end of this chapter, and to the vast research literature on the subject.

11.3.6 Algebraic Models

The simplest closure models are the algebraic models, in which no additional differential equations need to be solved. The turbulent eddy viscosity is approximated by an algebraic function of the characteristics of the mean flow. The models follow the assumption that the effect of transport and diffusion of turbulent fluctuation fields can be ignored, which is only true for a small class of simple flows, most importantly for parallel shear flows such as attached boundary layers, mixing layers, jets, and wakes. In these flows, there is one dominant mean velocity component $\langle u \rangle$, which depends only on the cross-flow coordinate y. The models are based on the Prandtl's mixing length theory.[3] It can be hypothesized that the length scale ℓ and velocity scale q of (11.36) are estimated as

$$\ell \approx \ell_m, \quad q \approx \ell_m \left| \frac{d\langle u \rangle}{dy} \right|, \tag{11.37}$$

where ℓ_m is the Prandtl mixing length. Substituting into (11.36) and setting the proportionality constant to one, we obtain

$$\mu_t = \rho \ell_m^2 \left| \frac{d\langle u \rangle}{dy} \right|. \tag{11.38}$$

---

[3]The mixing length theory was proposed by L. Prandtl (1925). The theory introduces the so-called mixing length ℓ_m, an analog of the mean free path in the kinetic theory of gases. The mixing length can be viewed as the typical maximum distance over which a lump of fluid particles moving in a turbulent flow retains its momentum.

This expression for the eddy viscosity can be used in (11.32). The turbulent kinetic energy $k = q^2$ can be estimated as in (11.37) or, for an incompressible fluid, included into the modified pressure field. To fully close the RANS equations, we need an approximation of the mixing length ℓ_m. Such approximations were proposed and fine-tuned for various types of parallel shear flows. For example, for a wake, mixing layer, or jet, the approximation is $\ell_m \approx \alpha\delta(x)$, where x is the coordinate in the direction of the flow, $\delta(x)$ is the flow width, and α is the nondimensional adjustment coefficient, which takes a special value in each case. In the attached boundary layer, the mixing length can be approximated on the basis of properties of the logarithmic layer, which we will discuss in Section 11.3.11, as $\ell_m \approx \kappa y$, where $\kappa = 0.41$ is the von Karman constant and y is the wall-normal coordinate.

There exist advanced versions of the mixing length model, which use more accurate approximations of ℓ_m (see, e.g. Wilcox 2006). The model can also be nominally extended to the general three-dimensional flow configuration. For example, the version proposed by Smagorinsky (1963) is

$$\mu_t = \rho\ell_m^2(2\langle S_{ij}\rangle\langle S_{ij}\rangle)^{1/2}. \tag{11.39}$$

Such extended models reduce to models of the form (11.38) in parallel shear flows.

The attraction of the algebraic models is in their simplicity and computational efficiency. We have to reiterate, however, that their applicability is almost exclusively limited to parallel shear flows. In more complex cases, an acceptably accurate prediction by the algebraic models is impossible due to their inability to account for the nonequilibrium nature of turbulence.

11.3.7 One-Equation Models

A simple first step toward avoiding the inherent limitations of the algebraic models is to introduce one partial differential equation that describes the transport and diffusion of some key turbulent fluctuation field. Such one-equation models are applied substantially more rarely than the two-equation models discussed in Section 11.3.8. Their applicability is largely limited to parallel flows, such as boundary layers of straight channels. Only a rudimentary discussion is, therefore, given here for the sake of completeness. As it is customary in this book, an interested reader is referred to specialized literature, for example, to the book of Wilcox (2006) listed at the end of the chapter.

In one type of the models, the transport equation is written directly for the turbulent eddy viscosity μ_t. The equation contains unknown terms, which correspond to higher-order statistical moments of turbulent fluctuations and have to be approximated on the basis of modeling assumptions. The better known examples are the models introduced by Sekundov (1971) (see also Vasiliev et al. (1997)), Baldwin and Barth (1990), and Spalart and Amaras (1992).

The other type of the one-equation models is based on the turbulent kinetic energy equation (11.30). As suggested by L. Prandtl we can employ the relation known from the fundamental theory of turbulence (essentially the Richardson–Kolmogorov energy cascade introduced in Section 11.1.1) that connects the total dissipation rate with the typical length scale of the largest eddies ℓ and the kinetic energy of turbulent fluctuations k:

$$\epsilon \sim \frac{k^{3/2}}{\ell}. \tag{11.40}$$

The model uses the approximation

$$\epsilon = C_D \frac{k^{3/2}}{\ell}, \tag{11.41}$$

where C_D is some constant. It also uses the eddy viscosity assumption (11.32). The eddy viscosity is approximated as in (11.36) with the constant $C_\mu = 1$, so (11.41) leads to

$$\mu_t = \rho q \ell = \rho k^{1/2} \ell = C_D \frac{k^2}{\epsilon}. \tag{11.42}$$

The turbulence diffusion terms in (11.30) are approximated as the diffusion of k with the variable diffusivity coefficient μ_t/σ_k, where σ_k is the new parameter, which is called the turbulent Prandtl number and usually assumed to be unity:

$$\frac{\rho}{2}\langle u_i' u_i' u_j' \rangle + \langle p' u_j' \rangle \approx -\frac{\mu_t}{\sigma_k}\frac{\partial k}{\partial x_j}. \tag{11.43}$$

The energy production term in the right-hand side of (11.30) can be approximated using the eddy viscosity hypothesis (11.32) as

$$P_k \equiv \tau_{ij}\frac{\partial \langle u_i \rangle}{\partial x_j} = -\rho\langle u_i' u_j' \rangle\frac{\partial \langle u_i \rangle}{\partial x_j} = 2\mu_t \langle S_{ij} \rangle\frac{\partial \langle u_i \rangle}{\partial x_j}. \tag{11.44}$$

Note that the contribution of the second term in the right-hand side of (11.32) disappears for an incompressible flow, since summation over i and j gives $\delta_{ij}(\partial\langle u_i\rangle/\partial x_j) = \partial\langle u_i\rangle/\partial x_i = 0$.

After all these approximations, the equation for the turbulent kinetic energy becomes

$$\frac{\partial(\rho k)}{\partial t} + \langle u_j\rangle\frac{\partial(\rho k)}{\partial x_j} = 2\mu_t\langle S_{ij}\rangle\frac{\partial\langle u_i\rangle}{\partial x_j} - C_D\rho\frac{k^{3/2}}{\ell} + \frac{\partial}{\partial x_j}\left[\left(\mu + \frac{\mu_t}{\sigma_k}\right)\frac{\partial k}{\partial x_j}\right].$$

(11.45)

The equation is still not closed, since it includes the unknown typical length scale ℓ. Following Prandtl's suggestion, it can be approximated using the mixing length theory. This is evidently the weakest spot of the entire model. It means that the model is accurate only in the case of simple parallel flows, for which the theory produces accurate results. This is also the reason why the models based on (11.45) are used much more rarely than the two-equation models presented in Section 11.3.8.

11.3.8 Two-Equation Models

The RANS models most commonly used in engineering CFD are based on two additional transport equations. As we have seen in Section 11.3.4, the eddy viscosity hypothesis (11.32) and the dimensionality analysis expressed by (11.36) provide a description of the effect of turbulent fluctuations on the mean flow by just two fields: the typical local velocity scale q (or the local turbulent kinetic energy $k = q^2$) and the typical local length scale ℓ. A model with two additional transport equations is, thus, the minimum model that allows us to describe turbulence in a way free from the arbitrary assumptions required by the zero-equation and one-equation models.

It must be stressed from the beginning that the two-equation RANS models are certainly also not free from strong and poorly justified assumptions. We have already mentioned in Section 11.3.4 that the eddy viscosity hypothesis is by itself unrealistic. Further departures from reality will become visible in the following discussion. Still, the extensive engineering experience shows that the two-equation models, if properly implemented and used with appropriate validation, produce reasonably accurate results for a wide range of flows.

As a typical representative, we will consider the widely used k–ϵ model. Similarly to other popular models, it includes the equation for the kinetic energy k. To derive it, we start with (11.30) and apply the same approximations as in Section 11.3.7 (the eddy viscosity formula for τ_{ij} in the energy production term leading to (11.44) and the approximation based on (11.43) for the diffusion of turbulent fluctuations). Unlike the approach taken in the one-equation models, the dissipation rate ϵ is not estimated according to (11.41) but considered as an independent scalar field $\epsilon(x, t)$.

The final form of the k-equation is

$$\rho\frac{\partial k}{\partial t} + \rho\langle u_j\rangle\frac{\partial k}{\partial x_j} = 2\mu_t\langle S_{ij}\rangle\frac{\partial\langle u_i\rangle}{\partial x_j} - \rho\epsilon + \frac{\partial}{\partial x_j}\left[\left(\mu + \frac{\mu_t}{\sigma_k}\right)\frac{\partial k}{\partial x_j}\right]. \quad (11.46)$$

We can now use (11.40) as a definition of the typical length scale

$$\ell = \frac{k^{3/2}}{\epsilon} \quad (11.47)$$

and determine the eddy viscosity as

$$\mu_t = C_\mu\rho\frac{k^2}{\epsilon}. \quad (11.48)$$

The equation for the dissipation rate ϵ is also derived from the Navier–Stokes equation and then transformed in a series of drastic simplifying assumptions. The assumptions are, in fact, strong and unfounded to the degree that the final equation can be considered only loosely connected to the dynamics of averaged fields determined by the Navier–Stokes equations. The final equation for ϵ is

$$\rho\frac{\partial\epsilon}{\partial t} + \rho\langle u_j\rangle\frac{\partial\epsilon}{\partial x_j} = C_{\epsilon1}P_k\frac{\epsilon}{k} - C_{\epsilon2}\rho\frac{\epsilon^2}{k} + \frac{\partial}{\partial x_j}\left[\left(\mu + \frac{\mu_t}{\sigma_\epsilon}\right)\frac{\partial\epsilon}{\partial x_j}\right], \quad (11.49)$$

where P_k is the rate of turbulent kinetic energy production determined by (11.44) and $C_{\epsilon1}$, $C_{\epsilon2}$, and σ_ϵ are model constants.

Let us summarize. The k–ϵ model consists of two partial differential equations (11.46) and (11.49), the eddy viscosity hypothesis (11.32), and the algebraic expression (11.48) for the eddy viscosity. The equations for k and ϵ have to be solved simultaneously with the momentum and mass conservation equations for the mean flow (11.23) as parts of one PDE system. The standard version of the model contains five constant parameters, which have little theoretical footing and have to be determined through comparison with DNS and experimental results in the process of fine-tuning the model. The most commonly used sets of values are those proposed by Launder and Sharma (1974)

$$C_\mu = 0.09, \ C_{\epsilon1} = 1.44, \ C_{\epsilon2} = 1.92, \ \sigma_k = 1.0, \ \sigma_\epsilon = 1.3 \quad (11.50)$$

and by Chien (1982)

$$C_\mu = 0.09, \ C_{\epsilon 1} = 1.35, \ C_{\epsilon 2} = 1.8, \ \sigma_k = 1.0, \ \sigma_\epsilon = 1.3. \tag{11.51}$$

The standard version of the model is not unique. Others have been developed and tested. For example, Yakhot and Orszag (1986) used the method of the renormalization group theory to develop the RNG k–ϵ model. It differs from the model described above in that the constant $C_{\epsilon 2}$ is replaced by the variable coefficient

$$C_{\epsilon 2} = \tilde{C}_{\epsilon 2} + \frac{C_\mu \lambda^3 (1 - \lambda/\lambda_0)}{1 + \beta \lambda^3}, \tag{11.52}$$

where $\lambda = (k/\epsilon)(2\langle S_{ij}\rangle\langle S_{ij}\rangle)^{1/2}$ and the model constants, as they are reported in open literature, are

$$C_\mu = 0.085, \ C_{\epsilon 1} = 1.42, \ \tilde{C}_{\epsilon 2} = 1.68, \ \sigma_k = 0.72,$$

$$\sigma_\epsilon = 0.72, \ \beta = 0.012, \ \lambda_0 = 4.38. \tag{11.53}$$

Yet another popular version is the so-called realizable k–ϵ model. Its derivation takes into account that not all values of the turbulent stress tensor are physically possible. The resulting model (see, e.g. Shih et al. 1995) uses variable coefficient C_μ and differs from other k–ϵ models by the values of constants and some other aspects. The model is shown to perform better than the standard model for flows with expanding jets and for three-dimensional flows with complex features, such as large recirculation vortices, boundary layer separation, rotation, etc.

Numerous other two-equation models have been proposed and tested during the long history of RANS development. Many of them use the k-equation to determine the velocity scale, while the choices of the second equations and of the modeling approximations of the equation terms vary. In particular, the k–ω model uses the equation for the field $\omega = \epsilon/k$ instead of the ϵ-equation. A detailed description of this and other models as well as an excellent in-depth discussion of various aspects of RANS modeling can be found in the book of Wilcox (2006), reference to which is provided at the end of the chapter.

11.3.9 RANS and URANS

The mean flow fields computed in RANS have to be viewed as ensemble- or time-averaged fields. The inherent unsteadiness of turbulent eddies is removed by the averaging, so the RANS modeling generally implies that steady-state problems with $\langle u \rangle$, $\langle p \rangle$, k, and ϵ being functions of x but not t are to be solved.

The steady-state problem approach is often followed in the engineering RANS analysis. There are, however, situations in which it is justified to solve full equations, such as (11.23), (11.46), and (11.49) for time-dependent fields $\langle u \rangle$, $\langle p \rangle$, k, and ϵ. This approach is called unsteady RANS, often abbreviated as URANS. It is evident that it has to be used when the mean flow is itself unsteady due to the unsteadiness of external features, such as boundary conditions, flow geometry, or body forces. The logic is more subtle in other cases, when there are no such external factors, and the mean flow can be considered steady, but we chose not to do so because the flow is strongly affected by large-scale unsteady structures. Such structures may appear as a result of vortex shedding (separation of boundary layers) or hydrodynamic instabilities of the mean flow arising on a turbulent background. Such structures can be periodic in time or chaotic. Apart from large typical scale, they are characterized by large (in comparison with the modeled turbulence) typical time of evolution T_V.

In URANS, the mean flow is viewed as a result of the ensemble averaging (11.13) or time averaging over the time period much smaller than T_V (see (11.15)), but, evidently, not as a result of the standard time averaging (11.14).

11.3.10 Models of Turbulent Scalar Transport

Evolution and transport of scalar fields, such as internal energy or concentration of an admixture, is also modeled in RANS. This requires solving one more transport equation and additional modeling assumptions. As an example, we will show a commonly used model for the convection heat transfer in incompressible fluid. In this case, the temperature field is substituted for the internal energy. The conservation of energy is expressed by Equation (2.29). We write the equation in the conservation form. For simplicity, we assume incompressible fluid, zero rate of internal heat generation Q, and constant specific heat C:

$$\frac{\partial T}{\partial t} + \frac{\partial}{\partial x_i}(u_i T) = \frac{\partial}{\partial x_i}\left(\chi \frac{\partial T}{\partial x_i}\right), \qquad (11.54)$$

where $\chi = \kappa/\rho C$ is the thermal diffusivity coefficient. Applying the RANS averaging, we obtain the equation for the mean temperature field $\langle T \rangle$:

$$\frac{\partial \langle T \rangle}{\partial t} + \frac{\partial}{\partial x_i} \langle u_i T \rangle = \frac{\partial}{\partial x_i} \left(\chi \frac{\partial \langle T \rangle}{\partial x_i} \right). \tag{11.55}$$

We transform the averaged nonlinear term using the linearity and properties (11.16) of the averaging operation as

$$\langle u_i T \rangle = \langle (\langle u_i \rangle + u_i')(\langle T \rangle + T') \rangle = \langle u_i \rangle \langle T \rangle + \langle u_i' T' \rangle = \langle u_i \rangle \langle T \rangle + q_i^t. \tag{11.56}$$

Here

$$q_i^t = \langle u_i' T' \rangle, \; i = 1, 2, 3 \tag{11.57}$$

are the components of the vector of heat flux (scaled by ρC) caused by turbulent fluctuations. The heat transfer equation can be rewritten as

$$\frac{\partial \langle T \rangle}{\partial t} + \frac{\partial}{\partial x_i} (\langle u_i \rangle \langle T \rangle) = \frac{\partial}{\partial x_i} \left(\chi \frac{\partial \langle T \rangle}{\partial x_i} - q_i^t \right). \tag{11.58}$$

We see the problem similar to that found earlier for the momentum equation. The turbulent heat flux q^t cannot be mathematically expressed through the mean fields and requires a closure model. The typical approach is to use an analogy between the transport of heat by turbulent eddies and the mechanism of molecular heat conduction and assume that q^t is proportional to the gradient of temperature:

$$q_i^t = -\Gamma_t \frac{\partial \langle T \rangle}{\partial x_i}, \; i = 1, 2, 3. \tag{11.59}$$

The eddy diffusivity coefficient $\Gamma^t(x, t)$ needs to be modeled. A simple and often used solution is to assume that the turbulent diffusion of a scalar field can be approximately described in the same way as the turbulent diffusion of turbulent kinetic energy $k(x, t)$ (see (11.43)), i.e. to assume that the diffusivity is proportional to the eddy viscosity determined by the RANS model:

$$\Gamma_t = \frac{\mu_t}{\rho \sigma_t}, \tag{11.60}$$

where σ_t is the constant coefficient called the turbulent Prandtl number (for temperature) or turbulent Schmidt number (for other scalars). The value of σ_t varies depending on the details of the problem. It can be the same as the value of σ_t in (11.43) or different from it.

11.3.11 Numerical Implementation of RANS Models

We will focus the discussion on implementation of the RANS models that require solution of additional PDEs. The two-equation k–ϵ model will serve as an example.

The RANS equations, such as (11.46) and (11.49), are solved on the same computational grid as the equations for the mean flow. The numerical methods presented in the previous chapters of the book are applied directly to the extended system. For example, a steady-state problem for incompressible fluid is likely to be solved using a projection algorithm, iteration approach, linearization, and either finite volume or finite difference method (see Chapters 4, 5, 8, and 10). Several modifications are recommended in the RANS case to make the solution more efficient and robust.

One modification is related to the fact that the fields describing turbulence properties, such as k or ϵ, have shorter response time than the mean flow variables. This means that the extended PDE system is numerically stiff (different parts of the solution evolve at strongly different time scales), which usually leads to slow convergence of an iteration procedure. The convergence can be accelerated if the iterations for the mean flow and turbulence properties are separated in the manner similar to the sequential iterations discussed in Section 8.4.3. Every iteration is divided into two substeps. On the first, an outer iteration for the mean flow is performed with the eddy viscosity taken from the previous iteration. On the second, the obtained approximation of the mean flow is used to conduct an outer iteration of the k and ϵ equations.

An important fact to be kept in mind is that, by definition, k and ϵ cannot be negative. Violation of this condition may lead to a bizarre situation, in which the eddy viscosity and the turbulent stress tensor have wrong signs and the work of turbulent stresses $\tau_{ij}\langle S_{ij}\rangle$ acts as a source of energy for the mean flow instead of a sink. Similarly, negative eddy diffusivity Γ_t results in incorrect direction of the turbulent flux of a scalar. Even if such problems occur locally and at an intermediate stage of the solution, for example, after an intermediate iteration, there is a danger of numerical instability. We should take into account that the small response time of turbulence properties makes an overshoot into negative values of k and ϵ a likely event. The natural and efficient way to avoid the troubles is to apply successive under-relaxation (see Section 8.3.4). The relaxation parameter ω about 0.6–0.8 is typically recommended.

The RANS equations require boundary conditions for computed turbulent fields, such as k and ϵ. Setting them and arranging near-wall treatment involve interesting and not always easy questions. We will briefly discuss some of them.

The first question concerns the actual formulation of the boundary conditions. At solid walls, velocity fluctuations must be zero, so we require

$$k_{wall} = 0. \tag{11.61}$$

The dissipation rate does not have to be zero. Exact boundary conditions on ϵ are impossible in the RANS framework, but we can use approximations such as

$$\epsilon_{wall} = \nu \left(\frac{\partial^2 k}{\partial n^2} \right)_{wall} \quad \text{or} \quad \epsilon_{wall} = 2\nu \left(\frac{\partial k^{1/2}}{\partial n} \right)_{wall}^2. \tag{11.62}$$

The conditions at symmetry and periodic boundaries and at exits are typically set in the same way as for other flow variables (see Section 2.7). The situation is more difficult at the inlets, where we often do not know the parameters of turbulence in the flow. The common approach is to assume a certain level of turbulence and prescribe the turbulent kinetic energy as a fraction of the kinetic energy of the mean flow. This is done in terms of *turbulence intensity* defined as the ratio between the root-mean-square velocity of turbulent fluctuations and the absolute value of mean velocity:

$$I \equiv \frac{(2k/3)^{1/2}}{(\langle u_x \rangle^2 + \langle u_y \rangle^2 + \langle u_z \rangle^2)^{1/2}}. \tag{11.63}$$

Recommended values of I vary, depending on the specific situation. $I \approx 0.01$ is used for weakly turbulent inlets – for example, in computations of flows past moving bodies. Strongly turbulent inlets, such as the inlets into segments of heat exchangers or turbomachinery, require $I \approx 0.1$ or even higher. To determine the inlet values of ϵ, we apply the relation (11.47). A plausible assumption is used to estimate the turbulence length scale ℓ, typically as a fraction of the inlet width.

A comment should be made about the initial conditions applied in a marching problem or the initial guess of the iteration procedure in a steady-state problem. The general advice to choose the initial fields as close to the actual state of the flow as possible fully applies to the turbulence fields k and ϵ. There is another important rule. The initial values of k and ϵ should never be zero. The danger is easy to see if we inspect the right-hand sides of the eddy viscosity equation (11.48) and k and ϵ equations (11.46) and (11.49). Taking $k = \epsilon = 0$ leads to $\mu_t = 0$ and to zero right-hand sides of (11.46) and (11.49). k and ϵ would remain zero. No turbulence would be generated in the flow. This is a reflection of the fact that, unlike nature, the RANS models are unable to produce turbulence from a laminar high-Re flow. Moreover, calculations starting from such initial conditions would, most likely, lead to numerical instability.

One may think that there should be no special resolution requirements in RANS computations. After all, we only compute averaged turbulence properties, which have approximately the same typical length scale as the mean flow. Certain requirements are, however, imposed by the behavior of turbulent flows near solid walls. As illustrated in Figure 11.1, the fields of averaged turbulent fluctuations have strong and narrow peaks near the wall. Mean flow velocity has strong gradient in the immediate vicinity of the wall, in the so-called viscous sublayer. The typical length scale of the flow characteristics computed by a RANS model, therefore, decreases near the walls. This requires reduction of grid steps, especially the step in the wall-normal direction.

There is another feature of near-wall turbulence that requires special treatment. The intensity of turbulent fluctuations and turbulent transport and the amplitudes of the components of the Reynolds stress tensor τ_{ij} decrease to zero as the wall is approached. The true turbulent eddy viscosity defined by (11.32) must decrease accordingly. This phenomenon is poorly reproduced by typical RANS models. Even at an adequate numerical resolution, the use of an unmodified model near the wall would result in overprediction of turbulent stresses. The error would not be limited to the near-wall zone. Inaccurate estimate of the near-wall flux of momentum would inevitably lead to an incorrect picture of the entire flow.

Various algorithms have been developed to deal with this so-called low Reynolds number effect[4] by appropriately reducing the turbulent stresses in the near-wall zone. A review and discussion can be found, for example, in Wilcox (2006).

In high Reynolds number flows, the viscous sublayer is so thin that it becomes inefficient or even unfeasible to resolve it by a sufficiently fine grid. An alternative approach can be used, in which computations are performed on a relatively coarse grid, and turbulence modeling is only applied up to a certain distance to the wall. Near the wall, the solution is "patched up" by the so-called *wall function*, which implies universal boundary layer behavior and imitates the effect of a solution corresponding to that behavior.

---

[4]The term refers to the fact that near a wall the behavior of a turbulent flow is dominated by molecular viscosity as in the flows at low Reynolds numbers.

The wall function concept is based on two assumptions: the flow is in a local equilibrium, so turbulence production and dissipation are nearly equal, and the wall-parallel mean velocity at the wall-nearest grid point satisfies the logarithmic law

$$\langle u \rangle_{\mathrm{P}} = u_\tau \left[\frac{1}{\kappa} \ln \left(\frac{u_\tau z_{\mathrm{P}}}{\nu} \right) + B \right]. \tag{11.64}$$

In this formula, u_τ is the *wall shear velocity* defined as

$$u_\tau = (|\tau_w| \rho^{-1})^{1/2}, \tag{11.65}$$

where τ_w is the viscous shear stress at the wall and z_{P} is the wall-normal coordinate of the wall-nearest grid point. The two constants are the von Karman constant $\kappa = 0.41$ and the empirical constant B, which is about 5.5 at a smooth flat plate but may take different values in other cases. For the logarithmic law to be correct, the distance to the first grid point z_{P} should satisfy $30 < u_\tau z_{\mathrm{P}}/\nu < 300$.

The logarithmic law can be used to find the wall shear stress as a function of $\langle u \rangle_{\mathrm{P}}$. After that, the local equilibrium assumption allows us to find the turbulence properties at z_{P}. For the k–ϵ model, we have

$$k_{\mathrm{P}} = \frac{u_\tau^2}{C_\mu^{1/2}}, \quad \epsilon_{\mathrm{P}} = \frac{C_\mu^{3/4} k_{\mathrm{P}}^{3/2}}{\kappa z_{\mathrm{P}}}. \tag{11.66}$$

The wall function approach is strictly valid for an attached unidirectional boundary layer. Its generalization to the case of a more complex pattern of wall-parallel mean flow is straightforward. The situation becomes more difficult when the boundary layer experiences separation. The logarithmic layer ceases to exist in the separation zone, and the wall functions become invalid. Other approaches must be used: RANS with fine near-wall resolution and low Reynolds number modification, or LES.

11.4 LARGE EDDY SIMULATION (LES)

In LES, we directly calculate the mean flow and the part of the fluctuation fields that correspond to motions at large and intermediate length scales. The small-scale fluctuations are not calculated, and their effect on the rest of the flow is modeled. This introduces a modeling error, which, although generally smaller than in RANS, can be significant. Avoiding the requirement of accurate resolution of small-scale fluctuations substantially reduces the computational cost in comparison with DNS and makes it possible to simulate flows at much higher Reynolds numbers and in more realistically complex geometries. At the same time, the computational cost of LES is much higher than the cost of the RANS methods. Current use of LES in practical engineering analysis is mostly limited to the problems in which representation of large- and intermediate-scale turbulent fluctuations, and not just the mean flow, is essential, although the area of applicability is expanding with the growth of available computing power.

In this section, we provide an outline of the LES approach and introduce some simple and popular models. A broader and more thorough discussion can be found in specialized texts and research literature.

11.4.1 Filtered Equations

The key concept of LES is that of filtering. Instead of the actual flow, we simulate the behavior of fields obtained in the result of application of a low-pass filter. In the formal way, the filtered velocity is

$$\overline{u}(x, t) \equiv \int G(r, x) u(x - r, t) dr, \tag{11.67}$$

where the integral is over the entire flow domain. G is a filter function, which is nearly zero at $r = |r| > \Delta/2$ and satisfies the normalization condition $\int G(r, x) dr = 1$. The parameter Δ is called the *filter width*.

Various specific forms of the filtering function have been suggested. We will present two popular versions of the simpler type, in which the filter is uniform (the function G does not depend on x) and isotropic (the dependence on r is limited to the dependence on its absolute value r). Figure 11.3 shows the box filter

$$G(r) = \begin{cases} 1/\Delta & \text{if } r < \Delta/2 \\ 0 & \text{if } r > \Delta/2 \end{cases} \tag{11.68}$$

and the Gaussian filter

$$G(r) = \left(\frac{6}{\pi \Delta^2} \right)^{1/2} \exp\left(-\frac{6r^2}{\Delta^2} \right). \tag{11.69}$$

Evidently, the isotropic filtering operation is simply a weighted and localized (in a sphere of diameter Δ) averaging. The fluctuations with the typical length scales significantly smaller than Δ are smoothed out by such an operation. This is illustrated in Figure 11.4 using a one-dimensional

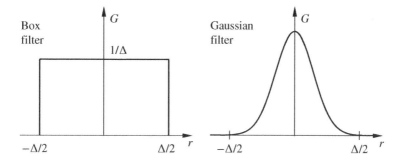

Figure 11.3 Box and Gaussian filters.

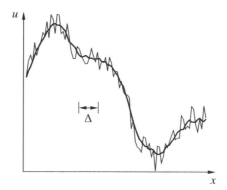

Figure 11.4 Effect of filtering on a turbulent signal. Bold curve shows the result of box filtering with filter width Δ.

example. The smallest scales of variation of the filtered signal are of the order of Δ. This explains the computational efficiency of LES in comparison with DNS. Since the fluctuations with significant variations of the length scales smaller than Δ are absent in the computed flow, it is sufficient to use a computational grid with steps approximately equal to Δ, which can be much larger than the steps needed to resolve the smallest-scale features of the flow in DNS.

Sometimes, the relation between the filter width and the size of the grid step is taken as far as assuming near identity between them. The filtered fields, such as $\overline{u}$ in (11.67), are called *resolved* fields in the sense that they are fully resolved by the computational grid. The fluctuations

$$u' = u - \overline{u} \tag{11.70}$$

are referred to as related to the *subgrid* (smaller than the grid step) scales.

We will follow the less restrictive approach, in which the possibility that the grid step and the filter width are not of the same size is not rejected. The appropriate terms in this case are the *filtered* fields for $\bar{u}$ and *residual* fields for u'.

Our next step is to apply the filtering operation to the Navier–Stokes equations. The filtering operation is linear and commutes with partial derivatives. For example, applying it to $\partial u / \partial t$, we obtain

$$\overline{\left(\frac{\partial u}{\partial t}\right)} = \int G(r,x)\frac{\partial u}{\partial t}(x-r,t)dr$$

$$= \frac{\partial}{\partial t}\int G(r,x)u(x-r,t)dr = \frac{\partial \bar{u}}{\partial t}.$$

For the Navier–Stokes system (11.19), the result of the filtering is

$$\rho\frac{\partial \bar{u}_i}{\partial t} + \rho\frac{\partial}{\partial x_j}(\overline{u_i u_j}) = -\frac{\partial \bar{p}}{\partial x_i} + \mu\nabla^2\bar{u}_i, \quad \frac{\partial \bar{u}_i}{\partial x_i} = 0. \qquad (11.71)$$

The main challenge of LES becomes visible now. It stems from the nonlinearity of the momentum equation. The filtered product $\overline{u_i u_j}$ cannot be expressed as a function of filtered velocity. In particular,

$$\overline{u_i u_j} \neq \bar{u}_i \bar{u}_j.$$

The filtered equations are usually rewritten as

$$\rho\frac{\partial \bar{u}_i}{\partial t} + \rho\frac{\partial}{\partial x_j}(\bar{u}_i \bar{u}_j) = -\frac{\partial \bar{p}}{\partial x_i} + \mu\nabla^2\bar{u}_i + \frac{\partial \tau_{ij}^R}{\partial x_j}, \quad \frac{\partial \bar{u}_i}{\partial x_i} = 0, \qquad (11.72)$$

where

$$\tau_{ij}^R \equiv -\rho\overline{u_i u_j} + \rho\bar{u}_i \bar{u}_j \qquad (11.73)$$

is the *residual stress tensor*.

Despite the similarity in appearance, there is a significant difference between this tensor and the Reynolds stress tensor (11.22) of RANS. The components of the Reynolds stress tensor describe the transport of momentum by *all* turbulent fluctuations, while the residual stresses only include the effect of fluctuations with the length scales smaller than the LES filter width.

For incompressible flows, further transformation is made by expanding τ_{ij}^R as a sum of anisotropic and trace parts

$$\tau_{ij}^R = \tau_{ij}^r + \frac{1}{3}\delta_{ij}\tau_{ii}^R \tag{11.74}$$

and introducing the modified pressure field

$$\bar{p}^* = \bar{p} - \frac{1}{3}\tau_{ii}^R. \tag{11.75}$$

The final system of LES equations is

$$\rho\frac{\partial \bar{u}_i}{\partial t} + \rho\frac{\partial}{\partial x_j}(\bar{u}_i\bar{u}_j) = -\frac{\partial \bar{p}^*}{\partial x_i} + \mu\nabla^2\bar{u}_i + \frac{\partial \tau_{ij}^r}{\partial x_j}, \quad \frac{\partial \bar{u}_i}{\partial x_i} = 0. \tag{11.76}$$

The trace part of the residual stress tensor is now included into the modified pressure field that can be found by a projection technique described in Chapter 10 (this would not work for LES of a compressible flow, in which case a special model of the trace part is required). The remaining anisotropic residual stress tensor τ_{ij}^r is unknown and cannot be mathematically expressed as a function of $\bar{u}$ and $\bar{p}^*$. The system of LES equations (11.76) is not closed. Its numerical solution is impossible unless we find a way to model τ_{ij}^r in terms of the filtered flow fields.

11.4.2   Closure Models

The models used to approximate τ_{ij}^r are called *closure* or *subgrid-scale* models. The first name is evidently due to the fact that the models close the LES equations (11.76) and allow them to be solved numerically. The second name reflects the fact that τ_{ij}^r represents the effect of residual (subgrid-scale) fluctuations on the computed part of the flow.

What requirements should we apply to a model? One obvious requirement is that, to be a true closure, an approximation of τ_{ij}^r should depend only on the filtered flow fields and known parameters, such as the filter width Δ. There is also a geometric requirement that a model is invariant with respect to the principal coordinate transformations. From the physical viewpoint, a model should represent the effect of residual fluctuations on the filtered flow fields as accurately as possible. Yet another view of the role of a closure model is illustrated in Figure 11.5. Since the small-scale fluctuations are not simulated in LES, the turbulent energy cascade cannot proceed in its

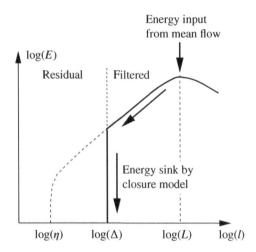

Figure 11.5 Schematic representation of the energy distribution and cascade in LES. L is the typical size of the largest and most energetic fluctuations. Δ is the filter width.

natural way, as shown in Figure 11.2. A closure model provides an energy sink at the length scale approximately equal to the filter width. The natural requirement is that the strength of this sink is appropriate, so that the closure model does not significantly affect the turbulent energy cascade at larger length scales.

Among the numerous existing closure models, we describe the simplest and oldest – the Smagorinsky model. Proposed in 1963, the model has become the most popular choice of LES analysis and served as a basis for many other, more advanced models.

The Smagorinsky model uses the *eddy viscosity hypothesis*

$$\tau_{ij}^r = 2\mu_t \overline{S}_{ij}, \tag{11.77}$$

where $\mu_t(\mathbf{x}, t)$ is the turbulent eddy viscosity and

$$\overline{S}_{ij} \equiv \frac{1}{2}\left(\frac{\partial \overline{u}_i}{\partial x_j} + \frac{\partial \overline{u}_j}{\partial x_i}\right)$$

is the rate of strain tensor of the filtered flow. The eddy viscosity is modeled, on the basis of the Prandtl's mixing length theory, as

$$\mu_t(\mathbf{x}, t) = \rho \ell_S^2 |\overline{S}| = \rho (C_S \Delta)^2 |\overline{S}|. \tag{11.78}$$

In this formula, $|\overline{S}| = (2\overline{S}_{ij}\overline{S}_{ij})^{1/2}$ (summation over $i = 1, \ldots, 3$ and $j = 1, \ldots, 3$ is assumed). The characteristic length scale ℓ_S is approximated as the product of the filter width Δ and the empirical Smagorinsky constant C_S. The rate of energy sink by this model is

$$P_t = \tau_{ij}^r \overline{S}_{ij} = 2\mu_t \overline{S}_{ij}\overline{S}_{ij} = \mu_t |\overline{S}|^2. \tag{11.79}$$

The numerical implementation of the Smagorinsky model is not difficult. We use the filtered velocity field computed at the previous time step or extrapolate from several previous time steps to evaluate $\overline{S}_{ij}$ and $|\overline{S}|$. The results are substituted into (11.78) and (11.77) to compute the eddy viscosity and residual stress tensor. Note that the filtering operation (11.67) is implied but not explicitly performed anywhere in this algorithm. The only information we need to know about the filter is its width Δ. The usual approach is to identify it with the typical grid step (local grid step in the case of a nonuniform grid) estimated as $\Delta = (\Delta x \Delta y \Delta z)^{1/3}$.

The popularity of the Smagorinsky model is justified by its simplicity and computational efficiency. The model, however, has significant drawbacks that make it less accurate and flexible than its more complex counterparts. The main problem is its reliance on the empirical Smagorinsky constant C_S. There is no theoretical derivation that would justify a single optimal value or a functional dependency of C_S. The Kolmogorov theory of isotropic homogeneous turbulence leads to the estimate $C_S \approx 0.17$. This becomes an overestimate in flows with strong mean shear – for example, in the channel flow shown in Figure 11.1. The optimal value of the channel flow C_S providing the closest agreement with DNS and experiments has been found to be about 0.1.

In general, the optimal value of C_S depends strongly and in a complex fashion on flow characteristics (strength of mean shear, Reynolds number, presence of density stratification, etc.). There is always the danger of getting inaccurate results because an incorrect value of C_S is used. One striking example is the behavior of the model in simulations of laminar flows. The residual stress τ_{ij}^r is practically zero in such flows, while the filtered strain rate $\overline{S}_{ij}$ is not. The Smagorinsky model applied to a laminar flow would predict nonzero residual stresses and add a false dissipation mechanism, leading to an utterly incorrect flow picture. This feature of the model makes its performance particularly poor in the case of the flows that evolve between laminar and turbulent states or have laminar and turbulent zones.

Advanced LES models provide higher accuracy and flexibility, although at the cost of higher complexity and larger amount of computations. Some of them are improved versions of the Smagorinsky model, while others follow different approaches (e.g. not based on the concept of eddy viscosity). A detailed discussion of these models is beyond the scope of this book but available in monographs and research papers, some of which are listed at the end of this chapter.

We only mention the dynamic Smagorinsky model proposed by Germano et al. (1991) and extended by Lilly (1992). The method remedies the drawbacks of the original Smagorinsky model by providing an elegant and effective algorithm of automatic adaptation of C_S to the local conditions of the flow. After every time step, the computed filtered flow field is subjected to an additional "test" filter, which has the width $\tilde{\Delta} = 2\Delta$. The Smagorinsky constant C_S, which now becomes a function of space and time, is estimated using the algebraic identities based on the assumption that the formulas (11.77) and (11.78) remain true at both the filter widths.

11.4.3 Implementation of LES in CFD Analysis: Numerical Resolution and Near-Wall Treatment

Discussing the numerical resolution requirements and, correspondingly, computational cost of LES analysis, we have to consider separately the inner part of the flow and the areas near the walls. For the inner part, the Kolmogorov description of turbulence can be adopted, at least in its most essential aspects: the energy of turbulent fluctuations decreases rapidly with the length scale, and the small-scale motions serve as a nearly passive receiver of the energy cascading from the large-scale motions (see Figures 11.2 and 11.5). A plausible and commonly accepted estimate is that an LES model gives an accurate prediction of the filtered flow dynamics if the filtered flow contains a large portion, say, 80 % of the total flow energy. The effect of the residual fluctuations on the filtered flow is inevitably predicted with some error by the closure model, but, since the residual fluctuations are weak, the error is not large.

Theoretical analysis and computational tests show that the requirement that 80 % of the flow energy is in the filtered velocity is not especially strict. The needed computational grids are, of course, much finer than the grids needed for laminar flows or for turbulent flows computed using RANS models. They are, however, much cruder than in DNS. Of the key importance is the fact that the size of LES grids for inner zones of fully turbulent flows *is practically independent of the Reynolds number*. The amount of computations, of course, varies with the complexity of the computational domain and large-scale features of the flow. In general, however, the computations are feasible on modern computers. As an example, we mention the simplest turbulent flow without walls, the isotropic turbulence in a periodic box. It can be shown (see, e.g. Pope (2000)) that the necessary grid consists of only about 40 points in each direction.

Simulation of near-wall zones is more difficult and, typically, requires special treatment. The difficulty is not in the implementation of the boundary conditions themselves. The natural and practically always used approach is to apply the conditions to the filtered flow fields. For example, the no-slip condition (2.49) becomes

$$\overline{u} = U_{wall} \text{ at the wall.} \tag{11.80}$$

The special treatment is necessary because of the nature of turbulent boundary layers developing near the solid walls. Here, the small-scale fluctuations cannot be treated as mere passive receivers of energy. On the contrary, the flow properties are strongly influenced by the dynamics of small-scale coherent structures. The flow is very anisotropic. It is also strongly inhomogeneous. As illustrated in Figure 11.1, the intensity of turbulent fluctuations has a peak at some short distance to the wall. Closer to the wall, the intensity decreases and approaches zero in the vicinity of the wall (in the viscous sublayer). The residual stress must, respectively, decrease or disappear. Finally, the mean velocity has strong wall-normal gradients in this area.

These changes of the flow in the boundary layers must be reflected by the implementation of an LES model. Simply ignoring the changes and extending the model and computational grid suitable for the internal flow into the near-wall zone would generate significant error. As we already discussed in the previous section, the error is not limited to the near-wall zone, but leads to an incorrect picture of the entire flow. For example, the inner part of the channel flow in Figure 11.1 could be accurately reproduced by the standard Smagorinsky model with $C_S \approx 0.1$ and the grid consisting of several tens of cells or points in every direction. The errors of the near-wall approximation would, however, lead to unacceptable inaccuracy in estimation of wall friction and momentum transport across the boundary layer, which would result in strongly distorted distributions of mean flow, turbulent fluctuation intensity, and other flow properties.

There are three principally different ways to treat the near-wall problem. One of them, schematically illustrated in Figure 11.6a, is to reduce the grid steps near the wall, thus providing a nearly DNS-like resolution. This guarantees an accurate approximation of boundary layers, but the computational

Near-wall resolution   Near-wall modeling

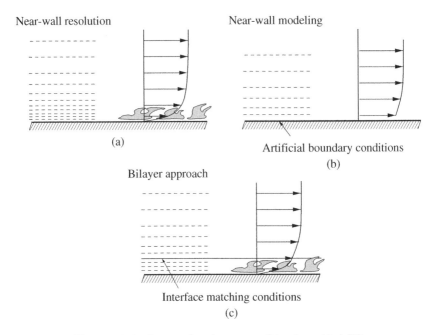

(a)

Artificial boundary conditions

(b)

Bilayer approach

Interface matching conditions

(c)

Figure 11.6   Approaches to near-wall treatment in LES.

cost goes up significantly. The required number of floating point operations increases with the Reynolds number as $\sim Re^k$, where the exponent k varies depending on the situation, but, typically, is around 2. Evidently, this approach is unfeasible in practical CFD analysis of high-Re flows.

The wall treatment based on increased resolution leaves us with one serious unsolved problem. Many closure models do not reflect the reduction of turbulent intensity near the wall. In particular, the Smagorinsky model (11.77) and (11.78) significantly overestimates the residual stresses. One possible remedy is to artificially reduce the turbulent eddy viscosity using the van Driest damping

$$C_S^2 = C_{S0}^2 [1 - \exp(-(u_\tau z/\nu)/25)]^2, \qquad (11.81)$$

where C_{S0} is the Smagorinsky constant used in the core flow and u_τ is the wall shear velocity defined in (11.65).

Another approach, illustrated by the scheme in Figure 11.6b, is to *model* the effect of boundary layers on the rest of the flow instead of simulating their internal dynamics. Artificial boundary conditions are imposed in the form of model approximations for the shear components of the residual stress. These conditions replace the no-slip conditions on tangential velocity components and imitate the momentum transport across the boundary layers. This approach works well when we can be certain that the flow has a unidirectional attached boundary layer. The artificial boundary conditions have the form of relations between the wall shear stress and the inner flow velocity at the grid point nearest to the wall. They are derived on the basis of the assumption of the logarithmic law behavior (11.64).

In the bilayer approach, illustrated in Figure 11.6c, the flow domain is divided into the core and the near-wall layer. The flow in the core is simulated using LES. The near-wall flow is found using a less accurate but computationally more efficient approach. Matching conditions are satisfied at the interface between the two zones. The popular and practical way to treat the near-wall layer is to apply a RANS model based on full Navier–Stokes or boundary layer equations. It is implemented in many modern engineering CFD codes.

The approach combining RANS and LES within one CFD analysis has the special name of *detached eddy simulation* (DES). It is often used for flows past moving bodies (airplanes, cars, etc.) in the manner discussed above, i.e. with RANS applied within the boundary layers and LES outside them (see, e.g. Spalart 2009). Both algebraic and more complex equation-based RANS models have been used. The approach is also applied in a more general sense and to a broader class of flows, as a hybrid LES–RANS method, in which the type of closure is determined locally on the basis of the local turbulence properties and parameters of the discretization grid.

BIBLIOGRAPHY

Books:

Davidson, P.A. (2004). *Turbulence*. Oxford, Oxford University Press.

Frisch, U. (1995). *Turbulence: The legacy of A. N. Kolmogorov*. Cambridge: Cambridge University Press.

Kolmogorov, A.N. (1941). The local structure of turbulence in incompressible viscous fluid for very large Reynolds numbers. *Dokl. Akad. Nauk SSSR* **3**: 299–303 (in Russian).

Pope, S.B. (2000). *Turbulent flows*. Cambridge: Cambridge University Press.

Richardson, L.F. (1922). *Weather Prediction by Numerical Process*. Cambridge, UK: Cambridge University press.

Sagaut, P. (2001). *Large Eddy Simulation for Incompressible Flows: An Introduction*. New York: Springer.

Tennekes, H. and Lumley, J.L. (1972). *A First Course in Turbulence*. Cambridge, MA: MIT Press.

Wilcox, D.C. (2006). *Turbulence Modeling for CFD*, 3e. DCW Industries, Inc.: La Cañada, CA.

Research Articles:

Baldwin, B.S. and Barth, T.J. (1990). A One-Equation Turbulence Transport Model for High Reynolds Number Wall-Bounded Flow. Technical report TM102847. NASA.

Chien, K.-Y. (1982). Predictions of channel and boundary-layer flows with a low-Reynolds-number turbulence model. *AIAA J.* **20** (1): 33–38.

Germano, M., Piomelli, U., Moin, P., and Cabot, W.H. (1991). A dynamic subgrid-scale eddy viscosity model. *Phys. Fluids A* **3**: 1760–1765.

Ishihara, T., Gotoh, T., and Kaneda, Y. (2009). Study of high Reynolds number isotropic turbulence by direct numerical simulation. *Annu. Rev. Fluid Mech.* **41**: 165–180.

Kim, J., Moin, P., and Moser, R. (1987). Turbulence statistics in fully developed channel flow at low Reynolds number. *J. Fluid Mech.* **177**: 133–166.

Launder, B.E. and Sharma, B.I. (1974). Application of the energy dissipation model of turbulence to the calculation of flow near a spinning disk. *Lett. Heat Mass Trans.* **1** (2): 131–138.

Launder, B.E. and Spalding, D.B. (1974). The numerical computation of turbulent flows. *Comput. Methods Appl. Mech. Eng.* **3**: 269–289.

Lilly, D.K. (1992). A proposed modification of the Germano subgrid-scale closuremethod. *Phys. Fluids A* **4**: 633–635.

Meneveau, C. and Katz, J. (2000). Scale-invariance and turbulence models for large-eddy simulation. *Annu. Rev. Fluid Mech.* **32**: 1–32.

Moin, P. and Mahesh, K. (1998). Direct numerical simulation: a tool in turbulence research. *Annu. Rev. Fluid Mech.* **30**: 539–578.

Orszag, S.A. and Patterson, G.S. (1972). Numerical simulation of three-dimensional homogeneous isotropic turbulence. *Phys. Rev. Lett.* **28**: 76–79.

Piomelli, U. and Balaras, E. (2002). Wall-layer models for large-eddy simulations. *Annu. Rev. Fluid Mech.* **34**: 349–374.

Prandtl, L. (1925). Bericht über die Entstehung der Turbulenz. *Z. Angew. Math. Mech.* **5**: 136–139.

Rogallo, R.S. (1981). Numerical Experiments in Homogeneous Turbulence. Technical report TM81315. NASA.

Sekundov, A.N. (1971). Application of a differential equation for turbulent viscosity to the analysis of plane non-self-similar flows. *Fluid Dyn.* **6** (5) 828–840.

Shih, T.H., Zhu, J., and Lumley, J.L. (1995). A new Reynolds stress algebraic equation model. *Comput. Methods Appl. Mech. Eng.* **125** (1–4): 287–302.

Smagorinsky, J. (1963). General circulation experiments with the primitive equations: I. The basic equations. *Mon. Weather Rev.* **91**: 99–164.

Spalart, P.R. (2009). Detached-eddy simulation. *Annu. Rev. Fluid Mech.* **41**: 181–202.

Spalart, P.R. and Almaras, S.R. (1992). A one-equation turbulence model for aerodynamic flows. AIAA paper, 92–439.

Vasiliev, V.I., Volkov, D.V., Zaitsev, S.A., and Lyubimov, D.A. (1997). Numerical simulation of channel flows by a one-equation turbulence model. *J. Fluids Eng.* **119**: 885–892.

Yakhot, V. and Orszag, S.A. (1986). Renormalization group analysis of turbulence. I. Basic theory. *J. Sci. Comput.* **1**: 3–51.

PROBLEMS

1. For the following hypothetical CFD tasks, compare the DNS, LES, and RANS approaches. Discuss which of them is feasible and which is likely to produce acceptable accuracy and level of description.

   a) Find flow rate and drag in a rectangular duct of a residential air-conditioning system. Assume that the duct cross section is a square of side 25 cm and the average velocity is 1 m/s.

   b) Find turbulent drag on a slender body of length 2 m moving at zero attack angle in the air with speed 10 m/s.

   c) Predict motion of a hurricane.

   d) Analyze flow around a swimming fish. Assume that the fish is 20 cm long and moves with speed of 40 cm/s.

2. For the following hypothetical CFD tasks, compare the DNS, LES, and RANS approaches. Discuss which of them is feasible and which is likely to produce acceptable accuracy and level of description.

   a) Try to understand how local properties of turbulent fluctuations created in a cup of coffee by spoon movements affect the rate of sugar dissolution.

   b) Analyze dynamics of a solar protuberance.

   c) Predict flow in a microfluidic device, essentially a network of channels of width 10 μm in which a water-based solution flows with the typical velocity 1 mm/s.

   d) Predict drag on a submarine moving through water with constant speed of 50 km/h. The length and the largest cross-sectional size of the submarine are, respectively, 100 and 15 m.

3. Does the steady-state formulation of a CFD problem make sense in DNS, LES, or RANS?

4. DNS of homogeneous turbulence is conducted by the Fourier spectral method (see section 11.2.1) applied to a flow in a periodic box. The expected value of the Reynolds number based on the typical size and velocity of large-scale eddies is 10^4. What are the resolution requirements? Is the analysis feasible?

5. Prove that the averaging operations (11.13)–(11.15) are linear and commute with space and time derivatives.

6. Prove the relations (11.16), (11.18), and (11.21).

7. $u(x, t)$ and $v(x, t)$ are velocity components of a turbulent flow. The RANS averaging $\langle v(au + b) \rangle$ is evaluated, where a and b are some constants. Rewrite the expression in the RANS form. i.e. as a sum of the terms containing $\langle u \rangle$ and $\langle v \rangle$ and the irreducable terms containing the averages of the products of the fluctuations u' and v'.

8. Do the same as in Problem 7 but for the expression $\langle u(au + v^2) \rangle$.

9. If the terms such as those in Problems 7 and 8 appear in the flow equations solved by an RANS method, do they need closure models? Explain your answer.

10. A steady-state turbulent flow through a long coiled duct is modeled using the standard k–ϵ model. Assuming the fluid is incompressible and isothermal, write the full system of equations and boundary conditions at the duct's walls, inlet, and exit.

11. For the flow analysis in Problem 10, consider the situation when temperature is not constant. There is an internal heat source with volumetric rate Q. The duct's walls are thermally insulated. Write the RANS heat equation and boundary conditions for temperature. Assume that the turbulent Prandtl number is 1.0.

12. Which of the three RANS model types – algebraic, k–ϵ RANS, or k–ϵ URANS – would you use in each of the following problems? Assume that the Reynolds number is sufficiently high to justify the RANS approach.

    a) Attached boundary layer at a long curved wall.
    b) Three-dimensional flow within a cube driven by one wall sliding parallel to itself with constant speed (the lid-driven cavity flow).
    c) Flow past a large truck moving with a constant speed.
    d) Propagation of an atmospheric smoke plume from a forest fire.

13. If your course involves exercises with CFD software, study the documentation to determine which RANS models are implemented. What approach is taken to the near-wall treatment?

14. Prove that the LES filtering operation (11.67) is linear and commutes with space and time derivatives. For the space derivatives, consider two cases: a uniform and a nonuniform filter.

15. What determines the adequate grid step in LES far from the walls?

16. For a flow past a moving car, would it be correct to apply the LES approach based on the standard Smagorinsky model (see Section 11.4.2) with constant C_S and a uniform finite volume grid covering the entire domain? Explain your answer.

17. LES is applied to a turbulent flow with convection heat transfer. Apply the filtering operation to (2.29) to derive the LES equation for temperature. Are there any terms that need modeling?

18. If your course involves exercises with CFD software, study the documentation to determine whether the LES option is implemented. Which closure models are used? What approach is taken to the near-wall treatment?

19. Consider the primitive LES and RANS models, in which the eddy viscosity formulas (11.32) and (11.77) are applied with constant eddy viscosity μ_t. What is the drawback of these models? Should we expect accurate results?

12

COMPUTATIONAL GRIDS

12.1 INTRODUCTION: NEED FOR IRREGULAR AND UNSTRUCTURED GRIDS

A well-designed computational grid is an essential ingredient of CFD analysis. Having such a grid is a necessary and significant step toward an accurate, efficient, and robust numerical solution. On the contrary, as any CFD practitioner can confirm and illustrate by spectacular examples, a poorly designed grid leads to low accuracy, slow convergence, and, sometimes, numerical instability.

Our discussion has so far largely ignored the questions of grid design. For the sake of simplicity and transparency, we have presented CFD techniques on the example of a structured, uniform, and orthogonal grid, in which grid points or finite volume cells are formed by intersections of coordinate lines of a Cartesian coordinate system. Non-Cartesian and unstructured grids have been addressed (for example, in the discussion of finite volume methods in Chapter 5), but only briefly.

Essential Computational Fluid Dynamics, Second Edition. Oleg Zikanov.
© 2019 John Wiley & Sons, Inc. Published 2019 by John Wiley & Sons, Inc.
Companion Website: www.wiley.com/go/zikanov/essential

The general principles of CFD approximation discussed in the previous chapters remain valid if the grid is nonorthogonal, nonuniform, or unstructured. At the same time, new questions arise. What is the effect of grid design on the order of approximation, accuracy, and stability of numerical schemes? Should the scheme itself and method of solution of the resulting discretization equations be changed if an irregular grid is used? How can irregular, in particular unstructured, grids be generated? The answers to these and other relevant questions are complex and require a detailed discussion going far beyond the scope of this book. For such a discussion, the reader is referred to specialized texts, some of which are listed at the end of the chapter. Here, we only present the most basic concepts and provide a few simple recommendations.

There are two often coexisting features of practical engineering flows and heat transfer processes that make the use of a uniform Cartesian grid a poor choice. One is illustrated in Figure 12.1. The solution domain may have complex, possibly curvilinear boundaries, which cannot be fitted with a Cartesian grid.

There are several ways to address this problem. The simplest one is to ignore it and employ a regular Cartesian grid anyway. This is illustrated in Figure 12.1a. The best available locations to set the boundary conditions are at the grid points nearest to the boundary. This means that the boundary itself is replaced by a staircase-like shape. The approach is not recommended, primarily because of the significant (generally, of the first order in terms of the grid step) error it introduces into the solution. Furthermore, implementation of Neumann and mixed boundary conditions is complicated.

The commonly used and recommendable approaches are illustrated in Figure 12.1b,c. As shown in Figure 12.1b, a curvilinear coordinated system can be introduced such that the boundaries coincide with certain coordinate lines. The grid points can be positioned directly at the boundary, so the boundary conditions are approximated with the scheme's accuracy. Another possible and commonly used approach is illustrated in Figure 12.1c. We use a boundary-fitting unstructured grid.

Sometimes, the boundaries of the computational domain are moving (imagine waves on the ocean surface or a flow around rotating blades of a wind turbine). In this case, a boundary-fitting *moving grid* can be designed and used.

The second common reason for the use of nonuniform grids is that gradients of solution variables may be of strongly different amplitudes

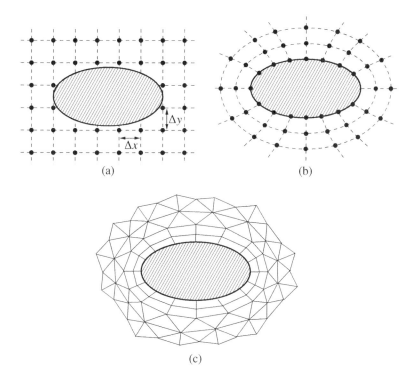

Figure 12.1   A case for irregular grids – complex geometry. (a) A nonrectangular geometry approximated poorly by a regular Cartesian grid. (b) The same geometry approximated by a boundary-fitting curvilinear structured grid. (c) The same geometry approximated by a boundary-fitting unstructured grid.

in different areas of solution domain. For example, mean velocity and turbulence variables experience large gradients in boundary layers near solid walls. On the contrary, far from walls, the gradients are small (see Figure 12.2). In all such cases, it seems reasonable to use smaller grid steps or cells in the areas, where the solution variables have strong gradients and reduce computational effort by using larger steps or cells in other areas, where the gradients are weak. This technique is called the grid *stretching* or *clustering*. As illustrated in Figure 12.2b, this can be achieved through the use of a structured grid designed on the lines of a coordinate system compressed and stretched as necessary. Alternatively, as illustrated in Figure 12.2c, we can build an unstructured grid refined in the areas of strong gradients.

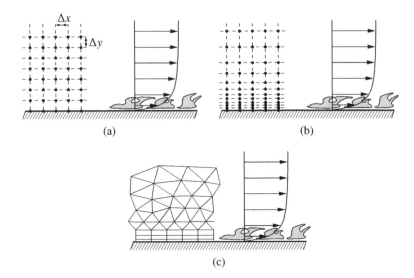

Figure 12.2 A case for irregular grids – strong spatial variability of gradients of solution. (a) A flow with strong wall-normal velocity gradient near the wall approximated poorly on a uniform grid. (b) The same flow approximated on a stretched grid adapted to solution gradients. (c) The same flow approximated on an unstructured grid refined in the area of strong gradients.

12.2 IRREGULAR STRUCTURED GRIDS

The structured grids are the only realistic choice for finite difference and spectral methods. They can also be used in combination with finite volume discretization, although unstructured grids discussed in the next section are more common in that case.

12.2.1 Generation by Coordinate Transformation

Grid points of a structured irregular grid can be generated as points of intersection of coordinate lines of a non-Cartesian (curvilinear) coordinate system. For simplicity, we will illustrate the method using the two-dimensional (2D) grid configurations shown in Figure 12.3. Generalization to the three-dimensional (3D) case is straightforward.

Two coordinate systems are considered: the original Cartesian system (x, y) and the specially designed curvilinear system (ξ, η). The two systems are related as

$$\xi = \xi(x, y, t), \quad \eta = \eta(x, y, t) \tag{12.1}$$

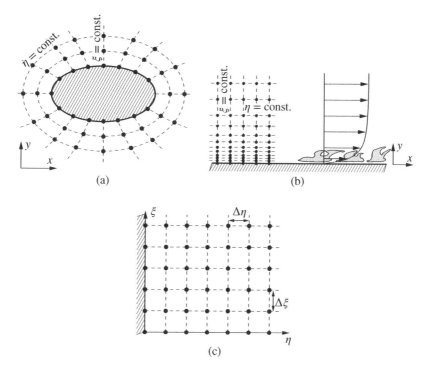

Figure 12.3 Coordinate transformation associated with irregular grids. (a, b) Irregular coordinate systems in the physical space. (c) Corresponding Cartesian coordinate system in the computational space.

and

$$x = x(\xi, \eta, t), \quad y = y(\xi, \eta, t). \tag{12.2}$$

The key is to select the curvilinear coordinates so that the boundary of the flow domain consists of one or several parts, each corresponding to the new coordinate lines ($\eta = $ const. in our example in Figure 12.3). If this is achieved, the relation between the two coordinate systems can be viewed as a mapping between the *physical* solution domain shown in Figure 12.3a,b and a rectangular (in 2D) or parallelepiped-shaped (in 3D) *computational* domain in the space of the transformed coordinates (see Figure 12.3c).

The transformation should satisfy certain mathematical properties that guarantee that the correspondence between the points of physical and computational spaces is one to one, the transformed lines of the same family (for example, lines $\eta = $ const. in Figure 12.3a or b) do not cross, and any two lines of different families do not cross more than once.

After identifying the proper coordinate transformation, we rewrite the governing equations and boundary conditions in terms of the transformed coordinates and discretize them. The discretization is not different from what we have done so far in this book, since it is performed in a rectangular computational domain on a rectangular uniform structured grid (see Figure 12.3c).

Rewriting the governing equations and boundary conditions is not difficult, although it sometimes involves tedious mathematical operations. We have to express everything in terms of functions of the transformed coordinates and their partial derivatives.

For example, let the unknown function be $u(x, y, t)$. In the transformed coordinates, it becomes $u = u(\xi(x, y, t), \eta(x, y, t), t)$. The first derivatives are replaced using the chain rule

$$u_x = u_\xi \xi_x + u_\eta \eta_x, \tag{12.3}$$

$$u_y = u_\xi \xi_y + u_\eta \eta_y, \tag{12.4}$$

$$u_t = u_\xi \xi_t + u_\eta \eta_t + u_t. \tag{12.5}$$

Formulas for second derivatives can be obtained by differentiating (12.3)–(12.5) and applying the chain rule again. For example,

$$
\begin{aligned}
u_{xx} &= \partial \left(u_\xi \xi_x + u_\eta \eta_x \right) / \partial x \\
&= \xi_x \partial \left(u_\xi \right) / \partial x + u_\xi \xi_{xx} + \eta_x \partial \left(u_\eta \right) / \partial x + u_\eta \eta_{xx} \\
&= \xi_x (u_{\xi\xi} \xi_x + u_{\xi\eta} \eta_x) + u_\xi \xi_{xx} + \eta_x (u_{\eta\xi} \xi_x + u_{\eta\eta} \eta_x) + u_\eta \eta_{xx} \\
&= u_{\xi\xi} \xi_x^2 + 2 u_{\xi\eta} \xi_x \eta_x + u_{\eta\eta} \eta_x^2 + u_\xi \xi_{xx} + u_\eta \eta_{xx}.
\end{aligned} \tag{12.6}
$$

After rewriting, the partial differential equations are discretized in the transformed coordinates. For example, if the finite difference scheme of the second order is designed, and we are approximating

$$u_x|_{(\xi_i, \eta_j, t^n)} = (u_\xi \xi_x + u_\eta \eta_x)_{(\xi_i, \eta_j, t^n)},$$

application of central differences leads to

$$u_x|_{(\xi_i, \eta_j, t^n)} = \frac{u_{i+1,j}^n - u_{i-1,j}^n}{2\Delta\xi} \xi_x + \frac{u_{i,j+1}^n - u_{i,j-1}^n}{2\Delta\eta} \eta_x, \tag{12.7}$$

where $u_{i,j}^n = u(\xi_i, \eta_j, t^n)$ and ξ_x, η_x are computed at the point (ξ_i, η_j).

The formula (12.7) illustrates the fact that the discretization equations do not require the explicit algebraic expressions (12.1) and (12.2). The only information needed is the set of values of the partial derivatives ξ_x, ξ_y, η_x, etc. at the grid points (ξ_i, η_j). They can be calculated using analytical derivatives of (12.1) or numerically. Another method useful when the only available analytical expressions for the coordinate transformation are (12.2) is to apply the Jacobian-based mathematical identities, such as

$$\xi_x = \frac{\partial(f,g)}{\partial(x,\eta)} \bigg/ \frac{\partial(f,g)}{\partial(\xi,\eta)}, \tag{12.8}$$

where the coordinate transformation is represented by

$$f = x - x(\xi,\eta,t) = 0 \quad \text{and} \quad g = y - y(\xi,\eta,t) = 0.$$

12.2.2   Examples

We will give two examples of the method of coordinate transformation. In the first example, the need for a non-Cartesian grid arises because of the nonrectangular shape of the boundary. Steady-state conduction heat transfer problem is solved in a domain that has the form of a circular ring with inner and outer radii R_a and R_b (see Figure 12.4). The temperature distribution is governed by the Laplace equation

$$T_{xx} + T_{yy} = 0.$$

The boundary conditions are of the von Neumann type (known heat flux) with the normal derivative $\partial T / \partial n$ equal to $g(\theta)$ and $f(\theta)$ at, respectively, $r = R_a$ and $r = R_b$.

The boundary-fitting coordinates are obviously polar coordinates r and θ, such that $x = r\cos\theta$ and $y = r\sin\theta$. The computational domain in these coordinates is the rectangle (see Figure 12.4)

$$R_a \leq r \leq R_b, \quad 0 \leq \theta \leq 2\pi.$$

It can be derived following the procedure described in Section 12.2.1 that the Laplace equation in the new coordinates is

$$\frac{\partial^2 T}{\partial r^2} + \frac{1}{r}\frac{\partial T}{\partial r} + \frac{1}{r^2}\frac{\partial^2 T}{\partial \theta^2} = 0. \tag{12.9}$$

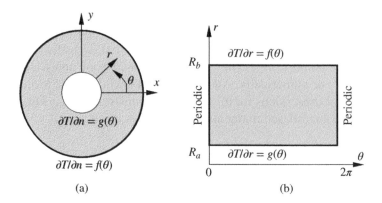

Figure 12.4 Heat transfer in a ring as an example of boundary-fitting coordinate transformation.

It is discretized as

$$\frac{T_{i+1,j} - 2T_{i,j} + T_{i-1,j}}{(\Delta r)^2} + \frac{1}{r_i}\frac{T_{i+1,j} - T_{i-1,j}}{2\Delta r}$$
$$+ \frac{1}{r_i^2}\frac{T_{i,j+1} - 2T_{i,j} + T_{i,j-1}}{(\Delta\theta)^2} = 0. \qquad (12.10)$$

Indices i and j refer to the grid point (r_i, θ_j).

The boundary conditions at $r = R_a$ and $r = R_b$ can be easily approximated since the new coordinate r is normal to the boundary. Using the one-sided second-order formulas, we obtain

$$\frac{-3T_{0,j} + 4T_{1,j} - T_{2,j}}{2\Delta r} \approx \left.\frac{\partial T}{\partial r}\right|_{(r_0,\theta_j)} = g(\theta_j), \qquad (12.11)$$

$$\frac{3T_{N-1,j} - 4T_{1,j} + T_{N-2,j}}{2\Delta r} \approx \left.\frac{\partial T}{\partial r}\right|_{(r_N,\theta_j)} = f(\theta_j), \qquad (12.12)$$

where $r_0 = R_a$ and $r_n = R_b$.

At $\theta = 0$ and $\theta = 2\pi$, periodic boundary conditions are imposed.

In the second example, the coordinate transformation is used to improve the numerical resolution in the zone of strong gradient. For the boundary layer flow illustrated in Figure 12.2, the grid should be stretched in the wall-normal y-direction so that the step Δy decreases near the wall and remains large far from it. There are many coordinate transformations that achieve this effect. For example, we can use exponential, trigonometric, logarithmic, polynomial, or hyperbolic mapping functions. The logarithmic transformation is (we assume that the wall is at $y = 0$)

$$\xi = x \quad \text{and} \quad \eta = \ln(y + 1). \tag{12.13}$$

Differentiating the second expression we find the relation between the grid steps

$$\Delta\eta = \frac{1}{y+1}\Delta y \quad \text{or} \quad \Delta y = (y+1)\Delta\eta.$$

For constant $\Delta\eta$, Δy has the minimum value equal to $\Delta\eta$ at the wall and increasing linearly with the distance to it.

12.2.3   Grid Quality

We have to mention that not any coordinate transformation produces an acceptable computational grid. Sometimes, numerical solution of the transformed equations leads to large discretization errors, slow convergence, or numerical instability even when the scheme performs well in the Cartesian coordinates. The reasons of this behavior and the mathematical apparatus needed to describe it are discussed in specialized texts. We will only list the properties that, taken together, can be called *grid quality* and give some rules of good computational practice.

Distortion   This term refers to the deviation from orthogonality between the intersecting lines of a non-Cartesian coordinate system. The best results are provided by orthogonal grids. If the grid needs to be nonorthogonal, the angles between the intersecting coordinate lines should be possibly close to $90°$. There are no universally valid limits, but grids with angles smaller than $45°$ or larger than $135°$ are usually considered dangerously distorted.

Ratio Between Sizes of Adjacent Cells   We have to avoid strong variations of grid steps. As a general guideline, it is usually accepted that the ratio between the typical sizes of any two adjacent grid cells should not exceed 2. This requirement often needs enforcing when we construct stretched grids for flows with boundary layers and other zones of strong gradient. The transition from the fine grid within such a zone to a crude grid in the outer flow should be gradual.

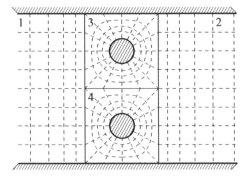

Figure 12.5   Example of a block-structured grid. A two-dimensional flow past two cylinders in a channel is computed using a grid consisting of four blocks marked by numbers 1–4.

Cell Aspect Ratio   We have to avoid strongly anisotropic grids. In general, it is recommended that the ratio between the grid steps in different directions is neither large nor small. The specific limits on the aspect ratio vary with the nature of the flow and the type of the computational scheme. One particular case when high aspect ratio cells are acceptable is that of the boundary layer flows. Such flows are characterized by high velocity in the streamwise direction (let it be x) and strong velocity gradients in the wall-normal direction (let it be y). In that case, it is possible to use Δx significantly larger than Δy. Even here, cells with aspect ratios higher than five have to be avoided.

Block-Structured Grid   Our last comment concerns the possibility of block-structured grids. They can be used in the geometries, which are too complex to be meshed by a single structured grid, but can be subdivided into several zones of relatively simple shapes. Within each block, a structured grid is constructed. As an example, Figure 12.5 shows a block-structured grid build for a 2D flow past an array of two cylinders in a channel. It uses Cartesian grids in the blocks 1 and 2 and grids built in curvilinear coordinate systems in the blocks 3 and 4.

12.3   UNSTRUCTURED GRIDS

One can easily imagine a complex geometry, in which a structured or block-structured grid is either impossible or very difficult to build. Our only practical choice in this case is to cover the domain with an unstructured grid and apply finite element or finite volume discretization.[1] As a typical example, Figure 12.6 shows the grids built to solve the problems of turbulent gas flow and heat transfer within the nozzles of two thermal plasma spray systems used to generate special coatings.

In this section, we discuss the unstructured grids in context of their application with the finite volume method. This type of solution has become the prevalent tool of practical CFD analysis implemented in many widely used commercial and noncommercial CFD codes. There are three main reasons for the popularity of unstructured grids:

1. Possibility of application to arbitrarily shaped computational domains.
2. High levels of flexibility and control of grid parameters such as cell shape and size.
3. Existence of efficient algorithms of grid generation and solution of discretized equations.

The combination of these factors creates an attractive situation for a CFD practitioner. The same code can be used to solve a wide range of problems with different geometries, flow types, discretization accuracies, and so on.

It may appear that the unstructured grid approach has disadvantages related to the complexity of the grid description. The grid points and cells cannot be identified through a simple system of two or three indices. Instead, the information on location, shape (typically determined through locations of all vertices), and connectivity to neighbors should be specified and stored for every cell. In fact, this is usually not a problem in practical CFD analysis, since the task of handling these data is taken care of by a CFD code.

There are, however, consequences of the unstructured nature of the grid that cannot be ignored so easily. Most importantly, the matrices of the systems of discretization equations no longer have band-diagonal or block-diagonal form. This leads to slower convergence and higher computational cost of iterations. In general, solutions on unstructured grids have lower computational efficiency than solutions on structured grids of the same size. It is, therefore, recommended to use a structured grid any time it is allowed by the geometry of the computational domain.

---

[1] Finite difference discretization on an unstructured grid is possible in principle, but has never been seriously tried because of the difficulties involved and the existence of a much simpler alternative – the finite volume method.

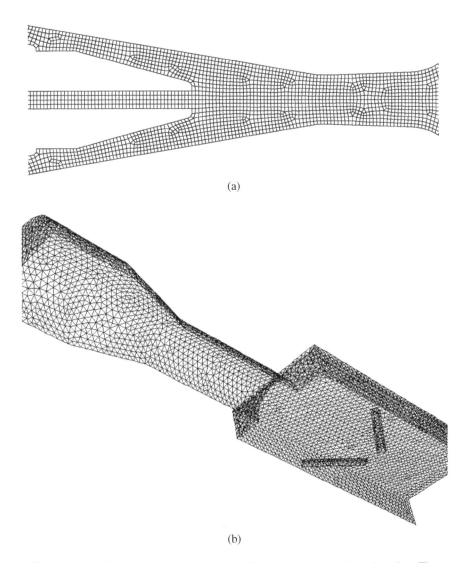

(a)

(b)

Figure 12.6 Examples of unstructured grids in complex solution domains. The grids are automatically generated using GAMBIT 2.3, part of FLUENT 6.3. Both pictures show the nozzles of plasma spray systems. (a) The grid used for preliminary analysis of axisymmetric flow within the nozzle of SG100 plasma torch by Praxair. (b) The three-dimensional grid for analysis of the novel torch design combining high-velocity oxy-fuel (HVOF) and twin wire arc spray systems. Source: Courtesy of Y. Wang and P. Mohanty, University of Michigan – Dearborn.

The issues involved in the solution of partial differential equations on unstructured grids and generation of such grids are complex. Taking into account that ready-to-use algorithms accomplishing these tasks are typically available in engineering CFD tools, we will limit our discussion to basic facts and a few practical recommendations.

12.3.1 Grid Generation

The common practice of modern engineering CFD is to use automated grid generation algorithms. Commercial codes usually offer such algorithms as parts of the software packages. The task of grid buildup, which used to require days or even weeks of CFD practitioner's time, is now accomplished with relatively low effort within hours or minutes. At the same time, the user retains substantial control over the grid properties. This includes setting the topology and typical size of the cells and determining the areas where grid refinement is needed. It is also the user's responsibility to verify that the grid contains no "bad" cells (the characteristics of grid quality are discussed in Section 12.3.3).

In the most common approach to grid generation, the process is initiated by the user, who specifies the mesh points at the boundaries of the domain. The user can exert control over the future distribution of cell sizes at this stage. The automated grid generator uses the mesh points to build the first layer of cells adjacent to the boundary. The second layer of cells is then built on the top of the first, and so on, converging toward the center of the computational domain in the manner of an advancing front.

After the grid has been built, the control is returned to the user, who checks the quality of the cells and modifies them, if necessary. Many CFD codes assist the user by providing information on essential cell characteristics and offering automated tools for fixing the cells.

12.3.2 Cell Topology

In theory, cells of arbitrary convex polygonal (in 2D) or polyhedral (in 3D) shape can be used in finite volume computations. In practice, however, CFD codes usually limit the choice to the few basic shapes shown in Figure 12.7. In the 2D case, we can use quadrilateral (Figure 12.7a) or triangular (Figure 12.7b) cells. In 3D grids, the most common elements are hexahedra (Figure 12.7c) and tetrahedra (Figure 12.7d), although prisms and pyramids (Figure 12.7e,f) are also used. Many CFD codes allow

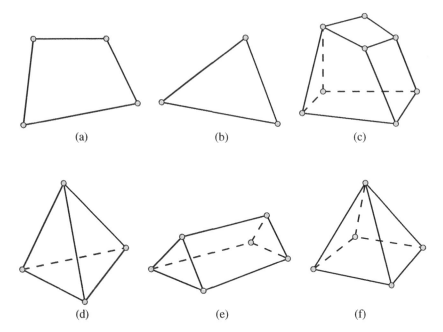

Figure 12.7 Typical shapes of unstructured grid cells. (a, b) Quadrilateral and triangular cells used in 2D grids. (c)–(f) Hexahedra, tetrahedra, prisms (wedges), and pyramids used in 3D grids.

combination of cells of different types (e.g. quadrilateral and triangular) within one grid.

The basic cell shapes are sufficient to build unstructured grids of great geometric flexibility. Any computational domain can be covered by such a grid. In many cases, this can be done relatively easily by applying an automated grid generation code. Figure 12.6 shows two examples of automatically generated grids: a 2D grid consisting of quadrilateral cells and a 3D grid in which the cells are tetrahedral.

12.3.3 Grid Quality

The grid quality is a complex characteristic that determines the overall accuracy of solution at a given number of cells and a given discretization scheme. We have already discussed the quality of structured grids in Section 12.2.3. In the case of unstructured grids, new aspects appear because of larger diversity and flexibility of cell shapes. In general, the matter of quality requires closer attention, since the grid generation algorithms occasionally create problematic cells. Typically, an unstructured grid needs quality control and optimization after it has been built.

Some parameters of the grid quality are evident. For example, the cells must be small enough to guarantee the desirably small discretization error. Other parameters are less evident and do not allow precise definitions. Furthermore, importance of various quality criteria varies with the flow type and the kind of analysis. In the result of all this, determining the grid quality is often a matter of experience and empirical knowledge rather than theoretical arguments.

We will briefly review the basic aspects of quality of unstructured grids. More detailed and specific information is available in specialized monographs, CFD research literature, and online resources.

Near Orthogonality: As we have discussed in Section 5.4, the best accuracy is provided for the second-order finite volume schemes if the unstructured grid is orthogonal. Building a perfectly orthogonal grid is not always feasible. The recommended strategy is to design a grid that is as close to orthogonal as possible.

Undesirable Cell Shapes: Certain cell shapes have to be avoided, since they may cause large approximation errors, spurious oscillations, slowdown of convergence, or even numerical instability. One common danger is the cells of unacceptably large aspect ratio. The recommended criterion is the same as in the case of structured grids. The ratio between cell dimensions in any two directions should not be smaller than 0.2 or larger than 5. Also, similarly to structured grids, we have to avoid strong (approximately larger than twofold) differences between the sizes of neighboring cells.

Undesirable topological distortions may occur to nontetrahedral cells of 3D unstructured grids. The list includes strongly warped and strongly sheared cells and cells with centroids located outside the cell body. The automated grid generation algorithms usually provide an opportunity of grid check operation, during which the topological quality of each cell is assessed and distortions are detected. It is a requirement of good CFD practice to always perform this operation and modify the grid in accordance to its results.

Choice of Cell Topology: We will consider the most common dilemma, the choice between quadrilateral and triangular cells in 2D and between tetrahedral and hexahedral cells in 3D. The triangular and tetrahedral cells provide certain geometric advantages. The automated generation of grids consisting of such cells is relatively easy. It can be accomplished for arbitrarily shaped domains at low risk of creating undesirable cell shapes. By contrast, generating a grid of quadrilateral or hexahedral elements often creates a number of topologically bad cells, especially in the cases of complex domains. Another advantage of triangular and tetrahedral cells is that we can generate a nearly orthogonal grid by simply requiring that the cell shape does not deviate too far from equilateral and that sizes of any two neighboring cells do not differ too much. Near orthogonality of grids consisting of quadrilateral or hexahedral cells requires that the cells are nearly rectangular, which cannot be achieved for all cells if the domain has a curvilinear boundary.

The arguments in favor of quadrilateral and hexahedral cells are related to the fact that the second-order finite volume schemes based on the midpoint approximation of cell and face integrals, linear interpolation, and central differences are most widely used with unstructured grids. The accuracy of these schemes is, in general, higher if the cells are quadrilateral or hexahedral. The reason is the partial cancelation of the errors of discretization of diffusive terms at opposite cell faces. The approximation of convective terms is also achieved with higher accuracy on quadrilateral and hexahedral cells, if the computed flow is predominantly in one direction and the cells are oriented with one set of opposite faces parallel or nearly parallel to the flow. An important and commonly observed situation of this kind is within the boundary layers at solid walls.

It is recommended for flows with attached boundary layers that the grid includes a layer of thin quadrilateral (in 2D) or hexahedral (in 3D) cells adjacent to the wall. The rest of the domain can be filled with cells of any kind chosen in accordance with domain geometry and accuracy considerations. Illustrations of such arrangements can be seen in Figures 12.1c and 12.2c.

Example To illustrate the discussion of grid quality, let us visually analyze the grids in Figure 12.6. Both grids are generated automatically using the code GAMBIT 2.3. The upper 2D and lower 3D grids are built of quadrilateral and tetrahedral cells, respectively. The grids do not contain topologically bad cells (it was confirmed to the author that the grids had passed the examination by a grid-checking algorithm). The near orthogonality is evident for the upper grid that consists of nearly rectangular cells. The lower grid can be assumed nearly orthogonal, since the tetrahedral cells are nearly equilateral and approximately of the same size.

Both grids in Figure 12.6 can be criticized.[2] The numerical resolution of the upper grid is inadequate. Four cells used across the channel are unlikely to be sufficient to accurately resolve the flow. In the lower grid, the boundary layer treatment, which would insert a layer of hexahedral cells between the wall and the tetrahedral mesh, is not applied.

12.4 ADAPTIVE GRIDS

In this section, we revisit the concept of local grid refinement (grid clustering) introduced in Section 12.1. As a reminder, the refinement is needed to assure sufficiently small discretization error in the area, where the solution variables have particularly strong gradients. In the rest of the domain, where the gradients are weak, a cruder grid, which provides the same magnitude of the error, has to be used for the sake of computational efficiency.

We have assumed so far that the location of the zone needing refinement is known before the problem is solved. A flow with an attached boundary layer is a good example of a situation, in which this approach is justified. There are, however, situations, in which the location of the refinement zone is a priori unknown. It can only be found as a part of the solution. In some cases, the solution is unsteady, and the strong gradient zone changes its location and shape with time.

We will mention just two among many such configurations. In flows with shock waves, refinement is required around the shocks, but the shock location and structure are only determined in the course of the solution. Another important and much researched type is that of flows with interfaces between two immiscible media: gas and liquid, two liquids, or liquid and solid. Examples are the unpremixed combustion and solidification. The interface dynamics is, as a rule, a complex, nonsteady, and nonlinear phenomenon coupled with flows and heat and mass transfer in surrounding media. The solution variables usually experience strong gradients near the interface. We need a moving and deformable refinement zone that follows the interface motion and deformation.

---

[2]It should be mentioned that the grids shown in Figure 12.6 exemplify the situation when the accuracy of the CFD solution is not of the highest priority. Figure 12.6 shows only parts of the solution domains. The parts outside the nozzle, which are not shown, play more important role in the analysis of the spraying process. The error of approximation of the flow within the nozzle, albeit possibly large by CFD standards, is definitely smaller than the error introduced by the physical model of electric discharge arc and plasma formation.

Evidently, in such situations, a pre-built clustered grid would be useless. We need a method, in which the refinement is made *in the course of computations in correspondence to the computed flow fields*. Numerous techniques have been developed for this purpose over the recent decades. The common names of *adaptive refinement methods* or *adaptive grid methods* can be applied to them. A thorough discussion of these complex methods would only be possible on the level significantly more advanced than appropriate for our book. The reader interested in the subject is invited to study the research literature. Some useful references are provided at the end of the chapter. Here, we only discuss the basic principles of the approach and briefly outline several established techniques.

The sequence of the key steps of an adaptive refinement procedure is obvious. We should:

1. Analyze the distribution of the discretization error and detect the zones where additional refinement is necessary.
2. Rearrange the grid, interpolate the solution onto the new grid, and complete the computations.

In steady-state solutions, the procedure is repeated in iterative manner until a desirably low level of discretization error is achieved throughout the domain. In unsteady solutions with moving refinement zones, the error analysis and grid rearrangement are done between the time steps.

It is the implementation of these obvious steps that makes the adaptive refinement a challenging and advanced technique. One serious issue is how to determine the new location of the refinement zone. We need a sufficiently accurate and robust tool to estimate the local discretization errors, the *error estimator*. This should be done by analyzing the computed solution, although, in some cases, understanding of the physics of the analyzed process can provide helpful indicators.

Perhaps the simplest, but not always reliable, error estimator is the normalized amplitude of the solution gradient. Many estimators have been proposed that evaluate the actual magnitude of the discretization error. Some of them are based on comparison between the solutions on differently refined grids and the Richardson extrapolation, which we present in Section 13.2.1. In others, the necessary information is derived comparing the values of the local solution properties, for example, convective and diffusive flux integrals obtained using schemes of different orders of approximation.

The second major issue is the actual grid refinement and handling solutions on the resulting complex grids. Several distinct approaches have been established. The classification popularized in the finite element computations separates p-, r-, and h-methods. In the p-methods, which are used exclusively in combination with finite elements, the grid remains unmodified, but the order of approximation of the scheme (the number of terms in the Galerkin expansion) is increased within the refinement zone. The r-methods keep the constant total number of grid points or cells but redistribute them so as to achieve finer resolution within the refinement zone and cruder resolution without it.

In the h-method, each cell within the refinement zone is split into a number of smaller cells. For example, a 2D quadrilateral cell is usually split into two or four. Several levels of refinement can be used. The traditional finite volume algorithms have to be modified to handle the complexity of the resulting grids. In particular, some faces of larger cells are in contact with faces of two smaller cells. In order to maintain the conservation properties of the method, the face of the larger cell should be treated as a combination of two faces, each representing an interface to a smaller cell. This is possible, but the method should allow a grid consisting of cells with the different numbers of faces.

If the h-method is used with a block-structured grid, the refinement is performed blockwise.

A completely original method has been actively applied for solution of hyperbolic systems, for example, in supersonic gas dynamics or in astrophysics. The refinement is accomplished through a sequence of overlapping patches. Each patch is covered by a structured and, typically, orthogonal grid. In the solution, each patch is treated quasi-independently. Coupling between the patches and the main grid is achieved using interpolation and other operations.

BIBLIOGRAPHY

Ferziger, J.H. and Perić, M. (2001). *Computational Methods for Fluid Dynamics*, 3e. New York, NY: Springer.

Liseikin, V.D. (1999). *Grid Generation Methods*. New York, NY: Springer.

Plewa, T., Linde, T., and Weirs, V.G. (eds.) (2005). *Adaptive Mesh Refinement–Theory and Applications*. New York, NY: Springer.

Thompson, J.F., Soni, B., and Weatherrill, N.P. (1998). *Handbook of Grid Generation*. CRC Press.

PROBLEMS

1. For each of the following situations, determine whether it is better to use a regular Cartesian, irregular structured, or unstructured grid. Provide arguments supporting your answer. If a grid with varying step or cell size is justified, explain how the size must vary and why.
    a) Flow in a gap between two concentric spheres rotating with different velocities around a common axis (the spherical Couette flow).
    b) Flow of air past a car moving in a tunnel (see Figure 2.5).
    c) Flow of ocean water around the supporting structure of an off-shore oil drilling platform.

2. For each of the following situations, determine whether it is better to use a regular Cartesian, irregular structured, or unstructured grid. Provide arguments supporting your answer. If a grid with varying step or cell size is justified, explain how the size must vary and why.
    a) Fully developed turbulent flow in a channel between two parallel walls (see Figure 11.1).
    b) Flow in a confluence of two rivers.
    c) Flow and conduction and convection heat transfer within a pack of air-cooled lithium-ion batteries.

3. For each of the flows in Problems 1 and 2, for which an irregular structured grid has been found necessary and possible, suggest a coordinate transformation.

4. For the coordinate transformation (12.1) and (12.2), derive the formula for u_{xy}.

5. For the coordinate transformation (12.1) and (12.2), develop the second-order finite difference approximations of u_y, u_{xx}, and u_{xy} at (ξ_i, η_j). Apply central differences for discretization of derivatives with respect to ξ and η.

6. If your course involves exercises with a finite volume CFD software, study the software manual to determine which types of cell shapes are available for unstructured grids. Does the software allow combination of cells of different types within the same grid? Which characteristics of grid quality are measured by the software?

7. If your course involves exercises with a CFD software equipped with automated grid generation algorithm, try to generate grids for several simple shapes: a pipe of circular cross section, a spherical ball, a duct of rectangular cross section, a 2D channel with a backward-facing step, and so on. In each case, create a grid with clustering near the walls. Try different cell shapes and different algorithms of grid generation, if available. Analyze the quality of each grid.

13

CONDUCTING CFD ANALYSIS

In this chapter, we consider the issues that belong to the realm of common sense as much as to that of CFD as a scientific discipline. The main question is formulated as follows: *given the available tools (discretization methods and algorithms for solution of discretization equations and grid generation), how should we conduct the CFD analysis to obtain reliable and meaningful results?*

13.1 OVERVIEW: SETTING AND SOLVING A CFD PROBLEM

A CFD analysis usually involves much more than simply generating a grid and running an available CFD code. As illustrated in Figure 13.1, there are other essential steps. The actual path of the analysis is usually quite convoluted, with multiple loops and bifurcations. Nevertheless, the following key stages are always present in some form.

Setting the Physical Model: With the exception of few simple cases, which are rarely encountered outside the classroom, the subject of the

Essential Computational Fluid Dynamics, Second Edition. Oleg Zikanov.
© 2019 John Wiley & Sons, Inc. Published 2019 by John Wiley & Sons, Inc.
Companion Website: www.wiley.com/go/zikanov/essential

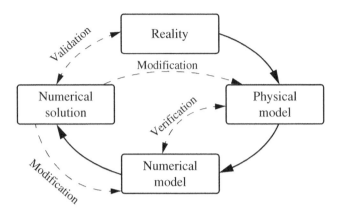

Figure 13.1 Key stages of a CFD analysis.

analysis is usually too complex for a direct CFD approach. Let us, for example, consider one specific area of application of CFD, the human blood circulation system. Parts of the system, such as the arteries, capillaries, and heart muscles, are all connected to each other and with the other organs within the body. The system's behavior involves complex and not fully understood effects: elastic and moving walls, complex blood rheology, varying oxygen concentration, and so on. Also, since no two human bodies are the same, each blood circulation system is unique.

Let us now take the position of a researcher trying to understand a certain pathological condition. A numerical simulation of the entire blood circulation system of a sick person is technically impossible. It would also be not very useful, because the critical information on the pathology would likely be lost in the sea of data produced by such a model. Furthermore, the results would correspond to the state of a specific body within a specific period of time. It is not at all certain that this state would reveal the behavior typical of the pathology.

The proper approach to the analysis is to set a *simplified and generalized physical model*. In our example, setting the model starts with locating the particular part of the system, where the pathology – for example, a partially blocked artery – occurs, and limiting the analysis to this part. The influence of the rest of the body is reproduced by, hopefully, plausible models implemented in the form of boundary conditions.

The next step is to decide which physical effects have to be included into consideration and which have to be neglected. For example, we can include the complex rheology of blood and the dynamics of plaque buildup or consider a simple flow of an incompressible fluid with constant viscosity and smooth solid walls. In CFD, where the physical model is usually set in terms of partial differential equations (PDE) expressing the conservation laws, the selection of physical effects is equivalent to the selection of a certain set of the equations.

Quite often, the physical phenomena are important, but cannot be directly addressed in the computations. The most obvious example is turbulence. We can also name multiphase flows and flows with chemical reactions. The approximations used in such cases, for example, the large eddy simulation (LES) and Reynolds-Averaged Navier–Stokes (RANS) closure models of turbulence considered in Chapter 11, are parts of the general physical model.

The boundary conditions have to be set in conjunction with the governing equations, so as to form a well-posed PDE problem. The conditions are a critical part of the physical model. Ideally, they accurately reproduce the actual impact of the surroundings on the solution within the flow domain. Since, however, sufficiently detailed information on the processes in the surroundings is typically unavailable, the reproduction is often just an approximation based on some simplifying assumptions (e.g. an adiabatic wall or prescribed mean velocity and turbulence intensity at an inlet).

One has to assign the values of physical parameters, such as the fluid viscosity and density. Furthermore, the CFD analysis has to be conducted for certain "typical conditions." In our example, this may mean generalizing the geometry and choosing the regime of operation of the blood system (pulsation frequency and flow rate) critical for the pathology.

The main point of the discussion above is that setting a physical model is unavoidable and often difficult, requires good understanding of the subject of analysis, and can be a significant source of error.

Selecting Numerical Approach: On this stage, we select components and features of the numerical model. This includes discretization method (finite volume, finite difference, etc.), order of approximation, discretization scheme, type of the problem (transient or steady state), and type and parameters of the grid (typical space and time steps, clustering, cell shape, etc.). Evidently, it is important to make correct choices, since they determine the accuracy of the future solution and the amount of computational work this solution will require.

Developing Numerical Model: In the earlier years of CFD, this was a difficult and time-consuming part of the analysis. Many thousand lines of code had to be written, tested, and debugged, which took, depending on the qualification of the author and difficulty of the task, from several days to several years. Development of a new algorithm to solve a particular problem still occurs nowadays, but mostly in fundamental research. Practical CFD analysis is predominantly conducted using readily available general-purpose algorithms.

This still leaves serious responsibilities to the user. One of them is the choice of the method to be applied to solution of discretization equations. For example, educated choice of one of the many available iteration methods usually has to be made to achieve reliable and rapid convergence.

Another responsibility appears when the physical model contains features, for which no ready numerical approximation is available in the CFD code. For example, in the analysis of a blood circulation system, such feature can be the dependence of the apparent viscosity of blood on the diameter of the vessel and the flow rate. It is usually possible to implement a new physical feature in the form of a *user-defined subroutine*, an additional segment of the code, supplied by the user.

Conducting Computations: This is the easiest part of the process. We let computers do the work.

Verification and Validation: There are many possible sources of error in CFD analysis. The only way to increase confidence in the results is thorough testing (verification and validation) discussed in detail in the next section. For now, we only mention that the absence of proper testing is a common and serious mistake, leading to sloppy CFD work.

Modifications: The CFD analysis rarely succeeds on the first try. Usually, the first results are either absent (convergence is not achieved or the computed fields look physically impossible) or do not pass the verification and validation tests. Modifications of the physical and numerical models have to be made, and the computations repeated. This is an iterative procedure, which, if the person conducting the analysis is skillful and lucky, eventually leads to acceptable results. Virtually the only kind of situation outside of a classroom in which the modifications can be unnecessary is when the same system is repeatedly analyzed for slightly different regimes of operation.

13.2   ERRORS AND UNCERTAINTY

In theory, we should separate between errors and uncertainty. For example, we can say that the error is a deficiency that occurs not because we do not know something, but because of limitations of our technical (e.g. computer power) capacity. The uncertainty is then characterized as a deficiency occurring because of lack of knowledge. In the following discussion, we use a simplified, albeit theoretically imperfect, approach. We ignore the difference and use the term *error* for all inaccuracies of a CFD analysis.

13.2.1   Errors in CFD Analysis

It is clear from our discussion throughout the book that a CFD solution inevitably contains errors. How can we be confident that the errors are not so large as to render the solution meaningless? Are there ways to estimate and reduce the errors? An attempt to answer these questions is provided next.

The errors appearing in a CFD analysis can be classified into the following four types:

1. Errors of physical model.
2. Discretization errors.
3. Errors of solution of discretization equations (iteration errors).
4. Programming (algorithmic) errors.

Let us look at these error types in more detail.

Errors of Physical Model: As we have discussed in Section 13.1, a CFD analysis generates an approximate description of the behavior of a *physical model*, rather than of a real physical system. This substitution is inevitably a source of error. We can symbolically write that the values of a property u in the real system and the exact solution of the model differ by

$$\epsilon_{model} = u_{real} - u_{model}.$$ (13.1)

There are many reasons why a physical model may behave differently from reality. We will name several of them, which are almost always present.

The PDE that usually constitute a physical model are themselves only imprecise models of real behavior. This does not concern the underlying principles of mass, momentum, and energy conservation, which are exact. The problem is that the Navier–Stokes and heat transfer equations contain the Newton's law (2.18) for viscous stresses and the Fourier law (2.25) for heat flux, both of which are empirical, albeit in many cases quite accurate, approximations. Another, typically larger source of error, is the modeling approximations we have to use in the equations when dealing with complex phenomena unsuitable for direct numerical analysis. The list is long. It certainly includes turbulence, multiphase flows, and flows with chemical reactions.

Properties of real liquids and gases (e.g. density or viscosity) depend in a complex and not always well-documented way on temperature, pressure, and concentration of admixtures. This is often neglected in CFD analysis, and the properties are assumed constant. In the cases where this assumption is abandoned, the error is still introduced, since the properties are approximated by imprecise empirical functions.

As we have already discussed in Section 13.1, the physical model typically reproduces only a part (in space and time) of the operation of a real system. The influence of the rest is imitated by boundary conditions. The imitation is unavoidably artificial and imperfect. One good example is a computational domain with an inlet. The inlet conditions are important but usually unknown except, perhaps, for general characteristics, such as the flow rate and average temperature. In the absence of a better alternative, we assume an idealized form of the inlet flow (e.g. constant mean flow and temperature, certain turbulence intensity, and dissipation rate) that probably has little in common with reality. Another example of approximate boundary conditions is the temperature conditions (2.51) and (2.53) at perfectly conducting and perfectly insulated walls. Neither of these conditions is realistic. They are only used to avoid solving the conjugate heat transfer problem in the wall.

There is no systematic way to fully predict the magnitude of the error produced by physical modeling. Preliminary estimates can, however, be made on the basis of understanding the physics of the process. For example, knowing the typical magnitudes of flow velocity and variations of temperature and pressure, we can estimate the order of magnitude of density variations and, thus, see if the incompressibility condition is justified. More comprehensive and conclusive estimates are obtained through *validation*, the process of comparison of computed results with data of real system behavior (see Section 13.2.2).

Discretization Errors: In the CFD approach, we do not solve the equations of the physical model exactly. Instead, an approximation is found as a solution of a system of algebraic discretization equations. The process introduces the *discretization error*. It is defined as the difference between the exact solution of the model PDE and the exact solution of the discretization equations

$$\epsilon_{discr} = u_{model} - u_{discr}. \tag{13.2}$$

The behavior of the error is determined by the order of the discretization scheme. Let the order of the discretization in the x-direction be p. If the solution of PDE is smooth and Δx is sufficiently small, so that the leading-order term of the truncation error is $O((\Delta x)^p)$, the discretization error scales as

$$\epsilon_{discr} = O((\Delta x)^p). \tag{13.3}$$

We can rewrite it as

$$\epsilon_{discr} = C(\Delta x)^p + O\left((\Delta x)^{p+1}\right), \tag{13.4}$$

where C is a generally unknown constant, which depends on the local properties of the solution (C contains its derivatives), properties of the discretization scheme, and design of the grid. Analogous formulas can be written for other coordinates and for time.

Unfortunately, the scaling formulas do not measure the amplitude of the discretization error. The only conclusion that can be made from (13.4) is that the error is reduced by approximately m^p times when the grid step is reduced m-fold.

Because of the variations of C, the absolute value of the error varies strongly, sometimes by orders of magnitude within the solution. Similar uncertainty exists for integral characteristics. The discretization errors of their evaluation may be very different for two discretization schemes of the same order or the same scheme implemented on differently structured grids of the same cell size, because of the different errors in the evaluation of C.

The only way to obtain reliable quantitative measure of the amplitude of the discretization error is to compare solutions on systematically refined grids. The logic is simple. Let us assume that a scheme of order p is applied on two grids, which are geometrically similar but have different steps Δx_1 and Δx_2. The results are the two approximate solutions that are related to the exact solution of the physical model as

$$u_{model} = u_{discr,1} + \epsilon_{discr,1} = u_{discr,1} + C(\Delta x_1)^p + O\left((\Delta x_1)^{p+1}\right)$$

$$u_{model} = u_{discr,2} + \epsilon_{discr,2} = u_{discr,2} + C(\Delta x_2)^p + O\left((\Delta x_2)^{p+1}\right).$$

Solving this as a system of linear equations for C and u_{model}, we obtain

$$C = \frac{u_{discr,1} - u_{discr,2}}{(\Delta x_2)^p - (\Delta x_1)^p} + O(\Delta x_1, \Delta x_2) \tag{13.5}$$

and

$$u_{model} = \frac{(\Delta x_2)^p u_{discr,1} - (\Delta x_1)^p u_{discr,2}}{(\Delta x_2)^p - (\Delta x_1)^p} + O\left((\Delta x_1)^{p+1}, (\Delta x_2)^{p+1}\right). \tag{13.6}$$

The discretization error can be estimated, for example, for the solution on the grid with Δx_1, as

$$\epsilon_{discr,1} = \frac{u_{discr,1} - u_{discr,2}}{(\Delta x_2)^p - (\Delta x_1)^p}(\Delta x_1)^p + O\left((\Delta x_1)^{p+1}, (\Delta x_2)^{p+1}\right)$$

$$\approx \frac{u_{discr,1} - u_{discr,2}}{(\Delta x_2)^p - (\Delta x_1)^p}(\Delta x_1)^p. \tag{13.7}$$

The first term in the right-hand side of (13.6) can be used as an approximation of the exact solution u_{model}, which is more accurate (of the higher order) than either $u_{discr,1}$ or $u_{discr,2}$. Together with (13.7), this constitutes the method known as the *Richardson extrapolation*.

If the order of discretization p is a priori unknown, it can be found by conducting computations on yet another grid with the step Δx_3 and solving the system of three equations to find C, p, and u_{model}.

Albeit simple and logical at first glance, the method based on the Richardson extrapolation is difficult to implement. The reason is that it requires computations on two or three increasingly fine grids. It is essential for the accuracy of the method that the refinement is significant. The preferred choice is $\Delta x_2 = \Delta x_1/2$. In 3D, this means that the second grid has eight times more points or cells than the first grid. At the same time Δx_1 should be small enough for the asymptotic formula (13.3) to be valid.

Although difficult and time consuming, computations on systematically refined grids are recommended every time the analysis is conducted of a new problem or using a new scheme or new type of grid. The main purpose is not so much to estimate the discretization error, but to determine the level of refinement on which this error becomes smaller than a given tolerance. This level is commonly (and somewhat incorrectly) referred to as the level of *grid independency* of solution. The meaning of the term is that further refinement would change the solution very little.

The grid independency is often determined using the simplified procedure that does not explicitly rely on the Richardson extrapolation. Solutions obtained on systematically refined grids are compared with each other using integral properties and plots of essential characteristics. The solution is declared grid independent when further refinement does not lead to visible changes.

Errors of Solution of Discretization Equations (Iteration Errors):
The algebraic discretization equations are unavoidably solved with errors, which constitute the difference between the exact and actually computed solutions

$$\epsilon_{iter} = u_{discr} - u_{comp}. \tag{13.8}$$

One component of ϵ_{iter} is the round-off error of computer operations. Fortunately, it only becomes important if the scheme is numerically unstable (see Chapter 6). If the scheme is stable, and the round-off errors do not accumulate, their magnitude is typically several orders lower than the magnitude of the errors of other types, for example, of the discretization errors.

Much more significant and constituting practically the entire ϵ_{iter} are the iteration errors that appear when a linear or linearized system of discretization equations is solved by an iterative method (see Section 8.3). The iterations cannot be continued indefinitely. They should be stopped when the estimated iteration error becomes smaller than a certain small but nonzero tolerance level.

The two important questions concerning the iteration error are: what tolerance threshold is acceptable, and how can we estimate the error? The answer to the first question is case specific, although we have to take into account that the iteration errors occur on the background of unavoidable discretization errors and, quite likely, physical model errors. In general, there is no need to continue the iterations to the round-off level of accuracy.[1] The commonly accepted criterion is that the iteration errors are at least one order of magnitude lower than the discretization errors.

---

[1] The only situation, in which such ultra-precision may be necessary, is the verification testing of an iteration algorithm when it is applied for the first time to a new type of problem. It is desirable in that case to gain confidence that the algorithm does not have internal faults that would prevent convergence below a certain level.

The second question is difficult because, as we have already discussed in Section 8.3.1, the error cannot be computed directly. Let us, for consistency, return to the notation of Chapter 8 and consider the error $\epsilon^{(k)} = v - v^{(k)}$ obtained after the kth iteration in the solution of the matrix equation $A \cdot v = c$. Evidently, the error is impossible to compute since the exact solution v is unknown. We can only compute indirect characteristics, such as the difference between the results of successive iterations $\delta^{(k)} = v^{(k+1)} - v^{(k)}$ or the residuals $r^{(k)} = c - A \cdot v^{(k)}$. The usual approach is to calculate the norm $\|\delta^{(k)}\|$ or $\|r^{(k)}\|$ (see (8.48) and (8.49) for definitions of possible norms) and stop the iterations when $\|\delta^{(k)}\|$ or $\|r^{(k)}\|$ becomes smaller than a predetermined tolerance limit ϵ_0.

How do we determine the value of ϵ_0 that secures the desirable level of iteration error? The question is nontrivial, since $\|\delta^{(k)}\|$ or $\|r^{(k)}\|$ do not represent the actual magnitude of $\|\epsilon^{(k)}\|$. Luckily for us, the magnitudes of $\|\epsilon^{(k)}\|$, $\|\delta^{(k)}\|$, and $\|r^{(k)}\|$ are connected. It can be shown theoretically and confirmed in computational experiments that, after several initial iterations, the reduction of the error with k follows

$$\|\epsilon^{(k)}\| \approx A\|\delta^{(k)}\| \approx B\|r^{(k)}\|, \tag{13.9}$$

where A and B are the constant coefficients related to the largest eigenvalue of the iteration matrix. Simply setting ϵ_0 to a small number, say, to 10^{-4}, does not guarantee small iteration error, because A and B can be large.

The only reliable fact at our disposal is that $\|\epsilon^{(k)}\|$, $\|\delta^{(k)}\|$, and $\|r^{(k)}\|$ are all reduced at approximately the same rate during the process of convergence. According to (13.9), we can write

$$\frac{\|\epsilon^{(k)}\|}{\|\epsilon^{(0)}\|} \approx \frac{\|\delta^{(k)}\|}{\|\delta^{(0)}\|} \approx \frac{\|r^{(k)}\|}{\|r^{(0)}\|}. \tag{13.10}$$

The initial error $\epsilon^{(0)} = v - v^{(0)}$ is related to the exact solution v. $\|\epsilon^{(0)}\|$ is equal to $\|v\|$ if zero initial conditions $v^{(0)} = 0$ are used. If less trivial initial conditions better approximating v are applied, $\|\epsilon^{(0)}\|$ is smaller but still likely to be of the same order of magnitude as $\|v\|$. Our final formula

$$\frac{\|\epsilon^{(k)}\|}{\|v\|} \approx \frac{\|\delta^{(k)}\|}{\|\delta^{(0)}\|} \approx \frac{\|r^{(k)}\|}{\|r^{(0)}\|} \tag{13.11}$$

shows that the norm of the iteration error relative to the amplitude of the solution is approximately equal to the observed total reduction of the norms of differences $\delta^{(k)}$ or residuals $r^{(k)}$. *We have to define ϵ_0 as the level, at which $\|\delta^{(k)}\|$ or $\|r^{(k)}\|$ becomes a predetermined fraction of $\|\delta^{(0)}\|$ or $\|r^{(0)}\|$.* For example, we can stop the iterations when the norms of residuals fall by four orders of magnitude and be confident that the norm of the iteration error is close to 10^{-4} and definitely smaller than 10^{-3} of the norm of the solution.

We will complete the discussion by two comments. The sentence "The residuals are changing only little from one iteration to the next, so I can stop the iterations" is meaningless. The residuals may be changing slowly because the convergence is slow while the iteration error is still high. The second comment is that the estimate $\|\epsilon^{(0)}\| \approx \|v\|$ becomes an overestimate when the initial conditions are close to v. This may happen, in particular, when the results of previous computation, either underconverged or conducted at slightly different parameters or on a slightly different grid, are used as initial conditions. Smaller reduction of $\|\delta^{(k)}\|$ or $\|r^{(k)}\|$ can be required in such cases.

Programming (Algorithmic) Errors: Programming errors are virtually inevitable and omnipresent. There are many opportunities for a CFD practitioner to make one. Even if a readily available general-purpose code is used, errors can be made while setting the problem (determining the geometry, boundary conditions, physical parameters, etc.) or preparing a user-defined code.

Another, somewhat disturbing fact should be kept in mind. Any CFD code, commercial or noncommercial, contains a significant number of algorithmic errors. The analysis of Hatton (1997) found on average about 10 faulty lines per 1,000 lines of code in more than 100 scientific and engineering codes reviewed. There is no ground to assume that the situation has significantly improved since then.

The programming errors can be roughly divided into two types. There are fatal errors that prevent the code from executing or generate evidently incorrect results. Another type includes "mild" or "sleeping" faults that generate incorrect answers only in certain circumstances that may or may not appear in the course of code execution. The errors hidden in the ready-to-use CFD codes usually belong to the second type. The resulting mild incorrectness is very dangerous, since it is usually not clearly visible (one may be tempted to declare that everything is fine, seeing that the iterations have converged and the plots of solution look plausible), but nevertheless real.

There is no systematic way to estimate and control the effect of programming errors. Rather, we have to give full effort to detect as many of them as possible and reach the maximum possible confidence in the results of computations. The methods of verification and validation used for this purpose are discussed in the following section.

13.2.2 Verification and Validation

The previous discussion shows that results of a CFD analysis unavoidably contain errors, the effect of which is either difficult or impossible to determine before or in the course of computations. In this section, we will talk about the additional tools available to detect the errors and evaluate their magnitude. The ultimate goal is to increase the level of confidence in the results.

We should say in the beginning that in CFD there is no universally reliable method of achieving absolute confidence. Rather, the process reminds a criminal investigation (how it appears in books and movies). We collect direct and indirect evidences and modify the assumed picture until a reasonably high level of confidence is achieved that the picture accurately represents reality.

There is a scientific discipline dealing with the general issues of accuracy and reliability of computational modeling. The methods of this discipline have been formalized and developed primarily for high-consequence systems, such as weapons or transportation systems, or nuclear power stations. Full-scale consistent application of these methods in CFD is a relatively recent phenomenon still largely limited to a few specific areas, such as military systems or weather prediction. However, the basic elements of the accuracy and reliability analysis, perhaps taken informally, have always been considered an essential ingredient of CFD.

The two main tools are verification and validation. The difference between them can be seen in the following definitions, which we quote from the AIAA guidelines published in 1998:

Verification: The process of determining that a model implementation accurately represents the developer's conceptual description of the model and the solution to the model.

Validation: The process of determining the degree to which a model is an accurate representation of the real world from the perspective of the intended uses of the model.

These definitions can be rephrased using our terminology, in which *model* becomes *physical model* and *implementation* becomes *implementation of numerical model*, basically, the code and the grid we are using. Figure 13.1 provides an illustration. We can describe verification as the way to ensure that the numerical model solves the physical model. Note that even a fully verified numerical model may produce unrealistic results if the physical model is incorrect. Correctness of the physical model is primarily tested in the course of validation, which we can describe as the way to test and ensure that the results of a numerically adequate CFD analysis agree with reality.

Considering the types of the CFD errors, we can say that the iteration errors are evaluated and controlled in the course of verification. Physical model errors are addressed by validation. The algorithm errors are detected by both processes. At last, the discretization errors are primarily assessed through verification, although their part related to the effect of numerical resolution on accuracy of turbulence and other closure models requires validation.

Verification: A systematic verification of numerical model should be conducted every time a new code is developed, an essentially new grid is built, or a significant new feature is added to the model. The methodology relies on availability of *reference solutions*, i.e. exact analytical or highly accurate numerical solutions of certain benchmark problems. The logic is simple. The best and often only way to find out where and how our numerical model solves the PDE incorrectly is to apply it to a problem for which an exact solution is known and compare the numerical and exact results.

To separate different sources of error, it is recommended to conduct the verification for separate modules of the model, starting at the most elementary level and working way up. For example, developing a new finite volume algorithm, we have to verify that each operation of integral approximation and interpolation is coded correctly and has the desired order of truncation error. This can be done by applying the operations to simple functions (e.g. polynomials), for which integrals and interpolations can be found analytically, and comparing these reference solutions with the numerical results obtained using the values of the same functions at grid points. Similarly, the iteration solver for matrix equations can be verified by substituting the matrix and the right-hand side, for which an exact solution is known.

After all modules are verified and assembled or if a ready-to-use code is applied, verification of the entire numerical model is conducted. Since a reference solution is usually unavailable for the system for which we plan to conduct the CFD analysis (the analysis would be unnecessary otherwise), we have to use simplified test cases, in which geometry and flow parameters are modified to make a reference solution possible. For example, if we study a blood flow in an artery with a pathological wall deformity, the good first test cases can be a steady and a pulsating flow in a segment of a circular pipe.

Verification of a numerical model is a complex process with loops and bifurcations, many of which are needed to detect algorithmic errors. The first of the major steps not related to the algorithmic errors is the analysis of the iteration errors. A test case with steady reference solution should be used for this purpose. After running the iterative procedure to round-off accuracy, thus assuring convergence, the iteration error is analyzed in its relation to residuals and differences between the results of successive iterations. This helps to establish useful, albeit not fully reliable, estimates of the convergence criteria.

Another important step is the analysis of the discretization errors. This is done by solving the test case equations on successively refined grids and comparing the results with the reference solution. As the outcome of this procedure, we learn whether the discretization errors are reduced at the rate corresponding to the intended order of approximation, estimate their amplitude as a function of the grid steps, and make conclusions regarding the grid quality.

Validation: The validation is performed after we have gained confidence that the numerical model sufficiently accurately reproduces the behavior of the physical model (this practically means that the numerical model has been successfully verified). The computed solution is then compared with the results of experiments. The main purpose of this comparison is to determine the degree of accuracy with which the physical model reproduces the real world. We make the conclusion whether or not the physical model is adequate and evaluate the errors introduced by the model assumptions and approximations, such as the choice of computational domain, artificial boundary conditions, turbulence models, and so on.

Quite often, experiments with the entire system considered in a CFD analysis are impossible or impractical. It is, therefore, necessary to replace the system by a simpler prototype or decompose its behavior into smaller and experimentally accessible units. Our example of a blood flow in an artery presents a good case for such *benchmark experiments*, since making detailed measurements within a functioning human body is a difficult and generally undesirable task. The validation experiments can be conducted in a laboratory using an artificial artery segment, in which the flow is generated by a pump. Information on the flow behavior in typical regimes of operation can be collected and compared with the results of computations.

Another difficulty of validation is that the experimental data are, themselves, not free from errors. Evidently, reliable validation is only possible if the experimental uncertainty is known and sufficiently small.

As an additional complication, grid properties (quality and step or cell size) affect the accuracy of certain aspects of the physical model. This concerns, in particular, LES and RANS models of turbulence. The effect cannot be analyzed in the framework of the verification procedure, where the physical model is usually simplified (e.g. assuming that the flow is laminar) to be able to obtain an exact reference solution. On the contrary, the accuracy of physical approximations on various grids can be assessed in the validation tests through direct comparison with experiments.

BIBLIOGRAPHY

AIAA (1998). *Guide for the Verification and Validation of Computational Fluid Dynamics Simulations*. AIAA-G-077-1998. Reston, VA: American Institute of Aeronautics and Astronautics.

Hatton, L. (1997). The T experiments: errors in scientific software. *IEEE Comput. Sci. Eng.* **4** (2): 27–38.

Oberkampf, W.L. and Trucano, T.G. (2002). Verification and validation in computational fluid dynamics. *Prog. Aerosp. Sci.* **38**: 209–272.

Roache, P.J. (1998). *Verification and Validation in Computational Science and Engineering*. Hermoza Publishers.

PROBLEMS

1. In each of the following situations, determine which type of the error (physical model, discretization, iteration, or programming) is most likely responsible for poor performance. In each case, suggest a course of action (e.g. refining the grid, testing the code on a benchmark problem, comparing with experiment, etc.).

    a) Steady-state flow and heat transfer in a small-scale (research prototype) glass melting facility is simulated. The flow is known to be laminar. The simulations are performed using a commercial general-purpose software based on finite volume discretization.

The verification procedure, in which we compute a laminar thermal convection flow in a box, shows good agreement with the known benchmark solution. The simulation results show large discrepancy with the experimental data. Refining the grid and decreasing the tolerance of the iteration solution does not help.

b) A code is developed for simulation of unsteady conduction heat transfer. As a verification, the problem of heat conduction in a two-dimensional rectangular plate with fixed temperature at the boundaries is solved. The solution is utterly incorrect and remains such as we refine the grid and decrease the time step.

c) Three-dimensional flow and heat transfer in a counterflow heat exchanger is solved using a commercial CFD software. A well-tested model specially designed for analysis of heat exchangers is used. The simulations conducted on a grid consisting of 9301 finite volume cells produce incorrect results.

d) URANS solution for turbulent unsteady flows within a chemical reactor shows good agreement with experimental data. It also shows that the mean flow fields always converge to an asymptotic steady state. We decide to focus on them and solve the steady-state version of the RANS equations. The results are disappointing.

2. For the following problems, propose a verification and validation tests of the CFD solution.

   a) Flow in confluence of two rivers.

   b) Wind flow around a smokestack of a coal power station and the resulting distribution of smoke particles in the atmosphere.

   c) Flow of air past a car moving in a tunnel (see Figure 2.5).

   d) Turbulent flow in an oil pipeline.

   e) Flow and surface waves around a performance swimmer.

Index

A

Adams–Bashforth method, 163
Adams–Moulton method, 163
Adaptive grids (Adaptive refinement), 330
ADI method, 227–228
amplification factor, 135
approximate factorization
  for Beam–Warming scheme, 214–218
  for heat equation, 225–228
artificial compressibility, 261

B

Beam–Warming scheme for compressible flows, 214–218
body forces, 17
boundary conditions, 27–31, 40
  Dirichlet, 41, 85
  exit, 29–30
  heat flux, 28
  impermeable wall, 28
  inlet, 29
  Neumann, 41, 86
  Newton's cooling law, 29
  no-slip, 28
  periodic (cyclic), 30–31, 41, 276
  Robin (mixed), 41
  symmetry axis, 30
  wall temperature, 28
box filter, 298–299
Burgers equation, 39, 50, 84, 161–162

C

CFD
  analysis, 335–338
  definition, 1
  history, 4–5
  types of errors, 339–346
CFL condition, 152, 213, 235
closure models, LES, 301–304
colocated grid, 238–243
compact schemes, 71
compressible flows, 208–222
  properties, 211–212
  schemes for, 212–222
computational grid, 56, 63–65
  block-structured, 322
  boundary fitting, 314, 319–321

Essential Computational Fluid Dynamics, Second Edition. Oleg Zikanov.
© 2019 John Wiley & Sons, Inc. Published 2019 by John Wiley & Sons, Inc.
Companion Website: www.wiley.com/go/zikanov/essential

Printed and bound by CPI Group (UK) Ltd, Croydon, CR0 4YY

16/04/2025

14658581-0003